Miomir Vukobratović
Dragan Stokić

Applied Control of Manipulation Robots

Analysis, Synthesis and Exercises

With 100 Figures

Springer-Verlag Berlin Heidelberg New York
London Paris Tokyo Hong Kong

Professor Miomir Vukobratović, Ph. D., D. Sc.
Corr. member of Serbian Academy of Sciences and Arts
Foreign member of Soviet Academy of Sciences

Assoc. professor Dragan Stokić, Ph. D.

Institute Mihailo Pupin
Volgina 15
P. O. Box 15
11000 Beograd
Yugoslavia

Based on the original Upravljanje Manipulacionim Robotima
published by NIRO "Tehnička Knjiga", Beograd, Yugoslavia

ISBN-13:978-3-642-83871-2   e-ISBN-13:978-3-642-83869-9
DOI: 10.1007/978-3-642-83869-9

© Springer-Verlag Berlin Heidelberg 1989
Softcover reprint of the hardcover 1st edition 1989

2161/3020 543210 – Printed on acid-free paper

Preface

The first book of the new, textbook series, entitled Applied Dynamics
of Manipulation Robots: Modelling, Analysis and Examples, by M.
Vukobratović, published by Springer-Verlag (1989) was devoted to the
problems of dynamic models and dynamic analysis of robots. The present
book, the second in the series, is concerned with the problems of the
robot control.

In conceiving this textbook, several dillemas arouse. The main issue
was the question on what should be incorporated in a textbook on such
a complex subject. Namely, the robot control comprises a wide range of
topics related to various aspects of robotics, starting from the synthesis of the lowest, executive, control level, through the synthesis
of trajectories (which is mainly related to kinematic models of robots)
and various algorithms for solving the problem of task and robot motion planning (including the solving of the problems by the methods of
artificial intelligence) to the aspects of processing the data obtained from sensors. The robot control is closely related to the robot programming (i.e. the development of highly-specialized programming languages for robot programming). Besides, numerous aspects of the control realization should be included here. It is obvious that all these
aspects of control cannot be treated in detail in the frame of a textbook. Therefore, we decided to confine ourselves only to the synthesis
of control at the executive level, while the other aspects of the control mentioned above, should be treated in the coming volumes of the
series. To facilitate the understanding of the control synthesis at
the executive level, the first two chapters of this book deal with the
problems concerning higher control levels and fundamental aspects of
the robot control realization.

After deciding to restrict our attention only to the synthesis of control at the executive level, we posed the following question: what should be the necessary background knowledge of the user of this textbook?
It is known that the control synthesis at the executive level is, on
the one hand, closely related to theory of automatic control (i.e. theory of large-scale technical systems) and on the other, to dynamics of

active mechanisms. We assume the user of this book has already read
volume one of this series, so that he is acquainted with the basic
knowledge on dynamic models of robots. Therefore, in this book (Chapter 3) we only briefly consider the dynamic model of a robot without
repeating many of the concepts related to the robot mechanics. However,
the problem of the relation to systems theory is much more complex. Many of the approaches and solutions developed in the general control systems theory have also been applied, after more or less adaptation, to
the synthesis of robot control. Someone who wants to study robot control should be familiar with some basic concepts of theory of automatic control and theory of large-scale systems. For this reason it was
necessary to "repeat" in this book some of the approaches of both classical theory of automatic control and theory of control of large-scale
systems (i.e. at least those approaches that found a wider application in robotics). In order to avoid the unnecessary broadening of the
book by including detailed considerations of the concepts and approaches that can be found in the related literature on automatic control
and systems theory we assume the user of this textbook is familiar with
basic concepts and techniques of automatic control (such as the models
of linear systems in s-domain, models in state space, classical methods
of control synthesis, etc.). However, as we have endeavoured to write
a book which should be as much as possible a self-contained unit, we
included in it a number of notions of theory of automatic control (e.g.
the elements of position servo systems, fundamental concepts of the methods of pole-placement, the synthesis of the optimal regulator, and
the like). In doing this we avoid all theoretical corroborations and
all those details which are not necessary for the understanding of robot control (and which the reader, if need be, can find in the cited
literature). We hope this approach will enable a wider circle of readers to study robot control, with no need to consult additional literature.

In planning this textbook, the next dilemma was: which approaches to
the robot control synthesis should be described and how extensively it
should be done? Presently, a great number of various approaches to robot control are being developed. Many of them have not been verified
in practice, as they are still at the stage of theoretical and experimental studies. It is difficult to decide whether these approaches should be included in a textbook of this kind or not. We decided first to
consider thoroughly the classical approach via the servo systems around
particular joints, which is involved in the majority of the present-day
commercial robots. We think that, from an educational viewpoint, this

approach provides a most convenient way to understanding the dynamic
behaviour of the robot as a whole. For this reason we consider in detail the methods of the analysis of robot's behaviour. From our experience, it is most advisable to choose those approaches which, starting
from the decentralized control, introduce the correctional signals to
compensate for the effect of the robot's dynamics. For this reason we
consider dynamic control via the global one. In this way, the reader
learns gradually, starting from the most simple servo control and going
to the complex dynamic control. Thus, an important educational aim is
achieved: the user of the textbook becomes aware of the crucial importance of choosing the simplest control law in the control synthesis and
also, that it is equally important to check if such control satisfies
the robotic system in different dynamic regimes of its operation, i.e. if
a satisfactory accuracy and desired operation speed of the system can
be achieved. Furthermore, we think it is worth to emphasize the importance of the problems related to variations of the robot parameters,
and indicate the circumstances when it is necessary to introduce adaptive control. As in the case of non-adaptive dynamic control, we think
it necessary for the reader to develop the criterion for introducing
adaptive control. However, we thought that reviewing of various schemes of adaptive control would burden the book very much and that a basic coursebook on robot control should not be stuffed with schemes of
adaptive control which are, mainly, at the stage of the laboratory application. We thought it also necessary to include in the book some
representative approaches to the dynamic robot control, such as the
"computed torque method" and the classical approach to systems theory
via the linear optimal regulator. In addition, we decided to present
the Cartesian control which, though still being at the experimental
stage, is very important for educational purposes as it provides a good insight into certain crucial aspects of robot control. Finally, we
thought it necessary to pay a special attention to the robot control
with constrained motion of the gripper and to so-called hybrid positi-
on/force control, as these are specific problems in the field of robotics which attract a great deal of attention of a number of researchers.

Having made such a choice of the subject matter to be included in our
textbook, we divided it into seven chapters.

Chapter 1 presents the fundamental control principles of the current
robotic systems. Without repeating basic concepts of robotic mechanisms
(which can be found in the first book of this series), the chapter briefly describes the classification of robots according to the control

and then deals with the usual control hierarchy of automatic robots. Furthermore, it gives a concise discussion of some characteristic robot tasks and points out that the different requirements are imposed before the robot, what should be borne in mind in the control synthesis.

The aim of Chapter 2 is to provide the reader with some basic concepts of the kinematic control level and thus enable him to follow the chapters to come. Here, we are not concerned with the methods of forming kinematic models of robots (which can be found in the appropriate literature) but we endeavour to indicate the problems related to solving the inverse kinematic problem.

Chapter 3, which occupies the central place in the book, is devoted to the synthesis of servo systems around particular robot joints. As we mentioned above, the servo system approach is prevailingly found with the present-day commercial robots. We think it extremely important in the teaching to emphasize the specificity of the synthesis of servo systems for robots. First part of the chapter presents in short the robot models: the model of the mechanical part and the model of the actuators, and their assemblying into a unique model. Then, the model of the behaviour of one robot joint is derived under the condition that all other joints are kept locked. Such a joint model is the basis for the servo systems synthesis. Section 3.3. describes the synthesis of the local servo systems in two ways: in s-domain and by pole-placement method. Throughout the chapter, the servo systems are consistently represented via the models in s-domain and via the models in state space. The elements of servo systems and their transfer functions are presented in detail, although their more precise definitions can be found in the literature on automatic control. We wanted the reader to refresh his knowledge on classical servo system and thus be prepared, without consulting additional literature, to follow the further consideration for which a full understanding of the concepts of a position servo system is needed. The synthesis in s-domain is given in detail and then the synthesis is generalized using the method of pole-placement. We endeavoured to simplify this method as much as possible because it is here applied on the simple, low-order subsystems. The synthesis of the optimal local regulator, as one of alternative approaches to the synthesis of servo systems, is given in Appendix 3.A. without entering detailed theoretical considerations. Section 3.3.4. explains in detail the specificity of the robot's servo systems: the effect of variation of the mechanism's inertiality and gravitational moments. The next sections deal with the nonlinear effects due to the constraints on the actuators, friction, and the like, as well as the application of the PID regulator. Finally, Section 3.5. is devoted to the problems of tracking of trajectories at the servo system level, and to the precompensator synthesis. We thought it advisable to explain the problems related to the delay compensation on a simple subsystem model (one joint and one actuator).

Chapter 4 analyzes the behaviour of the robot system as a whole when all its joints move simultaneously and the local servo systems are applied. As first, a qualitative analysis of the effects of particular dynamic forces is given, and then, the robot stability is analyzed using a linearized model. This analysis is carried out in a gradual way: first, the robot's behaviour during the position control (around a given position) is analyzed, then, the change of the given position is considered, and finally, the trajectory tracking is described. At this point, nominal centralized control is introduced with the aim to explain the phenomena of dynamic coupling between the joints, and not as a representative approach to solving the problem of dynamic control. Moreover, all the shortcomings of this approach to robot control are clearly stated. Special attention is paid to the effect of the robot's dynamics when the local servo systems are used to track trajectories, as this approach is most frequent in practice. At that, we wanted to make a clear distinction between the case when such simple control can serve the purpose and the case when it is necessary to introduce dynamic control.

An analysis of stability of a nonlinear model during the position control and tracking of trajectories, is the subject of Appendix 4.A. The stability analysis of the nonlinear robot model requires the knowledge of theory of large-scale systems. A method of analysis of complex systems which, to our opinion, can be with success applied on robotic systems, is presented in Appendix 4.A. in a most simple way so that it may be understood by the readers which did not have opportunity to become acquainted with the literature in this field. However, this appendix is intended for those readers which would like to study more deeply this aspect of the control synthesis and is not necessary for the understanding the basic approaches treated in the subsequent chapters.

Chapter 5 treats dynamic global robot control via two main forms: the force feedback and the on-line calculation of robot dynamics. The application of various approximate models of robot dynamics for dynamic global control received special attention. At this point, we thought it advisable to exemplify the computer-aided synthesis of control by presenting a characteristic programming package. In Appendix 5.A., the analysis of robot stability is extended to cover the case when, apart from the local servo systems and nominal control, the global dynamic control is introduced. However, as in the case of Appendix 4.A. this appendix is not necessary for the understanding of the rest of the book. A substantial part of this chapter is devoted to the approach via the "computed torque method" as the one of the most popular approaches to robot control. We clearly indicate the existence of different variants of this approach, as well as some of its shortcomings. As we have already mentioned, this chapter presents the possibility of Cartesian robot control. The centralized quadratic regulator is described in Appendix 5.B. (again without repeating the known theoretical statements which can be found in the literature) as an example of the possible application of the control systems theory on robots. Besides, we indicate all the shortcomings of this approach which are the reasons why such control has not been applied in practice.

Chapter 6 is devoted to the problems of the variation of robot's parameters and to the robustness of the control laws considered. To save space in the book we confined ourselves to only a qualitative analysis of the robustness without providing rigorous definitions and analyses. We want to help the reader to develop a critical attitude to possible introduction of adaptive control.

Chapter 7 deals with the most delicate tasks involving the constrained gripper motion, and especially in the assembly processes. Following the principle "from the simpler to the more complex", which has been implemented in the whole book, we first describe various "contact situations" occurring in the characteristic "peg-in-hole" task. Then, we deal with the conceptual scheme of the control synthesis for such a task, and finally, we treat the problem of the synthesis of hybrid position/ /force control in the tasks involving the constrained gripper motion.

In a separate appendix at the end of the book we give a short version of the programme package for the synthesis of robot control. We have chosen some characteristic programmes which can be used both for the synthesis of local servo systems and analysis of robot's behaviour for different control laws. Because of the limited space, we included only the analysis of the linearized model. It is our opinion that these programmes, though being presented in a reduced form, make an independent package which can serve in the educational practice. The main purpose of this appendix is to enable the reader to master the computer approach to the robot control synthesis highly important for the education of both the future robot designers and robot users. Moreover, we hope these programmes might inspire the reader to develop his own programmes which could yield the synthesis of some other control laws, as well as the adaptation of these programmes to the robot types not included in this package.

Practically each section of the book contains a numerical example illustrating the theoretical results given in the text. We endeavoured to present all the examples for the one and the same robot (with the exception of some sections where for the obvious reasons this was not possible). We chose a robot having a simple structure (one prismatic and two revolute joints), but which enables the demonstration of the majority of characteristic phenomena of both the non-dynamic and dynamic robot control (such as the mechanism's inertiality, coupling of joints via the cross - inertia terms, and the like, as well as the example of a joint totally decoupled from the others in a dynamic sense).

The majority of the sections end with a set of numerical problems. These problems are especially important for the use of this textbook. Our intention was to stimulate the reader to use actively the results presented in the preceeding section and thus increase his knowledge through an independent activity. For this purpose we composed a number of exercises of several types which can be conditionally divided into three groups:

a) The purpose of the first group of problems is to illustrate the particular subject that has been described in the preceeding section. Most often they are numerical exercises and for their solving the reader should use the presented results. These problems can be solved analytically (by hand) and they are usually similar to the worked-out examples, but for another robotic structure. In this way the reader independently learns how the presented theoretical statements can be applied on different robot types, and thus gain experience in dealing with different structures of robotic mechanisms, which to a great extent, affect the control synthesis. In addition, we wanted to emphasize the importance of making a proper choice of actuators in the control synthesis. As in the text itself, we exclusively dealt with DC electro-motors, it is left to the reader to apply, by solving the exercises, the same approaches, to hydraulic actuators.

b) The aim of the second group of exercises is to provoke the reader to think about the approaches presented and find, in an independent way, the explanations of certain statements which have been given in the text without proving or detailed explanation. Often, these problems illustrate the given statements (e.g. the statement on the complexity of implementation of the presented control law is illustrated by the exercises in which the user has to determine the number of numerical operations needed to calculate the given control, as well as to estimate the "microprocessor architecture" needed to implement the control law, and the like).

c) The problems of the third group (marked with the asterisk) assume the reader has a sound knowledge of the related scientific fields. These problems imply writing the programmes to synthesize or implement a control law using a microcomputer. Some problems of this group assume the previous knowledge of systems theory, and their aim is to encourage the reader to delve more deeply into the literature and thus prepare himself for independent work. For such a reader, solving these problems should be compulsory.

We think the thus structured textbook will enable the reader to gain a sufficient knowledge for his further work on the problems of control of robotic systems and for implementation of theoretical approaches into practice. We want the reader to develop an engineer's approach to the subject and to direct him to use computer approach to the synthesis of robot control, as this approach enables efficient linking of mathematical models and the practical requirements to be realized by current robots. How well we have succeeded it remains to be judged on the basis of the use of this book as a textbook in teaching practice as well as in the research-and-development units for applied robotics.

Authors express their thanks to Mr. Dj. Leković B.Sc., for his help in preparing an educational version of the software for the synthesis of control of manipulation robots. Also, they are indebted to Miss. V. Ćosić for her careful and excelent typing of the whole text. Finally, authors extend their thanks to Prof. L. Bjelica for his high professional contribution to the book translation into English.

March 1989   A u t h o r s
B e o g r a d

Contents

Chapter 1
Concepts of Manipulation Robot Control

1.1 Introduction

The tasks that are nowdays assigned to robots are becoming more varied
and more complex. More and more often robotic units are becoming parts
of flexible technological cells, lines and intelligent technological
systems. In view of these facts, the organization of robot control
should be based upon the principle of control hierarchy.

Hierarchical levels in control systems have appeared in modern development of the control techniques to meet requirements of more complex
functioning of some technical systems and their control systems. Robotic
systems often posses a certain degree of adaptability with respect to
both changed parameters of the workpiece and the changing environment
in which the processes take place.

The control systems of manipulation robots may be of diverse nature, as
can be seen from their classification presented in Table 1.1.

This introductory discussion shall give a more detailed description
only of those automatic systems of robotic control that are of practical importance for industrial application. Reference [1] was served in
some hand as the conceptual basis for writing Sections 1.1 and 1.2 of
this chapter. More informations about all types of manipulation robots
and classification of robots and their control systems, in a broader
sense more may be found in several books of which here will be mentioned only two [2, 3]. To give some basic information about the "non-in-
dustrial" robots, in this chapter we shall present briefly the biotechnical and interactive robotic manipulators.

Biotechnical robots are called the systems requiring permanent involvement of an operator to control the process of manipulator motion.
Unlike the automatic systems, biotechnical systems do not possess the
ability of generating autonomously a desired motion or attaining a certain goal. A manipulation system moves only when the operator acts upon
it. The simplest kind of these systems is the system of direct control
- by pressing a push-button or a lever the operator effectuates parti-

cular degrees of freedom. By releasing the push-button the motion is stopped at the desired moment. The positioning accuracy in this case in not high because of the lack of a position feedback. In a strict sense these systems cannot be called robots but manipulators with commanding control.

Another kind of biotechnical controlling systems are various *master--slave systems*. The conventional scheme of a copying system contains an assigning device which is in a kinematical sense quite similar to the robot's arm. The sensors placed at the joints of this device supply signals to the servo systems of the corresponding manipulator joint. In some cases, the operator's hand is part of the assigning device. However, this device is usually supplied with a joystick by means of which the operator controls the whole system and brings its tip to a desired point in space. The robot end-effector is automatically brought up to the corresponding point in the workspace.

T Y P E	V A R I A N T
I AUTOMATIC	I.1. Programmable (I generation) I.2. Adaptive (II generation) I.3. Intelligent (robots with elements of artificial intelligence) (III generation)
II BIOTECHNICAL	II.1. Direct controlled (control of individual degrees of freedom) II.2. Master-slave (of one and two-side action) II.3. Semi-automatic (joystick and computer)
III INTERACTIVE	III.1. Automated (combination of automatic and biotechnical regimes) III.2. Supervised (automatic with functional commands of goal) III.3. Dialogue (higher forms of interaction with operator)

Table 1.1. Classification of manipulation robots

Interactive control systems diminish the permanent operator's involvement in the control process of biotechnical systems. The primary interactive systems are the automated systems in which operations are partly automated and partly controlled by the operator. Another kind of interactive systems uniting the human role and the automatic regime is the system of *supervisory control*. On the basis of the information from the monitor screen about the situation in the workspace the operator sends (through a control computer) the commands to the robot, e.g. to bring the gripper (tool) to a certain position. The robot carries out the operation in automatic regime. When the operation is over, the operator instructs the robot (by the aid of the computer) to carry out the next operation, and the procedure is repeated.

1.2 Automatic Manipulation Robots

Basically, four hierarchical levels of the automatic robot control can be established, each of them involving a computer processing of information. On the basis of the results obtained, the computer forms the appropriate control signals. These control signals are then transferred from the upper to the lower control levels until they reach the realization of motion (executive system).

The highest level of the control system of a manipulation robot contains elements of artificial intelligence. We think here about those elements of artificial intellect that enable processing of the sensory information (visual, tactile, proximity, etc.) along with recognition the situation in which the robot's arms are acting, and modelling the scene in one of the appropriate forms. The information are compared to those stored in the memory as part of a global plan of the robot task, after which this (highest) level adopts the solutions that are necessary for execution of the required operation in the recognized situation. At that, the previous working experience of the robot is automatically taken into account, i.e. the robot possesses certain self-teaching abilities.

The highest control level receives the relevant information from all the lower levels and from various sensors controlling the motion of robot's arms and the state of the workspace. In this way the feedback loops are established that form the many-fold contures of a complete control system.

The next, lower level, is the *strategic level* at which the global task is divided in accordance with the solution generated at the upper level, into the elementary operations. The notion of an elementary operation is not strictly defined as it depends on the particular task and system. For example, elementary operations may be as follows: to find an object of given characteristics, to pick it up, to bring it to a point of given coordinates, or to bring it into the contact with another defined object or tool, to put the object down, to fix it, or to subject it to certain processing, etc. An elementary operation may be even simpler (more elementary). For example, the object searching can be split into several more elementary operations of the type: to reach a certain point in the workspace, to carry out searching accompanied by the appropriate surveying or measuring, to determine the smallest cross-sectional dimension of the object and to orient appropriately the manipulator gripper. The operation of transferring the object (if the mode of transfer or its trajectory are important too) can also be divided into several more elementary operations, for example: to lift the object to a predetermined height, to carry it along a circular trajectory, to place it on a surface of defined slope and with the defined spatial orientation of the gripper. The operation of joinning two objects can be split as follows: to put the objects on pins, to take a set of nuts, to fix the nuts according to a certain sequence, etc. The last operation can be divided into several even more elementary operations, e.g. after establishing the sequence of screwing, or after ascribing a slow screwing of all the nuts in the begining, carry out the final tightening according to a certain sequence.

At the strategic level, the degree of splitting the task obtained from the higher level determines the working algorithm of the strategic level. The modes of automatic planning at strategic level satisfy certain criteria of speed, quality and accuracy of execution of the task as a whole. However, apart from the information from the higher level, the strategic level uses also the feedbacks to get the information about all lower levels, as well as of the executive level and the workspace.

At the strategic level are most frequently planned the elementary actions to be realized by the robot's gripper (e.g. determination of the position in the workspace the gripper should be brought in order to pick the object, determination of the trajectory to be followed by the gripper carrying the object, etc.). The coordinates of the robot's gripper with respect to the absolute coordinate frame attached to the robot's base, are termed the *external coordinates* (a precise definition will be given in Chapter 2). Therefore, at the strategic level are planned the trajectories of external coordinates of the robot. However, the robot's motion is realized *via* the movements of its particular joints; most often, each robot's joint is powered by an appropriate actuator. Therefore, to realize the gripper trajectories planned at the strategic level, it is necessary to determine how the robot joints are to be driven.

At the lower control level, *tactical level*, the external gripper coordinates are mapped to the robot's joints coordinates (the so-called *internal coordinates* of the robot), i.e. the required motion is distributed to the particular degrees of freedom (d.o.f) (joints) of the robot. In other words, at the tactical level is determined the motion of each manipulator d.o.f. in such a way that their overall motion realizes all the elementary operations defined at the strategic level.

This would be the basic function of the tactical level. However, as already pointed out, the elementary operations can be defined in a more general way, e.g. to find the object of given characteristics (with no detailed searching and the accompanying surveying), or, to transfer an object to a certain point in space with no prescribing the trajectories of motion. In this case, a detailed task splitting is carried out at the tactical level.

The strategic level can be avoided provided the elements of artificial intelligence give the solution with a sufficiently deatiled sequence of operations. In such a case, the highest level renders the signals directly to the tactical level. In practice, depending on the nature of the task, the functions of levels are generally intertwined. Thus, the hierarchy should be understood rather in a conditional sense.

Like the higher levels, the tactical level too makes use of not only the signals supplied by the higher levels, but also all the necessary feedback information of the lower (executive) level, concerning the executive mechanisms and the workspace.

Finally, when the necessary movements of particular robot joints are determined at the tactical level, they are transferred to the lowest, *executive level*, for realization. The task of the executive level is to realize the given motion of joints in order to perform the desired functional movements of the robot mechanism, defined at the strategic level. In this way the executive level realizes all operations required (ordered) by the higher levels. The executive level realizes the prescribed movements of joints on the basis of the information supplied by sensors about the instantaneous positions, velocities,

accelerations, moments (forces) at the robot joints, i.e., at this level are realized the feedbacks with respect the position, velocity and
forces at joints. To achieve a high precision in performing the desired movements of joints, and thus to ensure an adequate realization
of the task given by the higher levels, the executive level should take
care of the dynamic characteristics of the robot, i.e. about the interaction of the movements of particular robot joints. From the point of
view of the executive control level, a robot is a complex (multidimen-
sional) system having more inputs and outputs, the system in which
generally exist strong interactions between particular d.o.f. Because
of that, the synthesis of control laws which would guarantee a precise
realization of the motion defined at the tactical level is a very complex task. In this book we shall deal with the problems which are primarily concerned with the synthesis of the executive control level. We
shall consider different control laws ensuring at the executive level,
the realization of operations prescribed by the higher control levels.

The hierarchical control of automatic robots is schematically presented in Fig. 1.1.

It should be noted that a sub-level is usually inserted between the
strategic and the tactical level. This is the sub-level of the *robotic
language*. Nowadays, the control and programming of robots is carried
out most frequently through specialized robotic languages. In the robotic language are given all the elementary operations to be performed
by the robot. The introduction of robotic languages is dictated by the
need of simplifying the procedure of giving instructions, which in itself may be a complex task. Robotic languages are also included into
the robot controllers. Because of that, the strategic control level is
often connected to the robot through the language, i.e. this level
should give the robot instructions (desired operations) in the form of
a predetermined set of instructions in the given robotic language. However, the sub-level of robotic language has also the task of checking
the consistency of the given elementary operations. In principle, this
sub-level can be avoided, i.e. it is possible to couple directly the
strategic level to the tactical control level. In such a case, however,
it is necessary to transfer some of the functions of this sub-level to
the strategic level.

Since the functions of the highest and strategic level are extremely
complex, they often cannot be realized in the course of robots motion
and task execution (on-line), but should be carried out before the

robot begin the work (off-line regime), as is indicated in Fig. 1.1.
This means that the highest control level defines the task, and the strategic control level, on the basis of the sensors' information, plans
particular robot operations and, by means of the robotic language,
gives the appropriate instructions. When the instruction on the desired motion is given (in the robotic language) to the tactical level,
the task itself is then realized in on-line regime. However, such a
design does not allow adaptation of the robot functioning to the changed conditions during the task execution. Because of that it is presently endeveured to realize the strategic level in on-line regime too,
i.e. to plan robot's operations during the task execution itself and in
accordance with the instantaneous situation in the work space. Since even
very powerful computers are hardly able to achieve a sufficiently fast

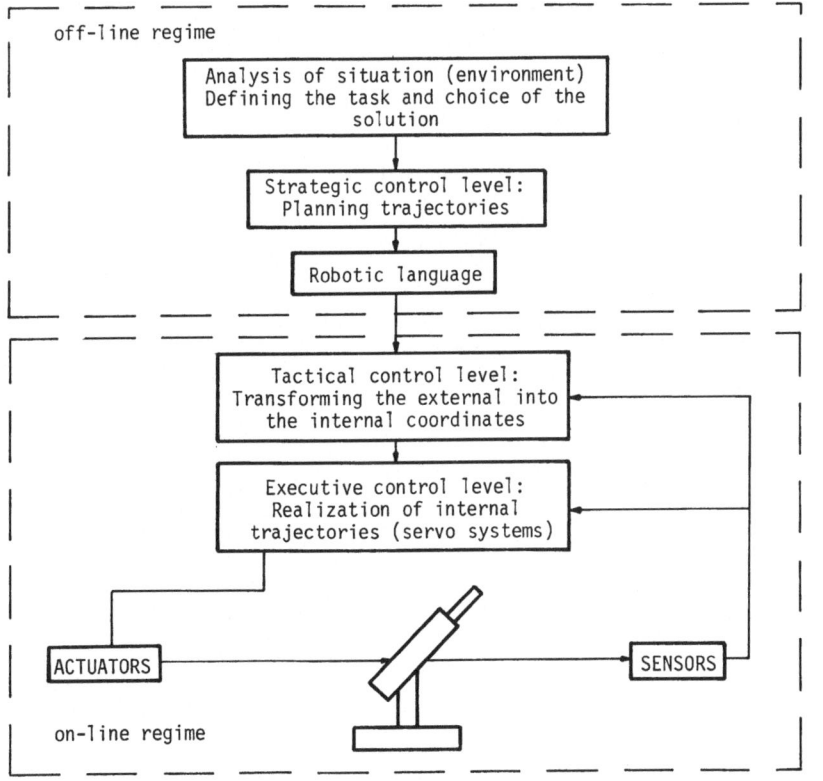

Fig. 1.1. Schematic diagram of hierarchical control of robotic systems

8

planning of elementary robot operations, only some simpler functions of
the strategic level can be realized in on-line regime. Because of that
the more complex operations are defined in the robotic language, whereas
the strategic level on the basis of the sensors information about the
actual situation in the workspace plans the elementary movements (e.g.
the motion of the robot gripper to avoid collision with some movable,
i.e. changeable obstacles in the work-space). The corresponding hierarchy of robot control is presented in Fig. 1.2.

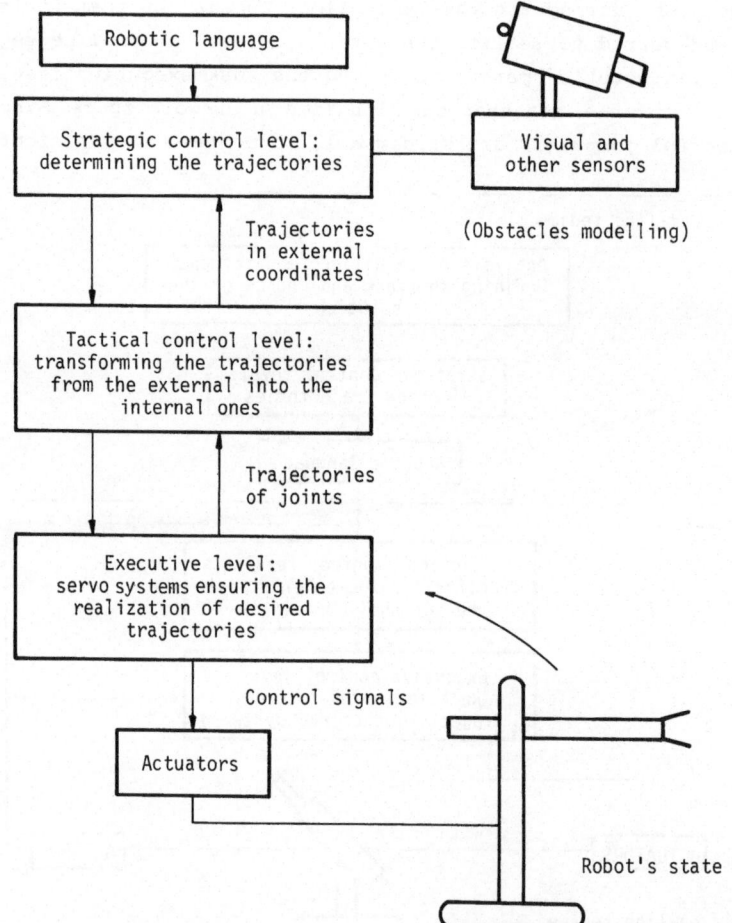

Fig. 1.2. Hierarchical structure of the robot control

It should be also noticed that the programmable robots (first-generation robots) or the automatically programmable manipulators have no
higher control levels, but only the tactical and executive level.

Adaptive robots (robots of second generation) too, have no the highest
control level, i.e. the level of artificial intelligence; they possess
the other three control levels.

With the adaptive robots, the strategic level receives the task from
the operator in the robotic language, as shown in Fig. 1.2., and the
robot controller plans automatically the gripper's trajectories. For
the first-generation robots, the operator has to prescribe the trajectories, i.e. positions of the gripper in a direct way and determine the
trajectories (for avoiding the obstacles, picking up and transferring
the object) through the appropriate programming language, or through a
teaching box (see Sect. 1.5).

1.3 Classification of Automatic Robots

There are four types of automatic manipulation robots: with fixed
(rigid) programmes, programmable, adaptive, and "intelligent".

The term "generation" is often used instead of "type". As manipulators
with fixed programmes in fact are not, robots, they can be considered
as the "zero" ("pre-robotic") generation. Hence, the programmable robots are considered as the first, the adaptive as the second, and the
intelligent ones as the third generation.

Manipulators with fixed programmes have a control system which is not
programmable. These are simply mechanical arms. They are firmly linked
to the technological equipment, subdueing themselves to a certain technological process as a whole. Their application is particularly characteristic for substituting manual work in mass production, e.g. on assembly lines of watch mechanisms, and similar. However, we are not
going to consider the problems concerning design and control of these
manipulators.

Programmable robots (first generation robots) have the controlled drives at all joints and their control system is easily adapted to various manual operations. However, after each adjustment these robots repeat one and the same fixed programme under strictly defined conditions and with fixed arrangement of objects. The majority of contempora-
ry industrial robots is of such type and they are applied for performing some auxiliary operations in pressing, welding, casting, machine
tools servising and similar. These robots demand technological arrangement in the work environment and position of parts.

Adaptive robots (second generation robots) are such robots that can, to a higher or less degree, orient themselves independently in the environment which is not fully determined, and to which they can adapt. They are equiped with sensors reacting to the situation and with an information data processing system aimed at generating adaptive data signals, i.e. flexible changes in the manipulator motion programme according to the real situation. In their modern versions, compact microprocessor systems are widely used. Adaptive industrial robots are needed in all cases when it is difficult to ensure a strictly defined situation, when avoiding obstacles, working with parts on conveyers in assembly operations, in arc welding, painting, applying protective layers, and other operations.

Intelligent robots (third generation robots) are equiped with a more varied range of sensors with the microcomputer processing of information, recognition of situations, automatic generation of the solutions for further actions by the robot itself, aimed at performing the necessary technological operations in an undetermined environment. These robots possess elements of artificial intelligence.

Transducers (sensors) are used with the robots of second and third generation. With the second generation robots, the force transducers, tactile, proximity (ultrasonic) and similar sensors can be used. The third generation robots are characterized by the presence of a complex of sensory devices, including technical vision, which, together with the advanced microcomputer data processing, forms an artificial intellect itself, i.e. the behaviour of the robot is more self-contained, and to a certain degree corresponds to the rational human behaviour in the process of working activity. Besides, the complex of sensory devices may include the equipment for controlling quality of products and properties of the environment, in case this is demanded by the automatic regulation of the working regime.

1.4 Characteristic Tasks and Applications of Robots in Industry

There is hardly any industrial production in which application of industrial robots has not been, at least, attempted. Robots are increasingly used in the following operations: arc welding, spot welding, pressure casting, spray painting, tool machine servicing, material

handling and paletizing, treatment of castings, applying glue and seal
layers, assembling, etc.

It is estimated that presently in the U.S.A., for example, 50-60% of
robotic technology belongs to the car industry, mainly the jobs of welding and spray painting. A similar situation is also in the West Euro-
pe. In Japan, in the car industry is engaged below 30% of the inbuilt
robotic technology, as robots are prevailingly used in the electronic
and similar industries. A forcast for the future is a sharp increase in
the number of robots engaged in the assembly and production control
jobs. This prediction particularly holds for the electronic, car and
some other industries.

To get an insight into some characteristic applications of industrial
robots we shall briefly review some of pertinent technological processes.

1. Pressure casting

This procedure of producing high-quality parts from molten non-ferrous
metals and alloys, consists in injecting the necessary amount of metal
melt into a special mould, usually made of two steel parts.

Generally, these are two basic procedures of pressure cassting: casting
from a hot chamber and cold-chamber casting. The former procedure is
faster since it saves the time needed to transport the melt from the
melting pot to the mould. Hence, this more recent casting method is
more interesting for robotization.

However, the procedure of pressure casting involving no human may have
certain drawbacks. As first, it is the need for manufacturing the
moulds of better quality in order to minimize the possibility of ad-
hesion of castings.

For economy reasons, one industrial robot and one cooling equipment
usually serve two horizontal pressure casting machines. The Japanese
robotic industry has produced several types of simple industrial robots
that are directly installed onto the casting machines. Of the several
millions of working hours of the robotized machines employed in pressure casting in the U.S.A., less than 3% has been lost because of the
mis-functioning of the robots.

2. *Spot welding*

The technique of electric spot welding is presently widely used in the
technology of joining parts of metal sheets, especially in car industry for connecting the body parts, in the industry of electrical household
hold appliences, etc.

The application of industrial robots for this purpose began in 1969,
when General Motors introduced a line of 26 UNIMATE robots to do the
job of car body welding. Only a year after, a line of robots was installed for the same purpose in the Mercedes factory of Daimler Benz.
Today, more than 4000 robots is engaged in the jobs of spot welding in
car industry which has reamined the main field of this application.
The highest efficiency that has been achieved on such a line is about
100 car bodies per hour. Some of car manufacturers, e.g. Volkswagen
and Renault, have developed special industrial robots only for this
purpose. The most modern lines for the car body production, including
their painting, are ROBOGATE with Fiat, and VOLVO in Geteburg, with a
capacity up to 60 cars an hour.

2. *Arc welding*

The proces of arc welding is especially unhealthy to workers because of
the presence of a strong and harmful ultraviolet radiation and evolving
smokes that are rather poisonous. This is a typical example of a job
where the replacement of man by robot represents a genuine humanization of labour. To get a welding seam of a high quality, it is necessary
to ensure not only a precise positioning of the tool tip along the
seam, but also an appropriate speed. To ensure the weld gun follows
the seam and in each time instant has the right position with respect
to it, 5 active d.o.f. at least, are needed. In order to obtain a good
sem it is necessary to use also the external sensors which should ensure good welding parameters. The advanced robots for arc welding possess an increasing adaptability, i.e. the control systems of higher
performances. In the U.S.A. there are even today several hundred thou-
sand jobs where arc welding is done by hand. Robots are increasingly
used in a similar technological process of oxyacetylene cutting.

4. Forging

The present-day use of industrial robots for this technological operation is relatively modest. A very interesting example of such application is the production of punched links chains. To overcome the limitations of robots application for forging operations caused by diffi-
culties in controlling quality of forgings, the attempts have been
recently made to introduce suitable sensory devices into the robot
control system. Presently, use is widely made of photoelectric and
thermal sensors whose task is to ensure the continuity of the technological process itself, and increase its reliability and safety. These
sensors are most often in-built in the appropriate place of the robot
gripper.

5. Spray painting

The problem of spraying a paint from its solution to form a layer which
will dry fast and thus prevent the occurrence of a production bottle-
-neck due to the shortage of man-power, is especially present in car
industry. When the parts of more complicated shape are involved, as for
example, the car bodies of different type (processed on the same production line), the use of automatic cabins or tunnels would be very
complicated and expensive. Hence, the use of industrial robots with
their high capacity of adapting themselves to different kind of products, is the logical solution to the problem. The world leader in this
field is the Norvegian company TRALLFA Niles Underhaug; the number of
specialized robotic units installed by this company is higher than in
any other car company. Since spray painting is an attractive area of the
applied industrial robotics, a number of companies, apart from TRALLFA,
is engaged in developing robots for this purpose. In addition, the leading car factories and other similar companies are involved in the
projects of developing the fully-automated and robotized lines for
spray painting.

6. Application of robots for tool machines servicing

This is one of the most important application fields of industrial robots. They are mainly employed with those universal tool machines that
are used for production of larger series, but which are not yet large
enough to justify the investment into buying fully automated machines
for producing the same parts. However, the application of industrial
robots in manufacturing of smaller series products can be justified by

the possibility of quality standardization and increased capacity of the production. In such a case use is made of the properties of high flexibility of the robotic control system. Thus different production programmes covering a wide assortment of products, can be easily stored and used when needed.

In recent years, the industrial robots for small-series production have been more often integrated into the groups of numerically controlled (NC) machines, making thus a *flexible production cell* or, wider, a *flexible technological line*.

7. Application of robots in thermal treatment processes

Thermal treatment is an almost unavoidable phase in the modern technology of metallic products. It includes hardening, improving, tempering, browning, and all other treatment procedures involving the exposure of the work-piece either to an abrupt change of temperature or keeping it at a fixed temperature for a prolonged time. The application of industrial robots in these operations is expediential and is already today rather widespread. However, a serious problem arises as how to protect the robot's gripper from the frequent action of high temperatures of the work-piece. This is achieved either by using special materials, or by cooling the gripper parts with water or some other cooling agent.

8. Application of robots in glass industry

Glass making, and especially production of sheet glass, is a technological process which is rather hazardous to workman. The fact that glass in some production stages is at very high temperature makes its handling very difficult. In the case of the sheet shattering, small sharp pieces of glass scattered around can cause severe, and even mortal, injuries to people in the working environment. In the production of cathode tubes (TV screens), the products are heavy, hot and very expensive. Thus, apart from a genuine humanization of the labour, the application of robots can substantially decrease wast due to breakage.

Recently, industrial robots have found some unconventional applications. Thus, due to their high accuracy in positioning and especially due to the accuracy of the robot arm position sensors (encoders, resolvers, measuring laths) robots can be used instead of the so-called "measuring machines", which are the extremely expensive, motor-driven rectangular measuring systems of large dimensions, used to control large-size products (aircrafts, ships, vehicles, etc.). Fig. 1.3. shows a sketch of such a machine for measuring the airplane body dimensions by the aid of two measuring robots which are "fingering" the object at certain characteristic points; the measurement data are converted into dimensions of the working object in its own coordinate frame and stored in the memory of the control system and finally protocolled in a prin-ted form. Another, quite different, unconventional use of industrial robots is illustrated in Fig. 1.4: an industrial robot ("PUMA" type) mounted onto a movable robot ("ROBOCAR") with the purpose of servicing several working positions in the so-called "clean" rooms for electro-nics and other highly sofisticated technologies.

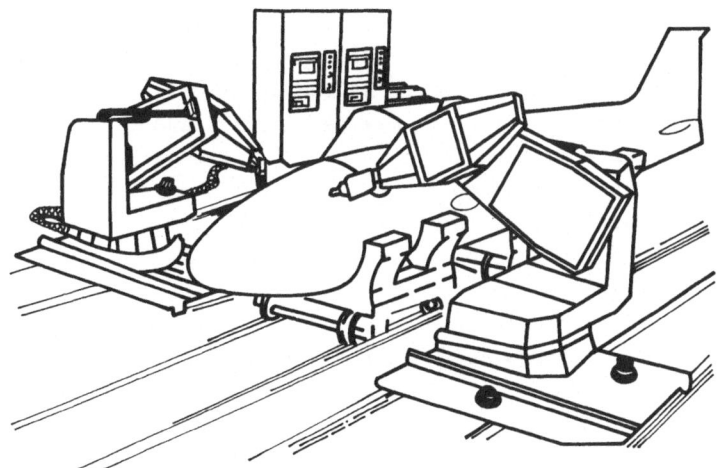

Fig. 1.3. A measuring robot

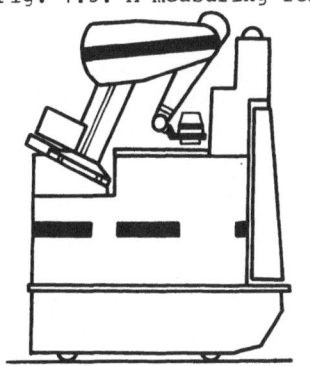

Fig. 1.4. A mobile robot with manipulator

1.5 Implementation of Control of Manipulation Robots

There are a number of factors which should be considered when the robot control synthesis is concerned, viz. the nature of tasks to be carried out, structural and dynamic characteristics of the robot, the equipment available for control realization (sensors and microcomputers), etc. In the preceding sections we presented a variety of tasks of different complexity that can be carried out by robots. The robot design, the choice of its geometrical characteristics (number and type of joints, mutual position of joint axes, link length, etc), and the choice of appropriate actuators are all determined by the class of designated task. The class of the task determines to a great measure the complexity of the robot control system, i.e. the choice and solution of the control. For example, if the robot is to perform simple operations (such as the transfer of an object from one place to another in an obstacle-free work-space and with the object position prescribed in advance) a simple robotic structure may be adopted, and the control system reduced to the executive level only. Thus, the desired robot positions can be given directly to the internal robot's angles, i.e. the positions to be attained by the particular robot joints in order to bring the robot to the desired position are prescribed in a direct way. However, in the case of more complex tasks (e.g. the transfer of work-pieces in the work-space with obstacles, etc) the robot control should be hierarchical, as it has been described above.

On the other hand, the chosen robotic structure and its dynamic characteristics determine to a great extent the complexity of laws of the robot control. In the chapters to follow we shall see that in case when the realization of a precise positioning of a robot and/or tracking of the desired gripper trajectories in the work-space is concerned, it is necessary for some robotic structures to take into account the dynamics of the manipulation robot, which substantially complicates the control system of the robot.

However, in the synthesis of the robot control special attention should be paid to the possibilities of implementation of a chosen control law. Obviously, the more complex control law is chosen, the more complex will be its implementation. When choosing a suitable robot control one should adopt the one which is as simple as possible from the viewpoint of its implementation, but which will, on the other hand, satisfy the given requirements. However, this book is not specially concerned with

the problems of robot control implementation [3]. We shall only briefly
point out some of the aspects of realization of robotic control units
which should be taken into account when the choice of appropriate control laws is concerned.

As with other systems, the robot control may be realized by *analogue*,
hybride or *microprocessor* technique. The analogue technique of robot
control is characteristic of the manipulators of the "zeroth" generation whose control system is reduced to the executive level only, i.e.
to the programming units controlling mechanical arms. Such a control is
practically abandoned in the present-day robots, so that we can say
that a microprocessor is an obligatory part of any robot control system. The use of microcomputers has enabled an easy changing of the task
to be performed (the so-called, *reprogramming* of robots), and this is
an important advantage of robots over classical manipulation mechanisms.
In order to be able to perform various complex tasks mentioned above,
the robot control systems should comprise both strategic and tactical
level. These control levels can be practically realized only by the aid
of computers.

The advanced robot control system should enable communication with
other computers, with the operator (through both the terminal and the
robot teaching unit), with the sensors in the work space, and with the
robot itself (Fig. 1.5).

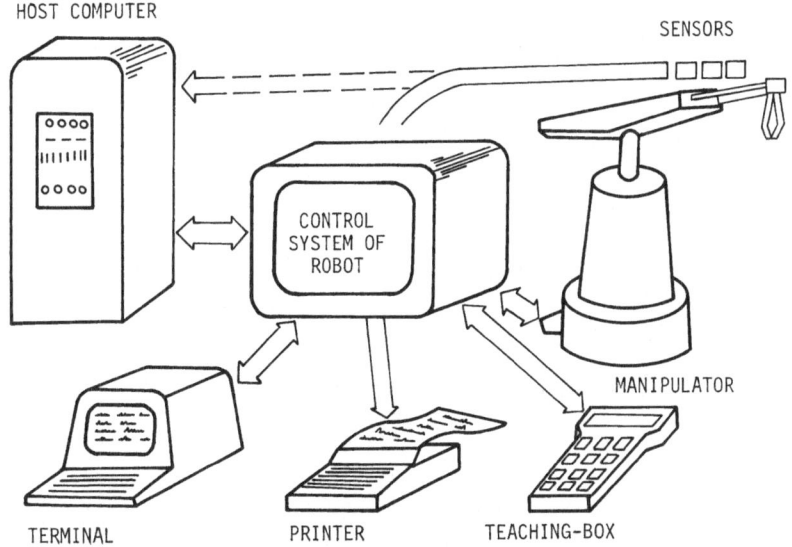

Fig. 1.5. Robot control system

The communication with another computer (host computer) should ensure the coordination and synchronization of the robot's work with its technological surroundings (other robots, NC machines, moving tracks, etc). Such a communication is of special importance for incorporating robots into the *flexible technological systems* in which one central computer is controlling several other computers and NC machines, conveyers, all being involved in a flexible technological cell or line. In some cases it is necessary to realize the communication between the robot control system and another computer for processing sensors data (e.g. signals obtained from the camera), since the information supplied by the sensors are necessary for solving the task at the strategic control level.

The communication between the robot control unit and the user is realized either through a terminal or a robot teching box. In communicating through a terminal the operator uses various high-level languages to programme the task the robot is to carry out (see Sect. 1.2). An easier way of instructing the robot is by using a *teaching box*. This unit enables an easy programming of simple tasks, and by combining them (using a robotic language) it is possible to handle more complex tasks. The teaching box usually has a certain number of keys for giving the robot proper instructions, e.g. to move the gripper to the desired direction, to move particular robot joints, etc. Often, the task assignement can be realized by means of the so-called "pilot-robot" which has the same kinematic structure as the robot itself and is supplied with the sensors for measuring the joint coordinates, but having no actuators. The operator drives the pilot along the certain paths that should be followed by the robot when executing the given task; the control system read out the sensors data about joints positions and memorizes them. When the pilot-robot motion is recorded, the robot can reproduce the given motion on the basis of the data stored in the memory.

The control system receives the information from the sensors in the work space and they are used in planning the robot trajectories at the strategic control level. If the sensor information are such that no substantial treatment is needed, they are introduced directly to the microcomputer of the robot control unit. On the other hand, if the sensors information needs further processing, this is done, as mentioned above, by means of separate computers. Often, the robot control unit has to follow the sensors information during the robot motion and on the basis of such information "change" both the motion and operation under execution (see Fig. 1.2).

Hence, an advanced robot control system should ensure all the above communications and, on the basis of the assignement received from the operator (through the programming language, teaching box, pilot) or from the higher control level (i.e. central computer), as well as on the basis of the sensory system information, plan the actions to be carried out by the robot. As we have explained above, the motion of the robot gripper is usually planned at the strategic level, i.e. the trajectories to be followed by the gripper are determined at this level and then transmitted to the tactical level.

At the tactical control level, the assigned gripper motion is mapped into the robot's joints trajectories (see Chapter 2). The execution of such mapping generally requires a lot of numerical calculation (additions, square roots, multiplications, calculating of trigonometric functions, etc). In order the robot could precisely realize the desired motion, the mapping of desired positions (and velocities) of the robot gripper into the corresponding positions (and velocities) of the robot joints should be carried out every 15 - 30 [ms]$^{*)}$, i.e. every 15 - 30 [ms] the gripper coordinates on the desired gripper trajectory should be mapped into the joints coordinates. This means that the microcomputer should complete each transformation within 15-30 [ms], i.e. it should be capable to carry out the necessary numerical calculations with a speed which allows the completion of this mapping during this time interval.

When the desired positions and velocities are determined, they are realized at the executive level. The executive control level should permanently generate the signals to the inputs of the actuators whose task is to drive the robot joints into the desired positions, i.e. to effectuate the joints to move with desired velocities and accelerations. The signals at the executive level are generated on the basis of both the trajectories received from the tactical level and the information obtained from the inner sensors (i.e. the robot's sensors supplying the informotion about the actual positions of joints, their velocities and accelerations).

The executive control level can be realized either by the analogue technique or by the aid of microprocessor. In the former case, the robot control unit is provided with a computer for implementation of

*) This is an average "speed" (sampling rate) needed to carry out the mapping at the tactical control level.

higher control levels and with the analogue servo systems with actuators to move the robot joints. Every 15-30 [ms] the tactical level in
the microcomputer calculates all the necessary positions and velocities of robot joints, and, by a D/A converter sends them to the analogu-
es control device for realization (Fig. 1.6). The analogue servo systems realize the prescribed positions (trajectories) of joints on the
basis of the information from the robot sensors. This means that,
starting from the prescribed trajectories (signals from the D/A converters) and signals from the sensors, the analogue controller generates the appropriate signals for the actuators at the robot joints.
However, because of the existence of strong dynamic interactions between the individual d.o.f., mentioned above, the control law at the
executive level has to be very complex for certain robot types and the
tasks imposing rigorous requirements with respect to the accuracy and
speed of their realization. The analogue technology is not suitable
for implementation of complex control laws. In the case the robot control system is realized by the aid of simple servo systems which independently control the individual joint actuators (see Chapters 3 and 4)
these laws are realizable by analogue technique. However, if the control has to include the compensation of complex dynamic interactions,
occurring between the motion of particular joints, the realization of
an analogue control device is much less suitable if compared to the
implementation of microprocessors. On the other hand, the digital
technique has substantial advantages as for the possibility of alte-
ring and adjusting the control laws, as well as from the viewpoint of
maintenance, reliability, and robustness under variable external conditions [4].

For these reasons, microcomputers are increasingly used in the implementation of executive control level (Fig. 1.7). Thus the microcomputer appears as a site for implementation of all control levels. The
microcomputer receives the information about the instantaneous states
of robot joints from the sensors through A/D converters, if the sensors used do not supply the information in a digital form. On the basis of these information and the robot trajectories determined at the
higher levels, the microcomputer computes the appropriate control signals, which are through D/A converters fed into the inputs of the actuators[*]. Calculation of the signals for actuators is carried out

_____

[*] It should be noticed that between the D/A converter and actuators
are inserted the appropriate amplifiers, not shown in Fig. 1.7

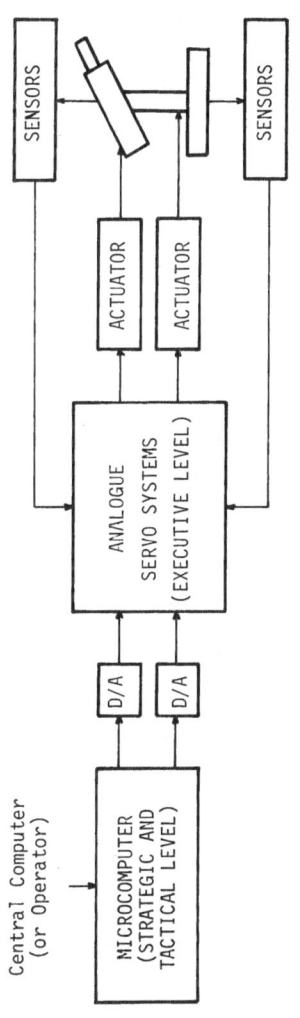

Fig. 1.6. The robot control unit with analogue implementation of the executive control level

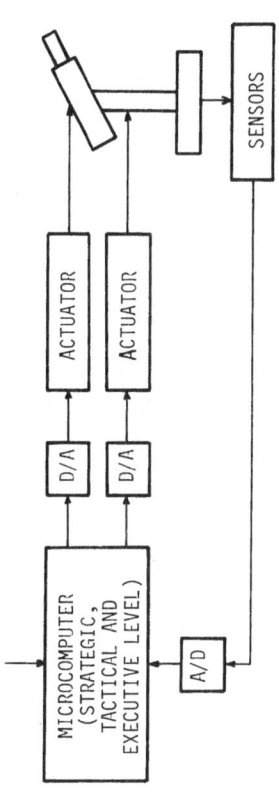

Fig. 1.7. The robot control unit with digital implementation of the executive control level

according to the selected control law and selected parameters, using a suitable programme module. As advanced microcomputers possess the translators for high-level programming languages, the programming on a microcomputer is relatively simple, which enables an easy change of the control law. The designer of the control unit should only change (in high-level language) the program module of "executive level". On the basis of the desired values for coordinates in a given time instant (calculated in the corresponding programme module which implements the tactical level) and the values of the actual coordinates (i.e. angles and velocities) as well as the moments (forces) at the manipulator joints (obtained through an A/D converter), this module calculates the signals to be supplied (through a D/A converter) to the actuators inputs. In this way, one and the same hardware (microcomputer) can be used for the implementation of different control laws. Moreover, such *direct digital control* of robots enables the development of a general purpose controller which would be (after minor alterations, mainly of the software) applicable for robots of different types and structure [5], which is practically unattainable by the analogue technique.

However, the realization of a digital robot control system is associated with certain problems. Namely, microprocessors require some period of time for execution of each mathematical operation. Thus, the control microcomputer needs certain amount of time to compute (on the basis of a given control law) the control signals for the actuators. The time interval between the moment the microcomputer had read the values for coordinates from the sensors (i.e. from the A/D converters) and the moment it computed the signals and sent them over (through D/A converters) to the actuators, is (conditionally) called the *sampling period*[*]. During this interval one and the same values of signals are sent by the microcomputer to the D/A converters, i.e. the actuators. To diminish the influence this microcomputer delay has on the robot's behaviour (i.e. on the precision of realization of the prescribed trajectories), the sampling period should be as short as possible in comparison to the time constants of the robot dynamics. It appeared that the sampling period with the majority of modern robotic systems should be below 10 [ms]. It means that in, let say 5 [ms], the microcomputer should (on the basis of the instantaneous values for robot coordinates and in accordance with the chosen control law) calculate the control signals and send them (through D/A converters) to the actuators.

---

[*] A more precise definition of the sampling period can be found in reference [4].

Different control laws require different number of numerical operations to be executed to compute the appropriate control signals. If more complex control laws are involved, the number of operations to be executed by the microcomputer during the sampling period is normally higher. Certain simpler control laws require about 30-40 numerical operations (additions and multiplications), whereas some others may require several hundred operations per sampling period. The microprocessor must be sufficiently fast to execute the necessary number of operations in 5-10 [ms]. The 8-bit processors can hardly achieve the calculation speed needed for the control of advanced robots[*]. Because of that the 16-bit microprocessors are used in advanced robotics, together with numerical co-processors which enable an accelerated execution of different numerical operations (addition, multiplication, square root, etc). Numerical co-processors ensure the realization of arithmetics with floating point which is necessary for calculating more complex control laws.

However, even the advanced 16-bit microprocessors with numerical co--processors, capable of doing multiplications in 20 to 30 [μs], cannot calculate more complex control laws during the allocated sampling period of 5-10[ms]. Besides, it should be borne in mind that the microprocessor has to realize not only the calculation of control signals according to the chosen law, but also to carry out the appropriate processing of the information obtained from sensors (their scaling, etc) as well as processing of the calculated control signals which are sent to the D/A converter, in addition to different checkouts, constraints, and so on. Apart from the executive level, the microprocessor should also implement the tactical level which includes, as we have already mentioned, the mapping of the external coordinates into the joint robot coordinates, which has to be done every 15-30 [ms]. This mapping may require several hundred numerical operations, so that the microprocessor should also accomplish these calculations in the allocated time. If the control unit includes the strategic control level (or, at least some of its elementary functions that has to be realized on-line, during the robot's motion), then, the microprocessor should ensure the execution of the necessary operations for this level too, and all that during the assigned time interval. The present-day commercial microprocessors cannot perform all the necessary calculations during the sampling period imposed by the robot dynamics (less than 10[ms]).

---

[*] The 8-bit processors are even today used with manipulators for some simple tasks; the executive level is realized either by the analogue technique or by the aid of a number of such microprocessors in parallel (see below).

For this reason, use is made of several processors working in a *paral-lel*, but which can exchange data between each other through a common memory. Usually, one of the processors is dedicated to the strategic control level which communicates with the user and the higher control level (in the host computer). The tactical control level is implemented by second processor, while the third one is dedicated to the executive control level. Thus, the robot control is realized by three processors. However, this number of processors is often insufficient. Hence, more complex control laws have to be realized at the executive level by means of several processors. Similarly, for the realization of complex algorithms and treatment of sensory data at the strategic control level, several processor should also be employed. The use of a large number of processors in parallel may substantially complicate the realization of the control unit and, on the other hand, affect the reliability of the work of the whole system. Therefore, the present--day 16-bit processors (together with numerical co-processors) are not yet capable of implementing some more complex algorithms at the strategic control level; most often one ought to be satisfied with some simpler solutions that enable performing certain specific tasks. Further advances in microprocessor technique and the enhancement of calculation speed will bring the progress in general problem solving at the strategic level.

On the other hand, the control law chosen at the executive level should be *as simple as possible* (requiring the smallest number of operations), but which can satisfy the desired requirements, in order to avoid the application of a large number of processors in parallel. However, implementation of many of the complex control laws requiring the calculation of robot's dynamic forces, demands application of several 16-bit processors in parallel. Such a realization is rather complicated, so that such control laws are not used with the majority of commerciall available robots. Therefore, one of the important criteria in choosing the robots control laws, imposed by the control realization, is the requirement for the lowest possible number of numerical operations to be carried out in calculating control signals during the sampling period.

In the realization of robot control unit, the requirements concerning the memory of the microcomputer should be also taken into account. The standard RAM (Random Access Memory) capacity of advanced robotic controllers is between 64 KB (of the 8-bit words) and 1 MB, which depends on the complexity of the algorithm for the control calculation and the amount of data to be stored. It should be pointed out that certain control laws may require a large memory capacity for storing the data calculated (off-line) in advance and which are used during the on-line robot control (see Chapter 4). Those robots which should perform complex motion for a prolonged time demand the storing of a large amount of data about the coordinates that are to be realized. The application of these robots is characterized by the use of peripheral magnetic memories, such as floppy and hard disc. The function of these peripheral memories is to store (either in the appropriate robotic language, or in the form of a set of coordinates which the robot has to realize during its motion) the programmes for various tasks to be assigned to the robot for execution when needed. Thus, various tasks can be programmed in advance, the programmes stored in the peripheral memory units, and used afterwards in the course of realization of the industrial process.

All these factors concerning the implementation of the robot control should be taken into account in the robot control synthesis. It should be mentioned that in the implementation of robot control by microprocessors, special attention should be paid to the accuracy of calculations. The data from the sensory system reaching the microcomputer, either directly in digital form or through an A/D converter, are most often 12 - or 16-bit. As we have already mentioned, calculation of the control is also carried out by 16-bit processors. In this way, a sufficient accuracy can be ensured for the majority of the present'day robots and their applications (i.e. the microcomputer truncation does not significantly affect the robot's behaviour). However, for some specific tasks and some robots, a higher calculation precision may be required. Besides, in the robot control synthesis, care should be taken of the reliability and robustness of the chosen control, i.e., it should be ensured the control is implemented in a reliable manner which is not sensitive to possible disturbances that might occur in the external conditions.

References

[1] Medvedov V.S., Leskov A.G., Juschenko, Control Systems of Manipulation Robots (in Russian), Nauka, Moscow, 1978.

[2] Vukobratović M., Applied Dynamics of Manipulation Robots: Modelling, Analysis and Examples, Springer-Verlag, 1989.

[3] Vukobratović M., (Ed.), Introduction to Robotics, Springer-Verlag, 1989.

[4] Cadzow A.J., Martens H.R., Discrete-Time and Computer Control Systems, Prentice-Hall, Englewood Cliffs, New Jersey, 1970.

[5] Vukobratović M., Stokić D., Kirćanski N., Non-Adaptive and Adaptive Control of Manipulation Robots, Scientific Fundamentals of Robotics 5, Springer-Verlag, 1985.

[6] Popov E.P., Vereschagin F.A., Zenkevich S.L., Manipulation Robots: Dynamics and Algorithms, Series: Scientific Fundamentals of Robotics, (in Russian), "Nauka", Moscow, 1978.

[7] Albus J.S., McCain H.G., Lumia R., NASA/NBS Standard Reference Model for Telerobot Control System Architecture (NASREM), NBS Technical Note 1235, 1987.

[8] Engelberger J.F., Robotics in Practice, Kogan Page, 1980.

Chapter 2
Kinematic Control Level

2.1 Introduction

As pointed out in Section 1.2, the tasks for which robots are applied
in industry and other fields are of very diverse complexity. The more
complex the task a robot has to perform and the more strict the requirements for its performing, the more complex should be the control system of the robot. The complexity of the robot control also depends,
as it will be shown in Chapter 4, on the robot's mechanical structure,
i.e. on the extent and mode the motion of one joint (mechanical degree
of freedom) of the robot influences the other joint. Because of that,
different "types" of control systems appear in practice, which in different ways solve the problems at both tactical and executive level
and enable accomplishment of tasks of different class. As will be shown
below, the "types" of control are most often related to different classes of tasks in robotics, which, on the other hand, have different requirements toward the executive control level. This chapter deals with
the problems concerning the tactical control level, while the subject
of the coming chapters will be the synthesis of control at the executive level.

2.2 Direct and Inverse Kinematic Problem-Determination of Robot Position

The most elementary task that appears in robot control is to bring the
robot to the desired position of the workspace. If we consider the
tasks encountered in robotics (Section 1.4) we realize that one of the
simplest tasks is the transfer of an object from one position to another (Figure 2.1). What is needed here is to bring first the robot
from the initial position (A) to the workpiece, at such a distance
from where the gripper can take hold of it (B). To execute the task,
the gripper has to be oriented in a proper way as to catch the workpiece. The position of the manipulator is, therefore, defined by the
desired position and orientation of the gripper in the workspace and
these are determined by the position and orientation of the workpiece

itself. The next step is to grip the workpiece[*]. After that, the workpiece should be transferred to a new position. This new desired position of the workpiece determines the new position of the gripper, and
thus, requires a new position of the manipulator (C). If the work space is not containing obstacles, and if there are no limitations upon
the mode of transferring the workpiece (e.g. if the orientation of the
workpiece is not important, in contrast to the case when a vessel containing liquid is to be transferred, etc), the manipulator can move
from B to C in an arbitrary way. When the workpiece is brought to position C, the gripper opens, and the workpiece (having acquired the
desired position in the workspace) is released. The robot is then positioned again at the initial position A, or it goes directly to position B.

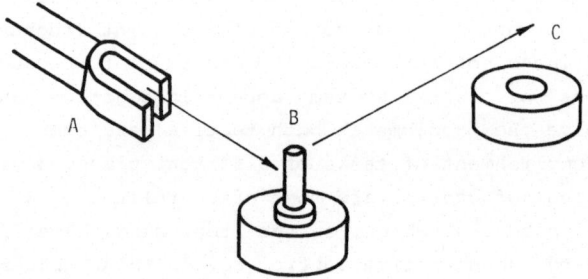

Fig. 2.1. The task of transferring a workpiece
from position B to position C

We have described this simple task to show that the only problem of
practical importance in its execution is the accurate positioning of
the robot and the workpiece which it is carring, i.e. the bringing of the
robot gripper to a desired position and acquiring the adequite orientation in the workspace. Such robot positioning is involved in many
tasks, described in Section 1.4, either as the only task, or as a sub-
task in a complex assignement. For example, spot welding requires a
precise robot positioning in regard the positions needed to carry out
the operation. In parts assemblying, the problem of accurate positioning arises at the state of bringing the workpiece to the site of parts
mateing, though the operation of assemblying is more complicated

---

[*] In a general case, the gripping phase can be complex too (in the
case the shape of the workpiece is not defined in advance), but
these problems are not going to be dealt with here.

(as it will be shown in Chapter 7). Thus, all robots have to ensure
the possibility of versatile positioning, and they differ with respect
to the mode the positioning is programmed and executed, and the accuracy the desired positioning is accomplished[*]. It should be noticed
that mechanical arms with open feedback have also the possibility of
positioning but their capability of reprogramming the desired positions and assigning new positions is very limited, if compared to robots
(as it was explained in Section 1.3).

The primary question arising in robot positioning is, how to assign a
desired position to the robot. In principle, there are two ways of assigning the robot's coordinates:

a) through the so-called *internal coordinates* of the robot (or the
   robot joints coordinates), or

b) through the so-called *external robot's coordinates*.

In the first book of this textbook series [1] we have explained some
basic concepts concerning the kinematics and dynamics of robots, such
as mechanical configuration of the manipulator, link, kinematic pair,
kinematic chain, etc. Internal robot coordinates are defined by the
scalar values describing relative position of the one link with respect to another link of the same kinematic pair. The internal coordinate for a revolute joint is the deflection angle at the joint, whereas, for a translational kinematic pair, it is the linear displacement
along the joint axis. In Fig. 2.2. the internal coordinates are denoted by q^i, and the vector of internal coordinates of a robot having n
simple joints (degrees of freedom) is $q = (q^1, q^2,...,q^n)^T$. (The definitions of a revolute and a linear joint are given in [1]).

The external robot coordinates are the Cartesian coordinates of a terminal robot link (gripper) with respect to the absolute coordinate
frame attached to the manipulator base, or, to any other point in the
workspace, as well as the gripper orientation with respect to the absolute coordinate frame (i.e. the angles between the terminal robot
link and the axes of the absolute coordinate frame). Therefore, the

---

[*] It should be mentioned that the task of robot positioning is the
most elementary one, though, as it will be shown later, the requirements put before modern robots may be much more complicated (to
follow a space trajectory, etc).

vector of external coordinates s consists of the coordinates x_c, y_c, z_c of a point on the gripper (e.g. its inertia centre, or the tip) with respect to the absolute coordinate frame and the angles θ, ϕ, ψ, which the axes of the coordinate frame, attached to the gripper, form with the axes of the absolute coordinate frame (Fig. 2.2).

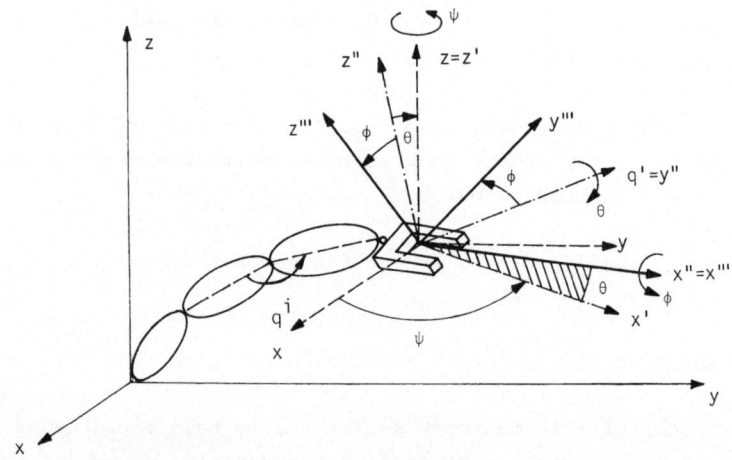

Fig. 2.2. Cartesian coordinates of a manipulator gripper

The yaw angle ψ corresponds to the rotation about axis z of the fixed coordinate frame, the peatch angle θ corresponds to the rotation about the newly formed axis y (after the rotation for angle ψ), whereas the roll angle corresponds to the rotation about the new axis x. Thus, the external coordinates describe the gripper position with respect to the fixed coordinate frame.

The vector of external coordinates s have, in a general case, m coordinates, where m is the number of coordinates needed to describe the gripper position in a particular class of manipulation tasks. Most often, it is accepted that m=6, i.e. the position and orientation of the gripper with respect to a fixed frame are fully described by the vector of external coordinates:

$$s = (x_c, y_c, z_c, \psi, \theta, \phi)^T$$

If a certain class of tasks can be described by a smaller number of external coordinates, such a vector of external coordinates is adopted which does not include these coordinates that are not of interest. For example, if the gripper positioning is only to be considered, the vector of external coordinates is of the form $x = (x_c, y_c, z_c)^T$.

It should be noticed that robot external coordinates can be defined in different ways. For example, the gripper position can be described by cylindrical or spherical coordinates, and the gripper orientation can be described by angles adopted in some other way.

As is known [2], robots are powered by means of actuators of various types. Most often, one manipulator joint is powered by one actuator. The motion of the actuator is transformed into the motion (either, rotation or translation) of the corresponding joint. As will be shown in the next chapter, each actuator is supplied with its own servo system which drives the actuator and brings it to the desired position. Therefore, to each actuator it should be assigned the desired position to be attained. This means that the executive level needs the robot position be defined in terms of internal coordinates. For each joints, it is necessary to define the angle (or displacement) q^{oi} to be attained, and this angle is the input to the corresponding servo system; the servosystem activates the actuator, which drives the joint until the desired joint position q^{oi}, is attained. In this way, the robot is brought to a desired position in the workspace.

If the manipulator position is assigned through the internal coordinates, then the robot positioning is reduced to assigning these coordinates at the executive control level, whose servo systems should realize the given joint positions. Therefore, the tactical control level is now trivial: the joints coordinates, assigned by higher levels (or by the operator), are transferred directly to the executive level. However, the question can be raised on how the joint coordinates q^{oi} are determined. In the task considered above (Fig. 2.1), the desired robot position is defined by the position of the workpiece, i.e. by the gripper position that should be attained, or in other words, by the Cartesian coordinates. One of possible ways of doing it is by teaching the robot: the operator, by means of either a joystick, a control panel, or by guiding the robot gripper manually (the robot is "loosened", so that the operator can easily guide its gripper)[*], or in some other way, brings the robot gripper to the desired position in the workspace; the joint coordinates q^{oi} corresponding to the desired gripper position are stored in the control system, and, when the

---

[*] The "teaching" method has been considered in [2].

robot works in an automatic mode, these coordinates are assigned to
the executive level for realization.

However, such a mode of teaching is not always acceptable. In some cases, it is not possible to realize a manual guiding of the robot because of its weight (size) and type of task (if the robot works in a
workspace inaccessible to the operator). On the other hand, the robot
teaching through the internal coordinates, i.e. by bringing the robot
gripper to a desired position, by assigning the joint coordinates
(either, via a terminal, a joystick, or a teching unit), can be a tedious and time consuming job. Because of that, modern robots should
have the possibility of assigning position in terms of external coordinates. This way of position assigning is incomparably more convenient from the point of view of the operator and the higher control levels (at which the task of planning trajectories is usually solved in
terms of Cartesian coordinates).

However, because the executive level (servo systems) requires that desired position is assigned in terms of internal coordinates, it is necessary to ensure conversion of the external coordinates into the corresponding internal coordinates. As we explained in Chapter 1, this
problem is solved at the tactical control level, whose task is to convert the given values of Cartesian coordinates s^o into the corresponding values of joint coordinates $q^o = (q^{o1}, q^{o2}, \ldots, q^{on})^T$.

It is obvious that there is a correspondence between the internal (q)
and external (s) robot coordinates. To each value of the vector of
internal coordinates q corresponds a value of the vector of external
coordinates s, that is:

$$s = f(q) \qquad\qquad (2.2.1)$$

where f represents the functional transformation of the vector of internal coordinates into the vector of external coordinates.

This functional transforming is called "direct kinematic problem" and
it involves the determination of external coordinates vector s for the
given internal coordinates (vector q). As the relevant vector of external coordinates is generally of dimension m = 6, and the vector of
internal coordinates is of dimension n, expression (2.2.1) represents
the transformation of the vector q of dimension n into the vector s of
dimension m.

Direct kinematic problem can be solved in different ways. There are several procedures enabling the correspondence between internal and external coordinates to be determined for different robot structures. The primary aim in all of these procedures is to develop an algorithm for systematic determination of functional relationship between the internal and external coordinates for an arbitrary robotic structure. As the robot control is presently realized by the aid of computers, a number of algorithms has been developed, enabling an easy calculation of external coordinates as a function of internal coordinates by using computers (microcomputers). The procedures differ in respect to a number of mathematical operations that are needed to calculate the external coordinates for the given values of internal coordinates, in respect of their generality (the applicability to different robotic structures), etc. These procedures have been reviewed in [3, 4]. It is clear that the transformation f represents a nonlinear trigonometric function, and its determination for complex robotic structures, having a large number of joints ($n \geq 6$), requires a large number of mathematical operations. The procedures for solving direct kinematic problem are generally reduced to the determination of relationship between the coordinate frames attached to the particular links. Thus, the relationship between the coordinate frame attached to the terminal robot link (gripper) and the fixed coordinate frame, represents the solution to direct kinematic problem. It should be emphasized that the functional transformation (2.2.1) can always be determined in analytic form, i.e. it is always possible to determine external robot coordinates as a function of internal coordinates in the form of analytic expressions.

However, as we mentioned above, the task is most often given in terms of external coordinates, i.e. the desired robot position is defined by the values of external coordinates. Because of that, one of the major problems in robotics is to determine the values of internal coordinates q^o, corresponding to the given external coordinates s^o. Determination of internal coordinates q^o (which are assigned to the executive level - the servo systems at joints - where they are realized) for the given external coordinates s^o is carried out by *inverse transformation:*

$$q^o = f^{-1}(s^o) \tag{2.2.2}$$

Determination of the inverse function f^{-1} is called "inverse kinematic problem". To solve this problem is incomparably more difficult than to

solve direct kinematic problem. Determination of f^{-1} depends also on
the dimensions of vectors s and q. In relation to this problem, three
cases are possible:

a) m=n, when the number of external coordinates is equal to the number
   of internal coordinates, it is possible to determine a unique q
   corresponding to the given s (under the condition that the manipulator is not in the special, so-called *"singular"* position for which
   several q values correspond to one s)[*];

b) m>n, then, it is not possible (except in some special cases) to determine q satisfying s;

c) m<n, there are more solutions q satisfying the given s; this is the
   case of *redundant* manipulators (for details see ref. [3]).

In a general case, for non-redundant manipulators too, the solution of
inverse problem is not unique, but there is a set of solutions for internal coordinates corresponding to one and the same position and orientation of the gripper.

As the function f, connecting the external and internal coordinates, is
a nonlinear trigonometric function, solving the inverse problem is equivalent to search solution to a set of nonlinear equations. Two general
approaches are known. In the one of them, the vector function $f^{-1}(s^o)$
is obtained in an analytic form for each particular manipulator, while
in the other approach, the solution is sought by one of the known methods of numerical analysis. Both approaches have certain advantages
and disadvantages.

2.2.1. <u>Analytic solution to inverse kinematic problem</u>

As the vector function f in (2.2.1), is a complex nonlinear (trigonometric) function of n variables, the analytic determination of the inverse function f^{-1} is a complex problem which cannot be solved generally, i.e. for an arbitrary robotic structure. For some robotic struc-

---

[*] The problem of singular points will be treated briefly in Section
2.2.2; a detailed consideration of the problem can be found in [3].

tures, however, it is possible to obtain an analytic solution of the inverse kinematic problem, while for some others, this is not possible.

Thus, it has been shown, that for robotic structures having three d.o.f., two intersecting and two parallel joint axes (Fig. 2.3) is possible to obtain an analytic solution to the inverse kinematic problem, i.e. to determine explicitly the joint angles q^i as a function of external co-ordinates. However, this is not always possible, even for the robots with three d.o.f. In [5], it was shown that solving the inverse problem for robots having three d.o.f. can be, in a general case, reduced to finding the zeros of a polynom of the fourth power.

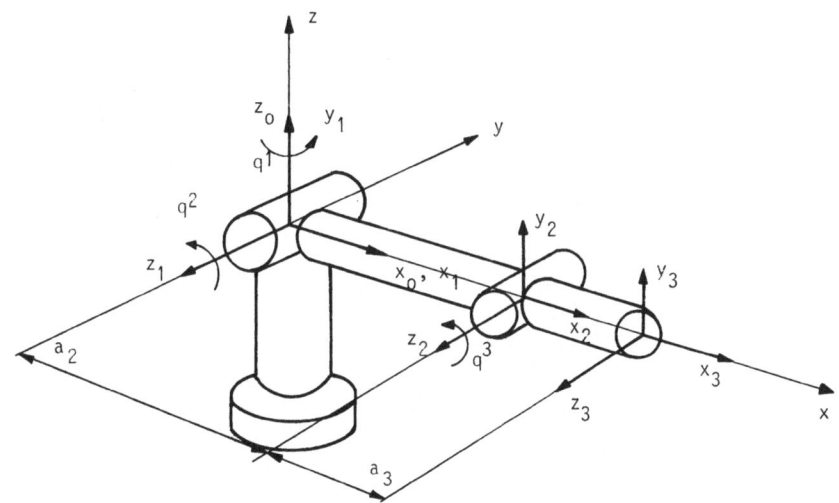

Fig. 2.3. A robot having three joints with two intersecting and two parallel axes (the robot for which is possible to solve the inverse problem)

To obtain an inverse solution in analytic form, for a manipulator having six d.o.f., it is necessary to split the problem into two independent subproblems: the determination of the solution for the first three joints q^1, q^2, q^3, and determination of the remaining three co-ordinates q^4, q^5, and q^6. The first three joints form the so-called *minimal configuration*, and the remaining three joints are usually associated with the robot gripper[*]. An inverse problem can be divided in two subproblems (for the minimal configuration and the gripper), provided the terminal three d.o.f. make a spheric joint, i.e. all

---

[*] It should be noticed that this holds for robots with n=6 d.o.f., so-called non-redundant robots.

three axes intersect at one point (Fig. 2.4), which in other words, means the gripper consists of one link only. In that case a simple calculation enables the position of the tip of the minimal configuration to be determined (on the basis of the given position of the gripper tip and the given orientation of the terminal link - gripper). Having the position of minimal configuration determined, it is possible to find an analytic solution to the inverse kinematic problem for the minimal configuration and thus, determine the internal coordinates q^1, q^2, q^3. Then, for the known values q^1, q^2, q^3, the orientation of the terminal link of the minimal configuration is determined. This enables the determination of internal coordinates q^4, q^5 and q^6 of the spherical gripper joint that satisfy the required gripper orientation.

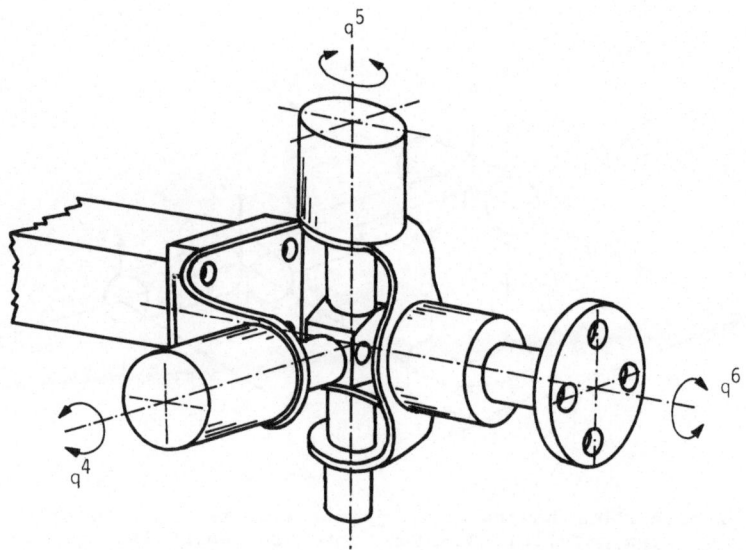

Fig. 2.4. The robot with a spherical joint (wrist)

Certain types of industrial robots possess such mechanical structure for which there is an analytic inverse solution. However, with some robotic structures, as well as with redundant manipulators, it is not possible to determine an analytic solution to the inverse kinematic problem, so that a numerical approach has to be used.

Example 2.2.1. We shall show an example of robotic structure for which the inverse function (2.2.2) can be obtained in an analytic form. For a cilindrical type of robot (Fig. 2.5) with n=4 d.o.f. the relation between the external and internal coordinates (2.1.1) (the vector of external coordinates is chosen as $s = (x_c, y_c, z_c, \phi)^T$, where x_c, y_c and z_c are the coordinates of the gripper gravity centre, and ϕ is the Euler angle shown in Fig. 2.5), is given as

$$x_c = (\ell_3^o + \ell_4^o + q^3)\sin q^1, \qquad y_c = (\ell_3^o + \ell_4^o + q^3)\cos q^1$$

$$z_c = q^2 + \ell_2^o, \qquad\qquad \phi = q^4$$

(2.2.3)

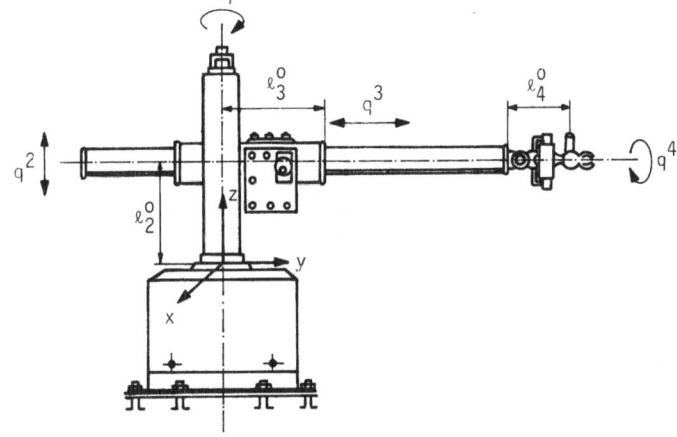

Fig. 2.5. A robot having cylindrical structure (with 4 d.o.f.)

where ℓ_i^o are the lengths of robot links (Fig. 2.5). The inverse function (2.2.2) can be now defined as:

$$q^1 = \text{arctg}(x_c/y_c), \qquad q^2 = z_c - \ell_2^o$$

$$q^3 = \sqrt{x_c^2 + y_c^2} - \ell_4^o - \ell_3^o, \qquad q^4 = \phi$$

(2.2.4)

Therefore, for such a manipulation robot, it is possible to determine the inverse kinematic model in an analytic form, i.e. to determine in an analytic way the coordinates q corresponding to the desired external coordinates s. However, as was explained above, this is not possible to achieve for all robotic structures.

Exercises

2.1. For the "spherical" manipulator having 3 d.o.f., presented in Fig. 2.6., determine the relation between the internal and external coordinates and the inverse kinematic model.

2.2. For the "anthropomorphic" manipulator having 3 d.o.f., presented in Fig. 2.7, determine the relation between the internal and external coordinates. Is it possible to determine the inverse kinematic model in an analytic form?

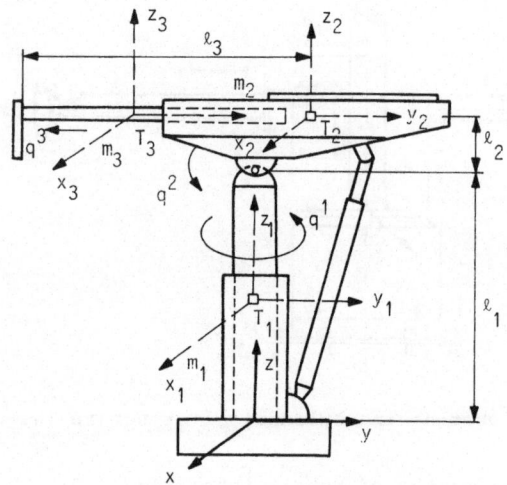

Fig. 2.6. A robot of spherical structure having 3 d.o.f.

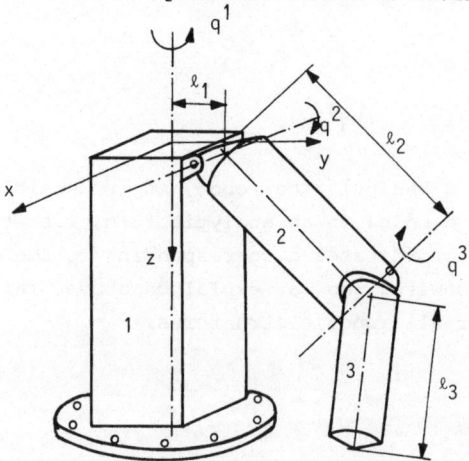

Fig. 2.7. A robot of anthropomorphic structure having 3 d.o.f.

2.2.2. <u>Numerical solutions to inverse</u>
       <u>kinematic problem</u>

In essence, an inverse kinematic problem represents the solving of a
set of nonlinear trigonometric equations with respect to internal coordinates q. A set of nonlinear equations can be solved by different
numerical procedures. Because of that, for those robots for which is
not possible to determine an inverse kinematic solution in analytic
form, this problem is solved by one of standard numerical methods.

We shall consider here a numerical method for solving inverse kinematic problem which is often used in practice. Let find the derivations
with respect to time for the functional relationship between the external and internal robot coordinates (2.2.1). Thus, we obtain:

$$\dot{s} = [\frac{\partial f}{\partial q}]\dot{q} = J(q)\dot{q} \tag{2.2.5}$$

where $J(q) = \partial f/\partial q$ denotes the *Jacobian matrix* of partial derivatives
of the function $f(q)$, having the dimension m×n.

Equation (2.2.5) represents the relationship between the vector of
Cartesian velocities $\dot{s}$ and the vector of joint velocities $\dot{q}$. If the
velocities of Cartesian coordinates $\dot{s}^o$ are given, the joint coordinates velocities should be determined from (2.2.5). Here, three cases
may arise:

a) m=n, in that case the Jacobian matrix is quadratic, and its inverse
   matrix can be determined:

$$\dot{q} = J^{-1}(q)\dot{s} \tag{2.2.6}$$

However, an inverse matrix exists if the matrix determinant $J(q)$ is
not equal to zero; the manipulator positions for which $\det(J(q)) = 0$
represent the so-called singular points for which inverse Jacobian
cannot be determined, i.e. the internal coordinates velocities cannot
be determined from (2.2.6) (Various procedures for solving the problem
of singular points have been described in the literature).

b) m>n, it is not possible to determine a $\dot{q}$ corresponding to the given
   $\dot{s}$ (except in some special cases);

c) m<n, this is the case of redundant manipulators, for which solution
is not unique, and some additional criteria should be introduced in
order to determine the unique $\dot{q}$ corresponding to the given $\dot{s}$; one
of possibilities is to adopt the so-called minimal inverse solution

$$\dot{q} = (J^TJ)^{-1}J^T\dot{s} \tag{2.2.7}$$

which gives a $\dot{q}$ which is, according to the criterion of minimum
quadratic error, closest to the "exact" solution.

In this way, the vector of internal coordinates velocities $\dot{q}^o$ can be
determined for the $\dot{s}^o$ values, assigned by the operator or by higher
control levels. The tactical level should determine the Jacobian $J(q)$
for an instantaneous value of internal coordinates q, carry out the
inversion of the matrix $J(q)$, and, on the basis of (2.2.6), calculate
the $\dot{q}^o$ for a given $\dot{s}^o$. The calculated $\dot{q}^o$ should be realized by the executive control level. This can be done in two ways: either to assign
the desired internal coordinates velocities $\dot{q}^o$ directly to the executive level (which means that the so-called speed controllers used at
the executive level can realize the given angular (translational) velocities of particular joints), or, to calculate the internal coordinates q^o and then assign them to the executive level (which now consists of position servo systems, as in the case of positioning control).
In the latter case, it is necessary to determine the internal angles
q^o corresponding to the calculated velocity $\dot{q}^o$, i.e. to find an integral of velocity:

$$q^o(t+\Delta t) = q^o(t) + \int_t^{t+\Delta t} \dot{q}^o(\tau)d\tau \tag{2.2.8}$$

The velocity integration may be carried out in different ways, depen-
dending on the mode of control realization. After the integration of
(2.2.8), the internal coordinates obtained should be assigned to the
executive level.

However, it has been supposed that the velocities of external coordinates $\dot{s}^o$ have been given. As the desired robot positions, assigned to
the tactical level, are often given in terms of external coordinates,
it is necessary to solve the inverse kinematic problem for the given
position s^o. An inverse Jacobian can be used to solve numerically the
inverse kinematic problem with respect to position, in the following
way. Let be given external coordinates s^o and let $q^{(k)}$ be an approxi-

mate (assumed) solution to the inverse kinematic problem, that is

$$q^{(k)} \approx f^{-1}(s^o) \qquad (2.2.9)$$

The difference between the exact values of external coordinates s^o and the values corresponding to the assumed solution $f(q^{(k)})$ may be approximately expressed in the following way:

$$\Delta s^{(k)} = s^o - f(q^{(k)}) \approx J(q^{(k)})(q^{(k+1)} - q^{(k)}) \qquad (2.2.10)$$

where $q^{(k+1)}$ is a more accurate solution to the inverse problem, corresponding to s^o. On the basis of (2.2.10), it is obtained that

$$q^{(k+1)} \approx q^{(k)} + J^{-1}(q^{(k)}) \Delta s^{(k)} \qquad (2.2.11)$$

Expression (2.2.11) can be used as an iterative procedure to determine solution to the inverse problem. The procedure is simple: for a given s^o, a solution to the inverse problem $q^{(o)}$ is assumed and corresponding values of external coordinates $f(q^{(o)})$ and error Δs^o determined; then, the Jacobian matrix at point $q^{(o)}$ is determined and also its inverse matrix (provided that the matrix is not a singular one); further, on the basis of (2.2.11), next approximate solution $q^{(1)}$, closer to the exact solution, is determined. The procedure is then repeated for the new $q^{(1)}$ and the repetition continued until the error for external coordinates satisfies the condition $|\Delta s^{(k)}| < \varepsilon$, where ε is a small positive constant defining the desired accuracy of the solution. This is, in fact, the classical Newton method for solving a set of nonlinear equations. The method is characterized by a quadratic convergence, which means that it has a high convergence rate toward the exact solution. Obviously, the Newton method gives only one solution to the inverse kinematic problem, and this is the one closest to the initial guess, $q^{(o)}$.

It is clear that equation (2.2.11) corresponds to equation (2.2.5) (if the velocities are replaced with finite increments, i.e. by the finite difference between two iterative solutions). Thus, the inverse Jacobian can be used for solving the inverse kinematic problem and for determining the internal coordinates for the given external coordinates (positions).

In practice, however, the desired gripper velocity (i.e. the external coordinates velocities) is often assigned in a direct way. If this is done by the "teaching" method, the operator should guide the manipulator to the desired positions, and it is obvious that it is easier for the operator to guide the robot by assigning the external than by assigning the internal coordinates (for this reason, the tactical control level is introduced). However, it appears inconvenient to the operator, guiding the robot by the aid of a joystick, to assign the external coordinates in a direct way. It is more convenient to the operator using a joystick to assign the velocity of the robot tip (gripper), i.e. the velocities of external robot coordinates. On the other hand, it appears (see the next section) that even when the desired external coordinates are determined (either by "teaching", or by direct assignement of the desired values by the operator or the higher control levels) a direct assignement of terminal values of the desired coordinates is not suitable; instead, it is necessary to ensure the robot "guiding" to the desired position. The robot guiding by assigning coordinates velocities is more suitable than by assigning the coordinates themselves (in time). For this reason, the desired velocity of the robot gripper is assigned in a direct way, and from (2.2.6) the velocities of internal coordinates are determined.

All that has been said above, reffers to the manipulation tasks requiring only the position control, i.e. bringing the robot to various positions in the workspace. Even nowadays, such tasks are most common in industry. However, a number of tasks appear in practice for which accurate robot positioning is not sufficient, but it is necessary to ensure that the robot moving from one to another position in the workspace, follows the more or less precisely defined paths. For example, in many manipulation tasks in the industrial practice, the robot moves in a workspace which is not empty, but contains different obstacles; in such a workspace the robot cannot move freely (i.e. along an arbitrary trajectory): hence, its trajectory should be defined in such a way the robot avoids collisions with the obstacles. In addition, in some tasks, it is necessary to realize the robot's tip moving along a particular path: such a task appears in paint spraying, where the robot has to spray paint uniformly over a particular surface. In some cases, a requirement may arise for a precise maintaining of the robot tip velocity along a given trajectory. A typical task of this kind is arc welding, where the robot tip should move along a defined seam at a strictly defined speed. In a great number of tasks, it is not required an explicit realization of a given velocity. Instead, a fixed motion time is required, so that the robot's velocity is defined in an indirect way.

In practice, the robot control is frequently accomplished through velocity: the gripper velocities are assigned in a direct way. When the joints velocities are determined by the aid of an inverse Jacobian, they are (as we mentioned above) sent directly to the executive level for realization, or integrated numerically according to (2.2.5) to determine the internal coordinates to be assigned to the executive level.

It can be seen that in the case of velocity control, the problem solving is generally simpler than in the position control: the Jacobian can always be determined and its inversion carried out if n=m (except at singular points). However, from the point of view of realization, a number of problems may arise. In a general case, a Jacobian is a complex function of all internal coordinates; the inversion of a Jacobian is not a simple task because it requires a relatively large number of mathematical operations (most often the matrix is of dimension 6×6, so that its inversion is a numerically complex problem); for this reason a powerful microcomputer (from the point of view of a speed of performing numerical operations) is needed in order to achieve a sufficiently fast calculation of velocities $\dot{q}^o$; besides, the problem of singular points and the problem of redundancy should be additionally solved. Similar conclusions hold also for numerical determination of internal coordinates by inversion of the appropriate Jacobian.

Here, we are not going to consider the problem of assigning external coordinates velocities $\dot{s}^o$, needed to guide the robot to a given position; the problem of calculating the trajectories to be followed by the robot in order to perform a given task, will be considered in the next section. The problems concerning the tactical control level (calculation and inversion of Jacobians, the problem of redundancy, etc) have been considered in [3].

Example 2.2.2. For the cylindrical manipulator shown in Fig. 2.5. the Jacobian is obtained by differentiating the right-hand sides of equations (2.2.3), that is

$$
J = \begin{bmatrix}
(\ell_3^o + \ell_4^o + q^3)\cos q^1 & 0 & \sin q^1 & 0 \\[2mm]
-(\ell_3^o + \ell_4^o + q^3)\sin q^1 & 0 & \cos q^1 & 0 \\[2mm]
0 & 1 & 0 & 0 \\[2mm]
0 & 0 & 0 & 1
\end{bmatrix}
\tag{2.2.12}
$$

Exercises

2.3. For the robot in Fig. 2.5. determine (in a analytic form) the inverse of (2.2.12), and check if this robot has singular points.

2.4. Draw a flow-chart of the algorithm for solving equation (2.2.2) by the Newton procedure.

2.5. For the manipulator in Fig. 2.6, determine the Jacobian and inverse Jacobian. Are there singular points?

2.6. Do the same, as in the previous exercise for the robot in Fig. 2.7.

2.7.* Write a programme (in one of high-level programming languages) for solving equation (2.2.2) for the manipulator in Fig. 2.7. by the Newton procedure.

2.8. Determine the number of numerical operations (multiplying and adding) to be carried out in order to determine an inverse matrix of the n×n matrix. Assume n=6.

2.3 Trajectory Synthesis for Manipulation Robots

In the preceeding two sections we showed how the problem of control is solved at the tactical level, i.e. how the joint coordinates or their velocities are determined on the basis of the given Cartesian coordinates or their velocities. Besides, we saw that in some tasks of practical importance, there is a need that the robot tip (gripper) follows a predetermined path at a given velocity (and acceleration). Here, we shall show, in short, how the trajectories are synthesized in order to

achieve desired motion of the gripper. This problem has been dealt
with in detail in [3]. To facilitate understanding of the further consideration of control at the executive level that will be presented in
the chapters to come, we shall discuss only some elementary notions of
trajectory synthesis.

A robot gripper trajectory may be assigned in different ways: one of
them is the robot teaching in which the operator guides the robot
along a desired trajectory and the control system stores the set of
points through which the robot has passed. However, this method of assigning trajectories is not convenient for several reasons. It is often required that the operator assign a desired trajectory numerically, through the terminal and in a suitable *programming language* (or,
the trajectory is assigned by the higher, strategic control level).
It is necessary to ensure that the assignement of a desired gripper
trajectory is realized in the simplest possible way, so that the operator (or the higher control level) supplies a minimal number of trajectory parameters; in this way the operator's job is made easier, and,
at the same time, the amount of data to be stored is minimized. Therefore, it should be determined the minimal and most suitable set of data enabling the assignement of a desired trajectory. For example, if
the motion is performed between the two given positions along a path
which represents a geometric figure, it is necessary to define the
initial and final position, the parameters describing the figure (for
example, if a circle is assumed between the two points, it is necessary to determine its centre, or the radius and the plane the circle
lies in), as well as the velocity profile, i.e. the change of the gripper velocity along the given path. On the basis of these parameters,
the control system should generate the gripper trajectory that should
be realized at the tactical and executive level.

Consider the simplest problem of synthesizing a trajectory between the
two given robot positions, defined by the vectors of external coordinates s_A and s_B, where the robot tip, when moving from one position to
the other, has to follow a straight line. As a straight line is defined by two points, no additional parameters are required to define the
gripper path. However, it is necessary to define the change of gripper
velocity along this straight line. If we suppose that the gripper velocity in the initial and final position is equal to zero, a triangular (Fig. 2.8), a trapezoidal (Fig. 2.9), a parabolic, or some other
kind of velocity profile, may be adopted. Let adopt a triangular velo-

city profile and let the time needed for robot passing from point A to point B be τ. In that case, the equations describing the change of external coordinates (the Cartesian coordinates of the gripper tip) are:

$$x_c^0(t) = \begin{cases} \dfrac{2(x_c^B - x_c^A) t^2}{\tau^2} + x_c^A, & \text{for} \quad 0 \le t \le \tau/2 \\[3mm] \dfrac{2(x_c^B - x_c^A)}{\tau^2}\left(2\tau t - t^2 - \dfrac{3\tau^2}{4}\right) + \dfrac{x_c^B + x_c^A}{2}, & \text{for} \quad \tau \ge t > \tau/2 \end{cases}$$

(2.3.1)

and the analogous for the other two coordinates, y_c and z_c.

Besides, it can be required that the (Eulerian) angles of the gripper change according to a triangular velocity profile, from the initial values θ_A, ϕ_A, ψ_A to the final values θ_B, ϕ_B, ψ_B. In that case, the change of external angles in time has the form:

$$\theta^0(t) = \begin{cases} \dfrac{2(\theta_B - \theta_A) t^2}{\tau^2} + \theta_A, & \text{for} \quad 0 \le t \le \tau/2 \\[3mm] \dfrac{2(\theta_B - \theta_A)}{\tau^2}\left(2\tau t - t^2 - \dfrac{3\tau^2}{2}\right) + \dfrac{\theta_A + \theta_B}{2}, & \text{for} \quad \tau \ge t > \tau/2 \end{cases}$$

(2.3.2)

and the analogous for the other two angles, ϕ and ψ.

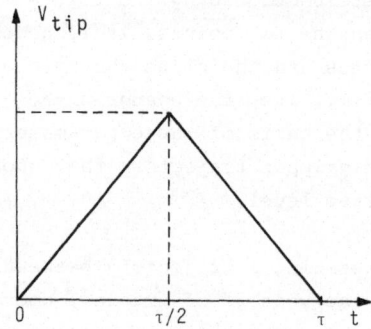

Fig. 2.8. Triangular velocity profile for the robot gripper

Fig. 2.9. Trapezoidal velocity profile for the robot gripper

In this way are obtained the trajectories of external coordinates (the gripper coordinates and angles) which should be then realized. It is easy to show that the external velocities are in this case defined as

$$
\dot{s}_i^O = \begin{cases} \dfrac{4\,(s_i^B-s_i^A)}{\tau^2}\,t & \text{for} \quad 0\leq t\leq \tau/2 \\[4mm] \dfrac{4\,(s_i^B-s_i^A)}{\tau^2}\,(\tau-t) & \text{for} \quad \tau/2<t\leq\tau \end{cases} \tag{2.3.3}
$$

for $i = 1, 2,\ldots,6$, $s^O = (s_1^O, s_2^O,\ldots,s_6^O)^T = (x_C^O, y_C^O, z_C^O, \theta^O, \psi^O, \phi^O)^T$, where s_i^A denotes the coordinate corresponding to position A and s_i^B corresponds to position B. Similarly, the accelerations are:

$$
\ddot{s}_i^O = \begin{cases} \dfrac{4\,(s_i^B-s_i^A)}{\tau^2} & \text{for} \quad 0\leq t\leq \tau/2 \\[4mm] \dfrac{-4\,(s_i^B-s_i^A)}{\tau^2} & \text{for} \quad \tau/2<t\leq\tau \end{cases} \tag{2.3.4}
$$

Therefore, the control system, in this case, generates easily the trajectories, as well as velocities and accelerations, in terms of external coordinates. The trajectories thus generated are converted at the tactical level into the trajectories of robot joints. This conversion can be accomplished in one of the following ways:

a) By a direct calculation of internal coordinates from the external coordinates on the basis of (2.2.2) (provided an inverse transformation, either analytic or numeric, is possible); in that case the joint trajectories $q^{oi}(t)$ are obtained, whose differentiation yields the velocities $\dot{q}^{oi}(t)$ and accelerations $\ddot{q}^{oi}(t)$ along the trajectories of robot joints;

b) By calculating joint velocities on the basis of relation (2.2.6), so that the values $\dot{q}^{oi}(t)$ are obtained (on the basis of the $\dot{s}_i^O(t)$ from (2.3.3)), whose integration yields the joint trajectories $q^{oi}(t)$ and differentiation yields the accelerations $\ddot{q}^{oi}(t)$ along the trajectories;

c) By calculating the joints accelerations $\ddot{q}^{oi}(t)$ along the trajectory on the basis of external accelerations $\ddot{s}^O(t)$, given by (2.3.4); the integration of accelerations $\ddot{q}^{oi}(t)$ yields the velocity $\dot{q}^{oi}(t)$ along the joint trajectory and the integration of the velocity provides the joint trajectory $q^{oi}(t)$. (It is assumed that there are no singular points along the chosen trajectory). When the accelerations of external coordinates $\ddot{s}^o$ are given, the joint accelerations are obtained on the basis of the relations between the accelerations of the external and internal coordinates. This relation is obtained by differentiating relation (2.2.5) with respect to time

$$\ddot{s} = J\ddot{q} + \frac{\partial J}{\partial q} \cdot \dot{q}^2 \qquad (2.3.5)$$

On the basis of this relation, we can determine the joint acceleration corresponding to the acceleration of external coordinates $\ddot{s}^o$:

$$\ddot{q}^o = J^{-1}(\ddot{s}^o - \frac{\partial J}{\partial q} \cdot \dot{q}^{o2}) \qquad (2.3.6)$$

In this way, the joint accelerations along a given trajectory are obtained, and then, the internal coordinates $q^o(t)$ are generated.

Generation of trajectories in terms of external and internal coordinates may be realized either off-line (at the stage of preparation, robot teaching, so that the trajectories are then stored), or on-line, during the robot motion along a desired trajectory, which depends on the type of task and the capabilities of the control system.

It can be seen that in this most elementary case, the trajectory generation is simple. However, if some more complex paths are involved, the problem becomes much more complex. Often, the operator, or the higher control levels, generate a set of points in space, through which the robot gripper should pass (without stopping), so that the gripper, moving from one given point to another, follows a straight line, as above, or a regular geometric figure[*]. In this case the problem of trajectories generation is much more complex, and it is especially complex if the motion time is not given, but it should be minimized [6]. The problem of minimization of the time of motion along a given path and the problem of optimal distribution of the robot tip velocity (from the point of view of time or energy), may be dealt with on the basis of a complete dynamic model of the robot; the problem is extremly complex, and the "optimal" solutions are difficult to realize in practice [7].

---

[*] This is a frequent case with the robots moving in the presence of obstacles, which are recognized by higher control levels, and which assign then a set of points through which the robot should pass in order to avoid obstacles.

Here, we are not goint to consider the problems concerning the synthesis of robot trajectories, because they have been dealt with in detail in [3]. For a better understanding of the further text it is essential to bear in mind that the tactical level generates either the joint positions q^{oi} to be realized, or the joint velocities to be realized, or the joint trajectories $q^{oi}(t)$ with corresponding velocities $\dot{q}^{oi}(t)$ and accelerations $\ddot{q}^{o1}(t)$, and the executive level should ensure the tracking of these trajectories.

Example 2.3. Consider the transfer of the manipulator shown in Fig. 2.5. from position A, defined by $s_A = (-0.064, 0.443, 0.624, 0)^T$ to position B, defined by $s_B = (0.2, 0.65, 0.68, 0.3)^T$. Assume the gripper moves along a straight line with the triangular velocity profile presented in Fig. 2.8, and let the required time of the movement be 1.5 [s]. The trajectories of external coordinates, based on relations (2.3.1) and (2.3.2) are presented in Fig. 2.10. Since the inverse transformation of the external into internal coordinates may be determined in the analytic form (2.2.4), it is easy to obtain the corresponding joint trajectories $q^o(t)$. These joint trajectories are presented in Fig. 2.11.

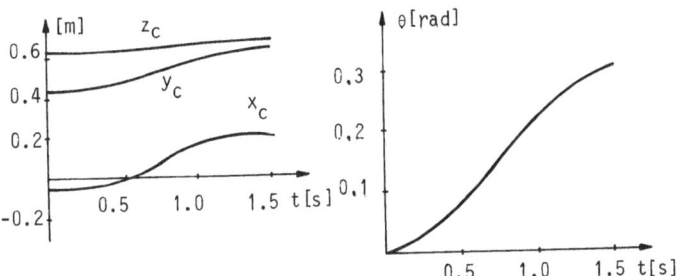

Fig. 2.10. Trajectories of external coordinates

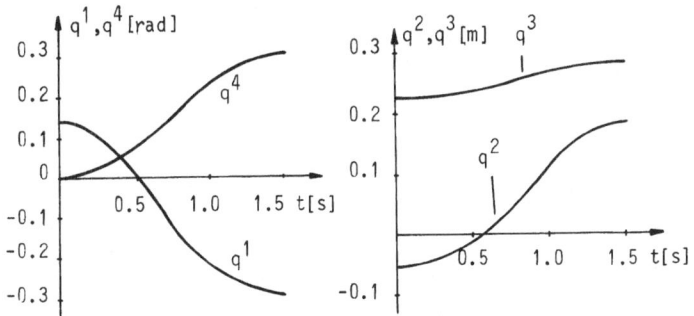

Fig. 2.11. Trajectories of joint coordinates

Exercises

2.9. Write the functions of the change of the manipulator gripper coordinates in time for the case the manipulator, while moving from
point s_A to point s_B, follows the trapezoidal velocity profile
(Fig. 2.9), for the time τ, and acceleration (i.e. deceleration)
time is $0.2\ \tau$.

2.10. Repeat the same as in Example 2.3. for the robot in Fig. 2.5, but
assuming a trapezoidal velocity profile (as in Exercise 2.9), instead of the triangular one.

2.11. Write the functions of the change of the manipulator gripper coordinates as a function of time for the gripper moving from the
point $s_A = (x_C^A, y_C^A, z_C^A)^T$ to $s_B = (x_C^B, y_C^B, z_C^B)^T$ and followsing a
circle line whose centre is at the point $s_O = (x^O, y^O, z^O)$. Assume a triangular velocity profile and a motion time τ.

2.12.* Write a programme for generating the joint trajectories for the
manipulator shown in Fig. 2.7. such that the robot tip moving
from one point to another tracks a straight line and a triangular
velocity profile for defined time τ. Use relation (2.2.6) and the
programme from Exercise 2.7.

2.13.* Repeat the task in Exercise 2.12 using relation (2.3.6) instead
of (2.2.6), i.e. calculate accelerations $\ddot{q}$ directly on the basis
of $\ddot{s}$. Compare the results. Which of the two variants of trajectory generation is more suitable?

References

[1] Vukobratović M., Applied Dynamics of Manipulation Robots: Modelling, Analysis and Examples, Springer-Verlag, 1989.

[2] Vukobratović M., (Ed.), Introduction to Robotics, Springer-Verlag,
1988.

[3] Vukobratović M., Kirćanski M., Kinematics and Trajectory Synthesis
of Manipulation Robots, Monograph Series, Vol. 3, Springer-Verlag,
1985.

[4] Craig J.J., Introduction to Robotics, Mechanics and Control, Addison-Wesley Publ. Company, 1986.

[5] Pieper D.L., Roth B., "The Kinematics of Manipulators under Computer Control", Proc. II Intern. Congress on Theory of Machines and Mechanisms, Vol. 2, Zakopane, 1969.

[6] Kahn M.E., Roth B., "The Near Minimum Time Control of Open Loop Articulated Kinematic Chains", Trans. of the ASME, Journal of Dyn. Systems, Measurement and Control, Sept., 164-172, 1971.

[7] Vukobratović M., Kirćanski M., "A Method for Optimal Synthesis of Manipulation Robot Trajectories", Trans. of ASME, Journal of Dyn. Systems, Measurement and Control, Vol. 104, No 2, 1982.

Chapter 3
Synthesis of Servo Systems for Robot Control

3.1 Introduction

In the previous chapter we have considered problems concerning the
synthesis of tactical control level. In the text to follow we shall
consider the synthesis of the executive control level. As we have already explained the executive control level has to ensure implementation of trajectories (or, only positions) of joint coordinates of a
robot. These trajectories are computed at the tactical control level.
The implementation of the trajectories (positions) directly involves
dynamic behaviour of the robot. Due to this, first we shall briefly
present dynamic model of the robot system. The dynamic model of the
robot "consists" of dynamic model of the mechanism and models of actuators which drive the joints of the mechanism. Next, we shall consider
synthesis of servo system around each joint (actuator) of the robot.
In practice, the executive control level is often implemented in the
form of independent controllers for each joint of robot. As we shall
explain, robot is a complex system with strong interactions between
the motions of its joints. However, the independent control of joints
is the simplest control law, and so it is the most appropriate approach from the standpoint of control implementation. In this chapter we
shall present various methods for synthesis of a local controller around
each joint, and in the next chapters we shall discuss certain disadvantages of this approach to control of robots. We shall also consider
the methods for improvement of this control in order to meet the requirements which are imposed before the robots in industry.

3.2 Dynamic Model of Robot

The robotic systems "consists" of mechanism (mechanical part of the
system), actuators which drive the joints of the mechanism and control
system (which includes various sensors and other equipment). The mathematical model of dynamics of the mechanical part of the system and the
models of actuators have been considered in detail in [1]. Here we
shall briefly present the mathematical model of robotic manipulators

only in a degree which is necessary to understand the synthesis of control at the executive control level.

3.2.1. Mathematical model of mechanical part of system

The mechanical part of the robotic system is a complex mechanism which consists of a number of bodies-links. The links are connected to each other by joints. This type of mechanism is called open kinematic chain[*]. We shall assume that each link of the robot mechanism is a rigid body, i.e. we shall neglect elasticity of the robot links. The validity of this assumption will be considered latter on. The joints between two neighboring links might be rotational or linear (and each joint might have one or more degrees of freedom)[**]. Let us consider the robotic system with n joints and n degrees of freedom (d.o.f.). Dynamic model of such mechanism might be extremely complex. Let us assume that each joint is driven by separate actuator which produce active force (moment) around the corresponding joint. The movement of the i-th joint is described by the corresponding internal (joint) coordinate q^i which represents angle between the two neighboring links (if rotational pair of links are considered) or, linear displacement between two neighboring links (if we consider linear joint). The i-th joint movement is also described by joint velocity $\dot{q}^i$ and acceleration $\ddot{q}^i$ (see Fig. 3.1. [1]). The active (driving) moment P_i of the actuator around the i-th joint causes the movement of this joint. The acceleration of the joint $\ddot{q}^i$ is proportional to driving torque P_i. The factor of proportionality between $\ddot{q}^i$ and P_i represents the moment of inertia of the complete moving part of the mechanism around the i-th joint. This moment of inertia is a complex function of the positions (coordinates) of all joints of the mechanism q^j which are in the kinematic chain "behind" the i-th joint, i.e. for j>i. However, besides the eigen moment of inertia of the mechanism around the i-th joint, the movement of the i-th joint is affected by the so-called cross-inertia terms which represent the influence of the accelerations of the rest of the joints in the mechanism $\ddot{q}^j$, j≠i upon the i-th joint. Namely, if the

---

[*] If the gripper of the robot is in a contact with some fixed object (with large mass) in the robot working environment, the robot becomes so-called closed kinematic chain (see Chapter 7).

[**] As it is explained in [1], the complex joints which have more than one degree of freedom can be considered as a set of several simple joints (the so-called pairs of the fifth class) with one degree of freedom each. Thus, in the text to follow we shall assume that all the joints of the mechanism are simple.

j-th joint is accelerated by $\ddot{q}^j$ the dynamic moment around the i-th
joint appear which is proportional to $\ddot{q}^j$. The factor of proportionality represents this cross-inertia member. The cross-inertia members are
also complex functions of the positions (coordinates) of the joints of
the mechanism q^j. Next, the moments due to joints velocities also influence the movement of the i-th joint. These moments are called Coriolis and centrifugal moments (forces) and they are proportional to
product of the joints velocities. At last, gravity forces (moments)
also affect the movement around the i-th joint, i.e. they also produce
moments around the i-th joint axis. The gravity force is also complex
function of all coordinates (positions of joints) q^j of the mechanism.

In accordance to the above considerations, the equation of equilibrium
of the dynamic moments which act around the i-th joint might be written in the following form:

$$P_i = H_{ii}(q)\ddot{q}^i + \sum_{\substack{j=1 \\ j \neq i}}^{n} H_{ij}(q)\ddot{q}^j + \sum_{j=1}^{n}\sum_{k=1}^{n} c_{jk}^i(q)\dot{q}^k\dot{q}^j + g_i(q) \qquad (3.2.1)$$

where H_{ii} denotes the moment of inertia of the mechanism around the
i-th joint which is the function of all coordinates of the mechanism q
(save for the coordinate of the i-th joint), H_{ij} denote cross-inertia
terms which also are the functions of all coordinates of the mechanism
q, c_{jk}^i represent Coriolis and centrifugal effects, g_i represents gravity moment around the axis of the i-th joint and it is also function
of all the coordinates of the mechanism q. The equation (3.2.1) describes the movement of the i-th joint for both rotational and linear
joint. (In the case of linear joint, P_i is the driving force of the
actuator, H_{ij} are masses, g_i is the gravity force, while in the case of
rotational joints we operate with driving torque of the actuator (P_i),
the moments of inertia (H_{ij}) and gravity moments (g_i)).

If we write down the moment equations (3.2.1) for all n joints of the
mechanism we obtain the complete model of dynamics of whole robotic
mechanism. The model of dynamics of the mechanism can be written in
the matrix form as:

$$P = H(q)\ddot{q} + h(q, \dot{q}) \qquad (3.2.2)$$

where $P = (P_1, P_2, \ldots, P_n)^T$ represents the n×1 vector of driving torques
(forces), $H(q) = [H_{ij}(q)]$ is n×n inertia matrix which is the function
of the vector of coordinates of the mechanism $q = (q^1, q^2, \ldots, q^n)^T$,

$h(q, \dot{q}) = \dot{q}^T C(q) \dot{q} + g(q)$ is the vector of dimension $n \times 1$, $C(q) = [C_{jk}^i(q)]$ is three dimensional $n \times n \times n$ matrix which is called matrix of Coriolis and centrifugal effects, or "C-matrix", $g(q)$ is the $n \times 1$ vector of gravity moments (forces) which is also the function of q.

As it can be seen, the dynamic model of the mechanism (3.2.2) represents the set of n nonlinear differential equations of the second order (the total order of this system is obviously 2n). It can also be seen that the movement of each joint is strongly interconnected with the movement of all the other joints and that the driving torques of actuators affects all the joints in the mechanism (i.e. the driving torque of the i-th actuator affects the movement all other joints in the system). We must emphasize that the matrix H and the vector h might be very complex nonlinear functions of all the coordinates of the mechanism, which makes the setting of differential equations (the mathematical model (3.2.2)) very difficult, in general case. The complexity of the model depends on the structure of the particular robot mechanism. Due to this complexity of robot dynamic model, there was developed a large number of various methods for automatic setting of the mathematical model of the mechanism at digital computer. These methods enable to generate automatically the mathematical model of the mechanism of arbitrary structure and of arbitrary number of d.o.f. The user has to impose just basic data about the mechanism (the mechanism structure, geometric data on the links lengths and positions of the centres of gravity of each link, masses, and moments of inertia of the links, etc.), and the computer automatically generates the equations (3.2.2). The generation of dynamic model of robot mechanism might be accomplished by various algorithms. These algorithms differs in the law of mechanics which is applied to form motion equations (3.2.1) [1]. Thus, the following methods are well known: the method for forming differential equations of the robot dynamics based on Lagrange's equations, methods based on Newton-Euler's dynamic equations, methods based on Appel's equations. Here we shall not present various methods for automatic generation of dynamic models of open kinematic chains since they can be found elsewhere [2]. It should be pointed out that the methods for generation of mathematical models of robots differs in number of multiplications and additions that they require to compute driving torque once the coordinates, velocities and acceleration of the joints are given (or, to compute inertia matrix H and vector h). This is a crucial point from the standpoint of application of this methods for computation of dynamic models required for control of robotic system. However, this fact will be discussed latter on when we shall consider dynamic control of robot (see Chapter 5).

EXAMPLE 3.2.1. The mathematical model of dynamics of the robot in Fig. 3.2. with n=3 d.o.f. might be written in the form (3.2.2) where the matrix H(q) and the vector h(q, $\dot{q}$) are given by:

$$H(q) = \begin{bmatrix} J_{1z}+m_1\ell_1^{*2}+m_3\ell_1^2+J_{3z}+m_3\ell_3^{*2}+2m_3\ell_3^*\ell_1\cos q^2, & J_{3z}+m_3\ell_3^{*2}+m_3\ell_3^*\ell_1\cos q^2, & 0 \\ J_{3z}+m_3\ell_3^{*2}+m_3\ell_3^*\ell_1\cos q^2, & J_{3z}+m_3\ell_3^{*2}, & 0 \\ 0 & 0 & m_3 \end{bmatrix}$$

$$h(q, \dot{q}) = \begin{bmatrix} -m_3\ell_1\ell_3^*\dot{q}^2(\dot{q}^2+2\dot{q}^1)\sin q^2 \\ m_3\ell_3^*\ell_1(\dot{q}^1)^2\sin q^2 \\ -m_3 g \end{bmatrix} \qquad (3.2.3)$$

It can be seen that there exist inertial and centrifugal coupling between the first and the second joint of this robot, while the third joint is not dynamically coupled to the first two joints.

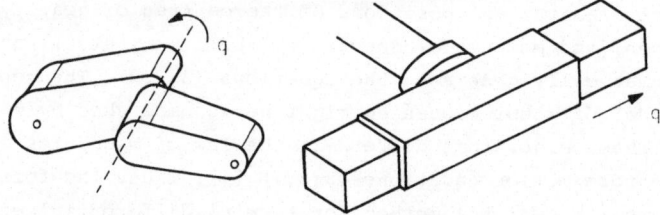

Fig. 3.1. Rotational and linear joint

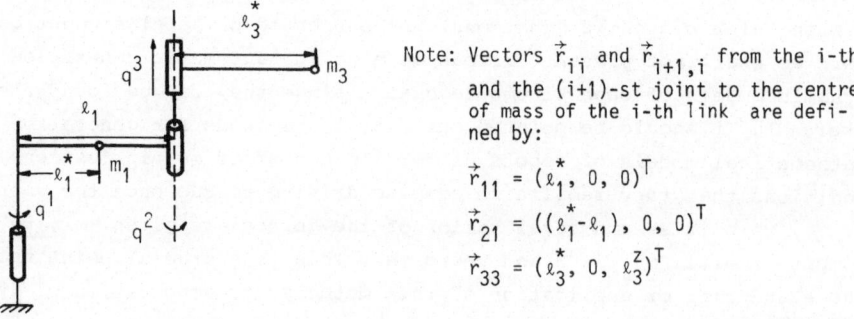

Note: Vectors $\vec{r}_{ii}$ and $\vec{r}_{i+1,i}$ from the i-th and the (i+1)-st joint to the centre of mass of the i-th link are defined by:

$$\vec{r}_{11} = (\ell_1^*, 0, 0)^T$$
$$\vec{r}_{21} = ((\ell_1^*-\ell_1), 0, 0)^T$$
$$\vec{r}_{33} = (\ell_3^*, 0, \ell_3^z)^T$$

Fig. 3.2. Robot with three d.o.f. (two rotational and one linear joint)

Exercises

3.1. Write the mathematical model of the dynamics of the robot in Fig.
2.5. Are there interconnections between the motions of the joints
of this particular robot (see [1])?

3.2. Repeat the Exercise 3.1. for the robot in Fig. 2.6.

3.3. Write the dynamic model for the two-joint robot presented in Fig.
3.3. What are interconnections between the joints?

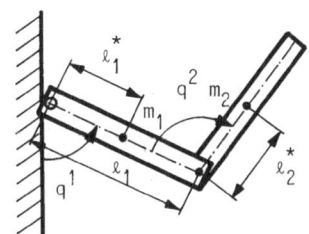

Fig. 3.3. Robot with two rotational joints

3.4. Determine the minimal number of additions and multiplications and
special functions (sinus and cosines) which is required to compute matrix $H(q)$ and vector $h(q, \dot{q})$ if values of vectors q and $\dot{q}$
are given, for robot presented in Example 3.2.1. (Fig. 3.2), and
for robots in previous three exercises.

3.5.* Write the programme, in some high programming language, for computation of matrix $H(q)$ and vector $h(q, \dot{q})$ if values of vectors q
and $\dot{q}$ are given (q and $\dot{q}$ are input variables for the programme,
and H and h are output variables) for the robot in Fig. 3.2. Repeat this for the robots in Figs. 2.5, 2.6. and 3.3. Try to minimize the number of additions and multiplications in each programme.

3.6.* Estimate the minimal sampling period (i.e. the time period required) for computation of matrix H and vector h (for given q and $\dot{q}$)
if we implement programme written in Exercise 3.5. at:

a) Microprocessor INTEL-80-80 (which requires 0.8 [ms] for one
floating-point addition and 1.5 [ms] for one floating-point
multiplication)

b) Microprocessor INTEL-80-87 (which requires 35 [μs] for one floating-point addition and 65 [μs] for one floating-point multiplication).

Assume that the processor time is consumed just for additions and multiplications.

3.2.2. Models of actuators

We assume that each joint of the robot is driven by the separate actuator. Majority of the robots at the market today are driven by D.C. permanent-magnet electro-motors. There are a lot of robots which are powered by electrohydraulical actuators, A.C. electro-motors and even electropneumatical actuators[*]. The mathematical models of these actuators have been considered in detail in [1]. Here, we shall briefly present just the model of D.C. electro-motors, since in this book we shall consider just the control of robots driven by D.C. motors. However, the reader can easily apply all considerations to other types of actuators (see exercises). We shall pay a special attention to direct-drive actuators, i.e. electrical actuator without reducer.

Let us assume that the i-th joint of the robot is driven by permanent magnet D.C. electro-motor with reducer. The scheme of such D.C. motor is given in Fig. 3.4. [1]. The differential equations which describe the behaviour of such D.C. motor, are:

- the equation of mechanical equilibrium of the driving torque and the equivalent moment of load at the output shaft of the reducer:

$$N_v^i N_m^i J_M^i \ddot{\theta}^i + F_v^i \dot{\theta}^i + M_i^* = C_M^i N_m^i i_R^i \qquad (3.2.4)$$

- the equation of electrical equilibrium in the rotor circuit:

---

[*] Pneumatic actuators have been applied for so-called mechanical hands, while with robots they were not used due to their disability to track continuous trajectories. However, appearance of pneumatic servo actuators announced their possible application in robotics. All considerations of control synthesis in the text to follow, could be extended to pneumatic actuators.

$$L_R^i \dot{i}_R^i + r_R^i i_R^i + C_E^i N_v^i \dot{\theta}^i = u^i \tag{3.2.5}$$

In the above equations the following notations are used: i_R^i is rotor current [A], u^i is rotor control voltage [V], r_R^i is rotor resistance [Ω], L_R^i is rotor winding inductance [H], θ^i is output angle of motor shaft after reducer [rad], C_M^i is torque constant (the ratio between the torque at the motor shaft and the current in the rotor circuit) [Nm/A], C_E^i is electromotor force constant (the ratio between the voltage in the rotor circuit caused by the rotation of the rotor in the magnetic field and the rotational velocity) [V/rad/s], F_v^i is viscous damping constant reduced to output shaft [Nm/rad/s], J_M^i is rotor moment of inertia, [kgm^2], J_R^i is rotor moment of inertia reduced to output shaft $J_R^i = J_M^i N_v^i N_m^i$ [kgm^2], N_v^i is speed reduction ratio (the ratio between the rotational velocity at the input shaft and the velocity of the output shaft of the reducer), N_m^i is torque reduction ratio (the ratio between the output and input torque at the reducer), M_i^* is load torque at the reducer output shaft [Nm]. The equation (3.2.4) and (3.2.5) represent mathematical model of dynamics of the D.C. motor. The order of the model is obviously $n_i = 3$. The index i denotes that the actuator is driving the i-th joint of the robot mechanism, and i might be i = 1,2,...
...,n.

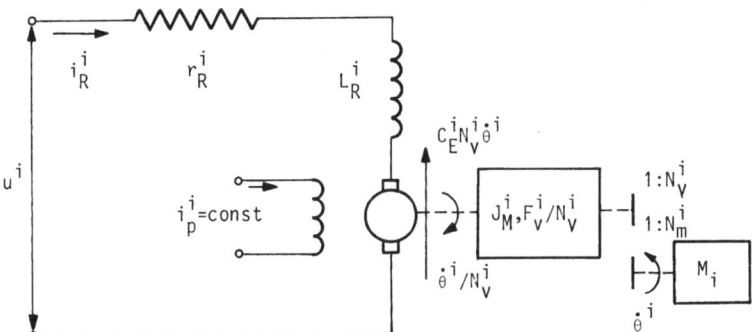

Fig. 3.4. Scheme of the permanent-magnet D.C. electro-motor

The model of the D.C. motor can be written in the state space. Let us adopt the state vector of the actuator model in the following form: $x^i = (\theta^i, \dot{\theta}^i, i_R^i)^T$ where the x^i is the $n_i \times 1$ vector. Then, we can write model of the D.C. motor (3.2.4), (3.2.5) in the state space as:

$$\dot{x}^i = A^i x^i + b^i u^i + f^i M_i^* \tag{3.2.6}$$

where A^i is $(n_i \times n_i)$ matrix of the system, b^i is $(n_i \times 1)$ input distribution vector, and f^i is $(n \times 1)$ load distribution vector. The matrix A^i and the vectors b^i and f^i are given by:

$$A^i = \begin{bmatrix} 0 & 1 & 0 \\ 0 & -\dfrac{F_v^i}{N_v^i N_m^i J_M^i} & \dfrac{C_M^i}{N_v^i J_M^i} \\ 0 & -\dfrac{C_E^i N_v^i}{L_R^i} & -\dfrac{r_R^i}{L_R^i} \end{bmatrix} , \quad b^i = \begin{bmatrix} 0 \\ 0 \\ \dfrac{1}{L_R^i} \end{bmatrix} , \quad f^i = \begin{bmatrix} 0 \\ -\dfrac{1}{N_v^i N_m^i J_M^i} \\ 0 \end{bmatrix}$$

$$(3.2.7)$$

In (3.2.7) u^i represents the scalar input to the system (i.e. the rotor circuit voltage). However, we have to take into account the fact that the input signal (voltage) is constrained by amplitude, i.e. the amplitude of the input must be below some maximal value u_m^i. This nonlinearity of the amplitude saturation type we have to introduce in the model (3.2.6) of D.C. motor, so it becomes:

$$\dot{x}^i = A^i x^i + b^i N(u^i) + f^i M_i^* .$$

$$(3.2.8)$$

where $N(u)$ is defined by:

$$N(u^i) = \begin{cases} -u_m^i & \text{for} \quad u^i < -u_m^i \\ u^i & \text{for} \quad -u_m^i \le u^i \le u_m^i \\ u_m^i & \text{for} \quad u^i > u_m^i \end{cases}$$

$$(3.2.9)$$

This nonlinearity is presented in Fig. 3.5.

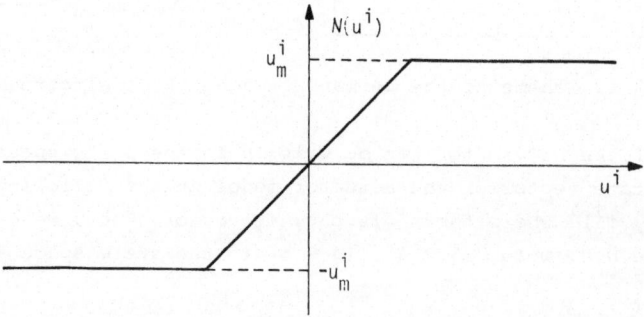

Fig. 3.5. Amplitude saturation constraint upon the actuator input

The model of D.C. motor is obtained as a system of linear differential
equations of the third order with the nonlinear amplitude constraint
upon the system input. However, the inductance of the rotor circuit can
be often neglected, and thus, the model of the actuator becomes system
of differential equations of the second order (but its form in the state space is given by (3.2.8), too).

The model of electrohydraulic actuator can be obtained in analogous
way. We obtain model in the form (3.2.8), too, and its order is $n_i = 3$.
However, the model of the electrohydraulic actuator can be written as
a higher order system, or, as the second order system (see [1]). In the
text to follow, we shall consider the model of the actuator in the form
(3.2.8) and the elements in matrices A^i, b^i, f^i are distributed as in
(3.2.7). Since the distribution of the elements in these matrices for
electrohydraulic actuators is the same as in (3.2.7), all considerations to follow are valid for both D.C. motors and electrohydraulic actuators. The only difference between the models of D.C. motors and
electrohydraulic actuators is in the physical meaning of the elements
in corresponding matrices. We must underline that for the electrohydraulic actuators these linear models are very approximate, since in the
hydraulic actuators nonlinear effects might be significant. Nevertheless, for the sake of symplicity first we shall consider linear models
of actuators and in Section 3.3.5. we shall consider the influence of
the nonlinear effects in actuators upon the servo system control.

EXAMPLE 3.2.2. For the permanent-magnet D.C. motor of the type IG2315-
-P20 the data taken from the data-sheets are given in Table 3.1.

Based on these data we can easily obtain the matrices A^i, b^i, f^i of the
actuator model acc. to (3.2.7) (for the first two joints of the robot
in Fig. 3.2., data are given in the first two rows of Table 3.1):

$$A^i = \begin{bmatrix} 0 & 1 & 0 \\ 0 & -0.201 & 53.03 \\ 0 & -621.74 & -695.65 \end{bmatrix}$$

$$b^i = \begin{bmatrix} 0 \\ 0 \\ 434.78 \end{bmatrix} \qquad f^i = \begin{bmatrix} 0 \\ -34.68 \\ 0 \end{bmatrix} \qquad (3.2.10)$$

ACTUATOR	1	2	3
$c_E^i \left[\dfrac{V}{rad/s}\right]$	0.0459	0.0459	0.0459
$c_M^i \left[\dfrac{Nm}{A}\right]$	0.0481	0.0481	0.0481
$J_M^i \ [kgm^2]$	0.00003	0.00003	0.00003
$N_v^i \ [-]$	31.	31.	2616.
$N_m^i \ [-]$	31.	31.	2616.
$r_R^i \ [\Omega]$	1.6	1.6	1.6
$F_v^i \left[\dfrac{Nm}{rad/s}\right]$	0.0058	0.0058	0.0154
$L_R^i \ [H]$	0.0023	0.0023	0.0023

Table 3.1. Data on actuators for the robot in Fig. 3.2.

Exercises

3.7. Show that if we assume that $L_R \approx 0$, the model of the D.C. permanent-
-magnet electro-motor might be obtained in the form (3.2.8) but as
the second order model $n_i = 2$, where the matrices are now given
by:

$$A^i = \begin{bmatrix} 0 & 1 \\ 0 & -\dfrac{1}{J_M^i}\left(\dfrac{F_v^i}{N_v^i N_m^i} + \dfrac{c_M^i c_E^i}{r_R^i}\right) \end{bmatrix} , \quad b^i = \begin{bmatrix} 0 \\ \dfrac{c_M^i}{N_v^i J_M^i r_R^i} \end{bmatrix} , \quad f^i = \begin{bmatrix} 0 \\ -\dfrac{1}{N_v^i N_m^i J_M^i} \end{bmatrix}$$

$$(3.2.11)$$

3.8. Compute the elements of the matrices A^i, b^i, f^i in (3.2.11) based
on data given in Table 3.1, for all three joints of the robot in
Fig. 3.2.

3.9.[*] See Appendix 6. in [1] for models of electrohydraulic actuators.
Write the state space equations and matrices of this model

for electrohydraulic actuator, for two cases: if we adopt the second order model ($n_i=2$) and for the third order model ($n_i=3$).

3.10.* See Appendix 6 in [1] for models of electropneumatic actuators, A.C. electro-motors, and direct-drive actuators.

3.2.3. Total model of the robotic system

As we have already explained, the robotic system consists of mechanical part of the system and actuators which drive the robot joints. The model of mechanical part of the system is given by (3.2.2), and the model of actuators is given by (3.2.8). Our next task is to combine these models into unique model of the robotic system [1].

The relation between the model of actuators and the model of the mechanism is by coordinates and by moments (loads). The movement of the actuator (output shaft of the reducer, if we consider D.C. motors with reducers) is the movement of the corresponding joint (rotational or linear displacement of the joint). Thus, the rotation of the actuator θ^i is transformed into the motion of the joint q^i. In the simplest case, the movement of the actuator shaft is equal to the movement of the corresponding joint, i.e.:

$$\theta^i = q^i \tag{3.2.12}$$

However, in general case, the relation between θ^i and q^i might be more complex. For example, if the linear displacement of the piston of the hydraulic cylinder is transformed into rotation of the joint (see Fig. 3.6), then the relation between the actuator coordinate θ^i and corresponding coordinate of the joint q^i is given by cosines formula:

$$\theta^i = \sqrt{a^{i2}+b^{i2}-2a^ib^i\cos(q^i+\alpha^i)} - \ell^{io} \tag{3.2.13}$$

In general case, the relation between the coordinate of actuator θ^i and coordinate of the corresponding joint q^i is given by:

$$\theta^i = g^i(q^i) \quad \text{or,} \quad q^i = \tilde{g}^i(x^i) \tag{3.2.14}$$

We can establish relation between the load M_i^* upon the actuator (output shaft of D.C. motor (reducer), or, at the piston of hydraulic actuator) and driving torque P_i which acts around corresponding joint of

the mechanism, in the similar way. In the simplest case (which corresponds to the relation between the coordinates (3.2.12)), the load moment upon actuator M_i^* is equal to the driving torque P_i (force) around
the joint axis:

$$M_i^* = P_i \qquad\qquad (3.2.15)$$

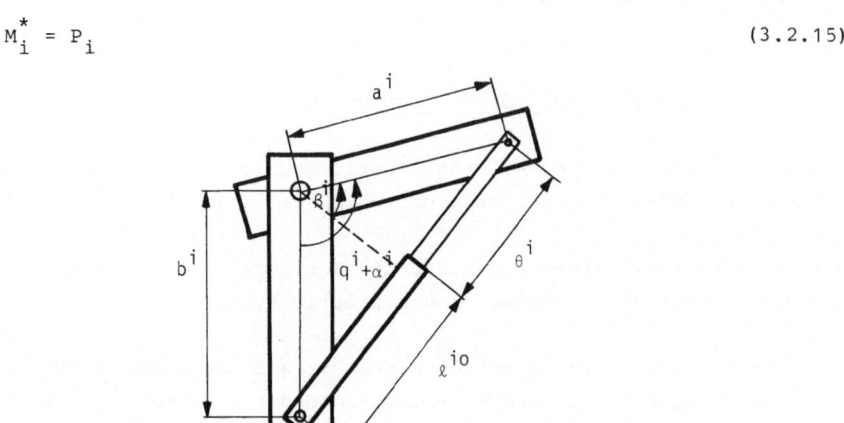

Fig. 3.6. Relation between actuator coordinate and joint coordinate

However, if the relation between the coordinate of the actuator and the
coordinate of the joint is not linear, then the relation between M_i^* and
P_i might also be complex. For example, in the case presented in Fig.
3.6, the relation between the load upon the piston of the hydraulic
cylinder and the driving torque which is produced around the joint axis
is given by:

$$M_i^* = \frac{P_i}{a_i \cos\beta^i} \qquad\qquad (3.2.16)$$

We may assume that the relation between the load moment upon the actuator M_i^* and the moment around joint axis P_i is given in the form:

$$M_i^* = z^i(q^i) P_i \qquad\qquad (3.2.17)$$

However, for the sake of simplicity we shall consider the case when the
relation between θ^i and q^i is given by (3.2.12) and the relation between M_i^* and P_i is given by (3.2.15).

The total model of the robotic system can be obtained by combining the actuators models (3.2.8) and the model of the mechanism dynamics (3.2.2). Since the relations between the actuator coordinates and the coordinates of the mechanism are given by (3.2.12), we may adopt the vector of the state of the total system in the form:

$$x = (x^{1T}, x^{2T}, \ldots, x^{nT})^T \qquad (3.2.18)$$

The state vector of the system is obviously of the order N which is given by:

$$N = \sum_{i=1}^{n} n_i \qquad (3.2.19)$$

Thus, the order of the total robotic system is N. Based on (3.2.12) we can write:

$$\dot{q}^i = T_i x^i \qquad (3.2.20)$$

where T_i is the $1 \times n_i$ vector. (For example, if x^i is of the order $n_i=2$, T_i is given by $T_i=(0, 1)$, and if $n_i=3$ then T_i is given by $T_i=(0, 1, 0)$). Based on (3.2.18) and (3.2.20) we may write:

$$\dot{q} = Tx \qquad (3.2.21)$$

where T is the matrix of dimensions $n \times N$ given by $T = \text{diag}(T_i)$, i.e.

$$T = \begin{bmatrix} T_1 & 0 & \cdots & 0 \\ 0 & T_2 & \cdots & 0 \\ \hline 0 & 0 & \cdots & T_n \end{bmatrix}$$

The model of the mechanical part of the system (3.2.2) might be expressed by the state vector of the total system in the following way:

$$P = H(x)T\dot{x} + h(x) \qquad (3.2.22)$$

We can combine the models of all actuators (3.2.8) of the robot in a unique system of differential equations in the matrix form:

$$\dot{x} = Ax + BN(u) + FP \qquad (3.2.23)$$

where we have used the relation (3.2.15) and where A is the $N \times N$ matrix given by $A=\text{diag}(A^i)$, B and F are the $N \times n$ matrices given by $B=\text{diag}(b^i)$,

$F = \text{diag}(f^i)$, while $N(u)$ denotes $n \times 1$ vector given by:

$$N(u) = (N(u^1), N(u^2), \dots, N(u^n))^T \qquad (3.2.24)$$

By u we have denoted the $n \times 1$ vector of inputs to the actuator system, which is defined by (the order of the system input is evidently n):

$$u = (u^1, u^2, \dots, u^n)^T \qquad (3.2.25)$$

If we substitute $\dot{x}$ from (3.2.23) into (3.2.22), and solve the system by P we get:

$$P = (I_n - H(x)TF)^{-1}[H(x)T(Ax + BN(u)) + h(x)] \qquad (3.2.26)$$

where I_n is the $n \times n$ unit matrix. Evidently the inverse matrix in (3.2.26) always exist, i.e. the corresponding matrix is regular. If we substitute P from (3.2.26) into (3.2.23) we obtain the total model of the robotic system in the following form:

$$\dot{x} = \hat{a}(x) + \hat{B}(x)N(u) \qquad (3.2.27)$$

where by $\hat{a}(x)$ is denoted the $N \times 1$ vector:

$$\hat{a}(x) = [A + F(I_n - H(x)TF)^{-1}H(x)TA]x + F(I_n - H(x)TF)^{-1}h(x)$$

and by $\hat{B}(x)$ is denoted the $N \times n$ matrix:

$$\hat{B}(x) = B + F(I_n - H(x)TF)^{-1}H(x)TB$$

In this way we obtain the total model of the robotic system in the state space as a system of N nonlinear differential equations. This form of model of the robotic system is called the *centralized model* while the model expressed by the models of actuators (3.2.8) and model of the mechanism dynamics might be called *decentralized form of the model*. We shall mostly use the decentralized form of the model of the robotic system. In this form of model, the robot is described by a set of n models of actuators. The actuators might be considered as subsystems which are interconnected by the mechanical part of the system (by robot mechanism). The "scheme" of the model of the robot is presented in Fig. 3.7.

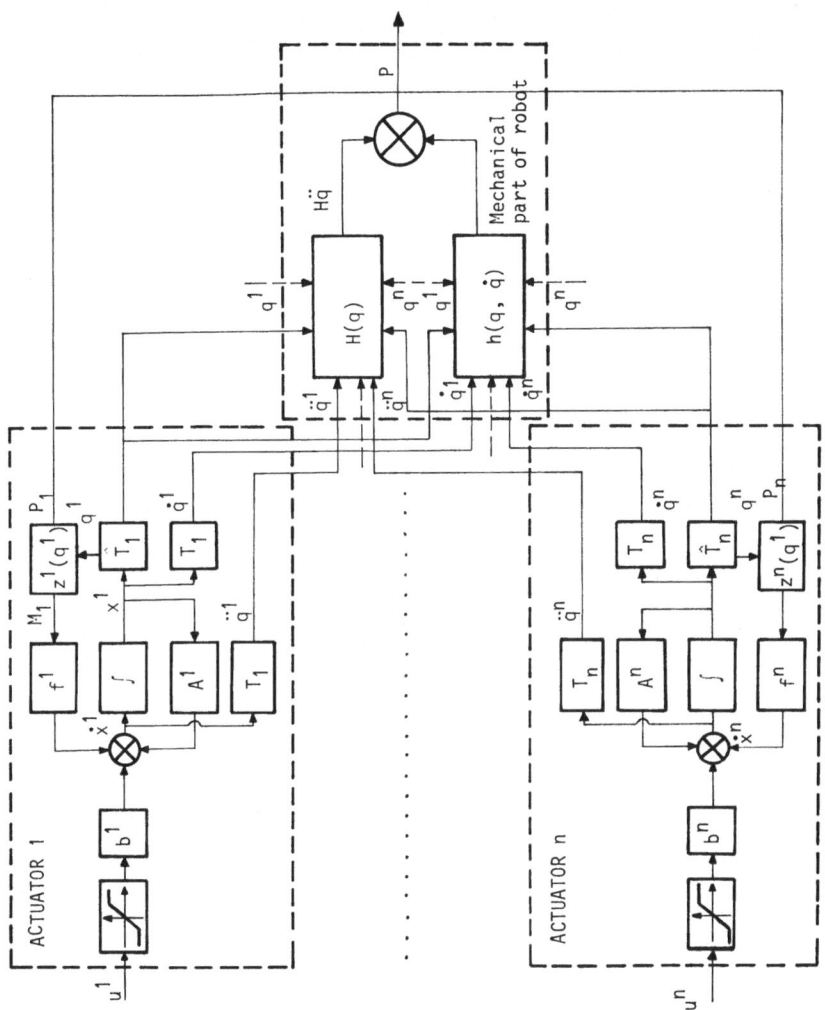

Fig. 3.7. "Scheme" of the model of the robot: mechanical part of the system and n actuators (in each joint one actuator); the vectors $\hat{T}_i$ are of dimensions $1 \times n_i$ and they are defined by $q^i = \hat{T}_i x^i$.

EXAMPLE 3.2.3. The model for the mechanical part of the robot in Fig. 3.2. is given by (3.2.3). The actuators for this robots are D.C. motors, the model of which are given by (3.2.8) and of the order $n_i=2$ for all three joints. We assume that the relation between θ^i and q^i is given by (3.2.12), and that the relation between M and P is given by (3.2.15). The model of the total system can be written in the form (3.2.27), where the state vector x is given by $x = (q^1, \dot{q}^1, q^2, \dot{q}^2, q^3, \dot{q}^3)^T$, and the input vector u is given by $u = (u^1, u^2, u^3)^T$, the order of the total system is N=6, the order of the system input is 3, and the vector $\hat{a}(x)$ is given by:

$$\hat{a}(x) = \begin{bmatrix} \dot{q}^1 \\ a_{22}^1 \dot{q}^1 + f_2^1(1-H_{22}f_2^2)d_1 + H_{21}(f_2^1)^2 d_2 \\ \dot{q}^2 \\ a_{22}^2 \dot{q}^2 + H_{12}(f_2^2)^2 d_2 + f_2^2(1-H_{11}f_2^1)d_2 \\ \dot{q}^3 \\ a_{22}^3 \dot{q}^3 + f_2^3(1-m_3 f_2^3)^{-1} m_3 (a_{22}^3 \dot{q}^3 - g) \end{bmatrix} \qquad (3.2.28)$$

The matrix $\hat{B}(x)$ in this particular case is given by:

$$\hat{B}(x) = \begin{bmatrix} 0 & 0 & 0 \\ \hat{b}_{21} & \hat{b}_{22} & 0 \\ 0 & 0 & 0 \\ \hat{b}_{41} & \hat{b}_{42} & 0 \\ 0 & 0 & 0 \\ 0 & 0 & \hat{b}_{63} \end{bmatrix} \qquad (3.2.29)$$

$$\hat{b}_{21} = b_2^1 + [f_2^1(1-H_{22}f_2^2)H_{11}b_2^1 + (f_2^1 H_{21})^2 b_2^1]/DET$$

$$\hat{b}_{22} = [f_2^1(1-H_{22}f_2^2)H_{12}b_2^2 + (f_2^1)^2 H_{21}H_{22}b_2^2]/DET$$

$$\hat{b}_{41} = [H_{12}(f_2^2)^2 H_{11}b_2^1 + f_2^2(1-H_{22}f_2^1)H_{21}b_2^1]/DET$$

$$\hat{b}_{42} = b_2^2 + [(f_2^2 H_{12})^2 b_2^2 + f_2^2(1-H_{11}f_2^1)H_{22}b_2^2]/DET$$

$$\hat{b}_{63} = b_2^3[1 + f_2^3(1-m_3 f_2^3)^{-1} m_3]$$

where a_{jk}^i, b_j^i, f_j^i are elements of the corresponding matrices and vectors $A^i = [a_{jk}^i]$, $b^i = [b_j^i]$, $f^i = [f_j^i]$, H_{jk} are elements of the matrix H (and they are functions of q^2, $\dot{q}^1$, $\dot{q}^2$) while d_1, d_2, DET denote:

$$d_1 = [H_{11}a_{22}^1\dot{q}^1 + H_{12}a_{22}^2\dot{q}^2 + h_1]/DET$$

$$d_2 = [H_{21}a_{22}^1\dot{q}^1 + H_{22}a_{22}^2\dot{q}^2 + h_2]/DET$$

$$DET = (1-H_{11}f_2^1)(1-H_{22}f_2^2) - H_{12}H_{21}f_2^1 f_2^2, \quad h = (h_1, h_2, h_3)^T.$$

Exercises

3.11. Data on parameters of the robot in Fig. 3.2. are given in Table 3.2. The joints of the robot are driven by D.C. motors the values of parameters of which are given in Table 3.1. Calculate the numerical values of the vector $\hat{a}(x)$ and matrix $\hat{B}(x)$ which are given by (3.2.28), (3.2.29), for the following values of the state vectors: $x = (0, 0, 0, 0, 0, 0)^T$, $x = (1.57, 0., 1.57, 0., 0.2, 0.)^T$ and $x = (1.57, 0.5, 1.57, 0.5, 0.2, 0.1)^T$.

LINK	1	2	3
MASS [kg]	7.	-	4.
r_{ii}^x [m]	0.3	0.	0.
r_{ii}^y [m]	0.	0.	0.2
r_{ii}^z [m]	0.	0.2	0.3
$r_{i+1,i}^x$ [m]	-0.3	0.	0.
$r_{i+1,i}^y$ [m]	0.	0.	0.2
$r_{i+1,i}^z$ [m]	0.	-0.3	0.
Moment of inertia around the z axis [kgm^2]	0.3	-	$\sim$0.

Table 3.2. Data on robot from Fig. 3.2. ($\vec{r}_{ii}$, $\vec{r}_{i+1,i}$ are vectors from the i-th and i+1-st joint to the centre of mass of the i-th link)

3.12. For the robot in Fig. 3.3. write the total model in the form
(3.2.27) assuming that each joint is driven by D.C. motor the
model of which is given by (3.2.8) (adopt the second order models of actuators). The link masses are m_1=2. [kg], m_2=1. [kg],
moments of inertia J_1 = 0.02 [kgm^2], J_2 = 0.01 [kgm^2], the
lengths of links are ℓ_1=0.3 [m], ℓ_2=0.15 [m] and the centers of
masses are at the middle-points of the links. Data on D.C. motors are given in the first two rows of the Table 3.1. Calculate
the values of the vector $\hat{a}(x)$ and the matrix $\hat{B}(x)$ for this particular robot and for the following particular values of the state vector $x = (q^1, \dot{q}^1, q^2, \dot{q}^2)^T = (0., 0., 0., 0.)^T$ and $x =$
$(0.3, 0.3, 0.2, 0.1)^T$.

3.13. Show that the vector $\hat{a}(x)$ and matrix $\hat{B}(x)$ might be written in
the form:

$$\hat{a}(x) = Ax+F[HT(I_N-FHT)^{-1}(Ax+Fh)+h]$$

$$\hat{B}(x) = B+FHT(I_N-FHT)^{-1}B$$

where I_N is the N×N unit matrix. Which of the two forms (this or
(3.2.27)) of matrices is numerically more convenient? (Instruction: substitute P from (3.2.22) into (3.2.23) and determine $\dot{x}$).

3.14. Show that if we adopt that the order of the model of actuator is
n_i=3 (for all actuators in robotic system) and if models of actuators are given by (3.2.7), (3.2.8), then the matrix $\hat{B}$ is not
function of system state i.e. $\hat{B}(x) = B$.

3.15. Show that if the relation between θ^i and q^i is given by (3.2.14)
and relation between M_i^* and P_i is given by (3.2.17) (for all joints of the robot), then the vector $\hat{a}(x)$ and matrix $\hat{B}(x)$ get the
following forms:

$$\hat{a}(x) = [I_N-Fz(G(x))\cdot H\cdot\tilde{G}(x)]^{-1}[Ax+Fz(G(x))(\hat{G}(x)x^T\cdot x+h)]$$

$$\hat{B}(x) = [I_N-Fz(G(x))\cdot H\cdot\tilde{G}(x)]^{-1}B$$

where $z(q) = diag(z^i(q^i)^{-1})$, $q = G(x) = (\tilde{g}^1(x^1), \tilde{g}^2(x^2),...$
$...,\tilde{g}^n(x^n))^T$ and $\ddot{q} = \tilde{G}(x)\dot{x} + \hat{G}(x)x^T\cdot x$.

3.16.[*] Write the programme (using some high programming language) to
compute the vector $\hat{a}(x)$ and matrix $\hat{B}(x)$ for an arbitrary robot
with n d.o.f. for which the matrix H and vector h are computed
in separate programme. Robot is powered by D.C. motors. The models of actuators are given by (3.2.8). The input to the programme is value of the vector of the system state x, the values of
the corresponding matrix H and vector h and the values of the
actuator matrices A^i, b^i and f^i. Combining this programme with
the programme written in Exercise 3.5. compute the total model
matrices for the robots in Figs. 3.2. and 3.3. for some particular values of the state vector.

3.3 Synthesis of Local Servo System

In the previous sections we have presented the model of dynamics of the
robotic system. In the text to follow we shall consider synthesis of
control at the executive control level. As we have explained in the
previous chapter, the tactical control level generates desired positions and/or trajectories of the so-called internal coordinates (or joint
coordinates) and sends them to the executive control level, which has to
realize them. Thus, the task of the executive control level is to ensure driving of the robot joints to the the desired positions, or their
driving along prescribed trajectories. The executive control level has
to ensure desired positions q^o of joint coordinates, or to ensure tracking of desired joint trajectories $q^o(t)$, velocities $\dot{q}^o(t)$ and accelerations $\ddot{q}^o(t)$.

In this chapter we shall consider synthesis of servo system for each
individual joint (and corresponding actuator) of the robot. We shall
consider one joint of the robot and we shall synthesize servo system
which has to realize any desired position q^{oi}, or desired trajectory
$q^{oi}(t)$ of that joint. Such a control law, which is synthesized for each
joint of the robot independently, is called *decentralized control of
robot*[*]. This control, which "neglects" dynamics of the total robotic
system, cannot be applied for all the tasks that are imposed to the
robots in industry and in other applications. This fact will be considered in the next chapter. However, this control law is the simplest

[*] In literature this control law is often called "naive" decentralized
control since in synthesis of servo systems for each joint independently we do not take care of dynamics of the total systems, nor of
interconnections between the joints.

one from the point of view of implementation, and this is the reason
why it is most frequently applied in robot controllers in practice.

We shall consider the i-th joint of the robot and its corresponding
actuator (see Fig. 3.8). Let us assume that all the other joints are
locked in some positions, i.e. that they keep constant values of their
angles (or linear displacements) q^{j^*} = const., for $j \neq i$, so that just
the i-th joint might be moved. Thus, the actuator in the i-th joint
moves that joint and the part of the mechanism which is behind the
i-th joint in the kinematic chain. The moment of inertia of the moving
part of the mechanism reduced to the axis of the i-th joint (for the
fixed values of the other joints q^{j^*}) will be denoted by $\bar{H}_{ii}$. This value is, obviously, fixed since the joints behind the i-th joint are
locked and H_{ii} does not depend on q^i. The mechanism can move just around
the i-th joint and the load moment around the axis of the i-th joint
includes two components only: inertia moment due to rotational or linear acceleration $\ddot{q}^i$ and gravity moment of the mechanism around the i-th
joint $g_i(q^i, q^*)$. Here q^* denotes vector $q^* = (q^{1^*}, q^{2^*}, \ldots, q^{i-1^*}, q^{i+1^*},$
$\ldots, q^{n^*})^T$. Thus, we can write that the moment around the i-th joint is:

$$P_i = \bar{H}_{ii}(q^*)\ddot{q}^i + g_i(q^i, q^*) \tag{3.3.1}$$

Gravity moment is in general case, nonlinear function of all joint coordinates q^j, $j=1,2,\ldots,n$. It is also function of q^i. First we shall
neglect this gravity term, but latter on we shall take it into account.

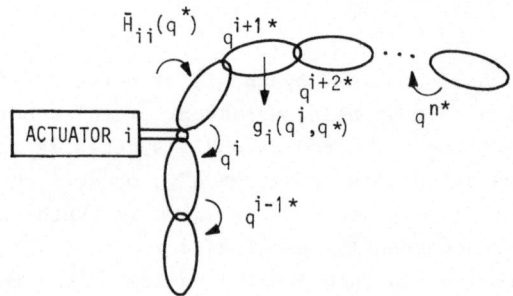

Fig. 3.8. Actuator in the i-th joint of the robot
(all the other joints are locked)

The actuator drives the i-th joint and the load M_i^* is acting upon that
actuator. This load is equal to moment P_i which is determined by

(3.3.1). Thus, the model of the considered system (if just the i-th joint is moving and all the other joints are locked) can be easily obtained by combining model of the actuator driving the mechanism around the i-th joint (3.2.8) and the model of the load moment (3.3.1):

$$\dot{x}^i = A^i x^i + b^i N(u^i) + f^i (\bar{H}_{ii} \ddot{q}^i + g_i (q^i, q^*)) \tag{3.3.2}$$

If we apply relation between q^i and x^i given by (3.2.20) we obtain

$$\dot{x}^i = (I_{ni} - f^i \bar{H}_{ii} T_i)^{-1} [A^i x^i + b^i N(u^i) + f^i g_i (q^i, q^*)] \tag{3.3.3}$$

where I_{ni} denotes the $n_i \times n_i$ unit matrix.

If we neglect gravity term in (3.3.3), we obtain model of the actuator and the load for this specific case in the following form:

$$\dot{x}^i = \hat{A}^i x^i + \hat{b}^i N(u^i) \tag{3.3.4}$$

where $\hat{A}^i$ is $n_i \times n_i$ matrix given by:

$$\hat{A}^i = (I_{ni} - f^i \bar{H}_{ii} T_i)^{-1} A^i$$

and $\hat{b}^i$ is $n_i \times 1$ vector given by:

$$\hat{b}^i = (I_{ni} - f^i \bar{H}_{ii} T_i)^{-1} b^i$$

In this way we obtain the model describing the behaviour of the individual actuator in the i-th joint if all the other joints are locked. Our task is to synthesize the control law which will control the actuator and the mechanism modelled by (3.3.4). We have to synthesize control which will ensure that the i-th joint will be driven towards the desired position q^{oi} imposed by the higher control level, or which will ensure tracking of desired trajectory $q^{oi}(t)$. First, we shall consider the case when the control system has to ensure just positioning of the joint, and afterwards the tracking problem will be addressed. First we shall neglect gravity term, and latter on we shall consider its effect upon the control of the actuator. We shall synthesize the so-called *static controller* first, and afterwards we shall consider *dynamic controller* (i.e. controller which includes feedback loop by integral of position error so that the order of the system model increases - see Section 3.4).

EXAMPLE 3.3. Consider the first joint of the robot in Fig. 3.2. Let us assume that the second and the third joints are locked in positions $q^{2*} = 0.$ and $q^{3*} = 0$. Then, the moment of the inertia of the mechanism around the axis of the first joint is given by:

$$\bar{H}_{11} = J_{1z} + m_1 \ell_1^{*2} + m_3 \ell_1^2 + J_{3z} + m_3 \ell_3^{*2} + 2m_3 \ell_3^* \ell_1 \qquad (3.3.5)$$

There is no gravity moment around the first joint. Thus, the model of the actuator and the mechanism in this case is given by (3.3.4) where the matrices $\hat{A}^i$ and $\hat{b}^i$ are (if we adopt the model of the actuator in the form (3.2.8) with matrices given by (3.2.7)):

$$
\hat{A}^1 =
\begin{bmatrix}
0 & 1 & 0 \\[2mm]
0 & -\dfrac{F_v^1}{N_v^1 N_m^1 J_M^1 + \bar{H}_{11}} & \dfrac{C_M^1 N_m^1}{N_v^1 N_m^1 J_M^1 + \bar{H}_{11}} \\[4mm]
0 & -\dfrac{C_E^1 N_v^1}{L_R^1} & -\dfrac{r_R^1}{L_R^1}
\end{bmatrix},
\quad
\hat{b}^1 =
\begin{bmatrix}
0 \\[2mm]
0 \\[4mm]
\dfrac{1}{L_R^1}
\end{bmatrix}
\qquad (3.3.6)
$$

If we apply the D.C. electro-motor the parameters of which are given in Table 3.1. (first row) and if the parameters of the robot mechanism are given in Table 3.2, then the matrices (3.3.6) get the following numerical values:

$$
\hat{A}^1 =
\begin{bmatrix}
0 & 1 & 0 \\
0 & -0.00165 & 0.42 \\
0 & -621.74 & -695.65
\end{bmatrix},
\quad
\hat{b}^1 =
\begin{bmatrix}
0 \\
0 \\
434.78
\end{bmatrix}
\qquad (3.3.7)
$$

If we compare (3.2.10) and (3.3.7), then it is clear that the model of actuator with the mechanism rotating around the axis of the i-th joint (if all other joints are locked) has the equal form as the model of the actuator itself (i.e. they are both linear time invariant system, their orders are equal and the distribution of the elements in matrices are equal), but the numerical values of the elements of the matrices are different (due to the fact that the moment of inertia of the mechanism is added to the actuator model). Namely, the only difference between these two models is in the moment of inertia: in the model (3.3.6) the equivalent moment of inertia of the motor rotor is increased by the moment of inertia of the mechanism $\bar{H}_{ii}$.

Exercises

3.17. Calculate the elements of the matrices $\hat{A}^i$, $\hat{b}^i$ in (3.3.6) for the first joint of the robot in Fig. 3.2. (Example 3.3) if the second joint is locked in position $q^{2*} = 1.573$ [rad] (data on parameters are given in Tables 3.1. and 3.2). How much have changed the numerical values of the elements of the matrices with respect to (3.3.7)?

3.18. Write the expressions for the elements of the matrices $\hat{A}^i$ and $\hat{b}^i$ in (3.3.4) for the first joint of the robot in Fig. 3.2, if we adopt the second order model of the actuator $n_i = 2$ and if the matrices of actuator are given by (3.2.11). Calculate these matrices for data given in Tables 3.1. and 3.2. for the following two cases: if the second joint is locked (a) in the position $q^{2*} = 0$, and (b) in the position $q^{2*} = 1.573$ [rad].

3.19. Determine the matrices $\hat{A}^i$ and $\hat{b}^i$ in (3.3.4) for the second and the third joints of the robot in Fig. 3.2, if we apply D.C. electro-motors which models are given by (3.2.8) and (3.2.7). Compute these matrices for data given in Tables 3.1. and 3.2.

3.20. Determine matrices $\hat{A}^i$ and $\hat{b}^i$ in (3.3.4) for the three joints of the robot in Fig. 2.5, if D.C. electro-motors are applied which models are given by (3.2.7) and (3.2.8). Data on parameters of the applied D.C. electro-motors are given in Table 3.3. and data on parameters of the robot mechanism are given in Table 3.4.

ACTUATOR	1	2	3
c_E [$\frac{V}{rad/s}$]	0.0459	0.0459	0.0459
c_M [$\frac{Nm}{A}$]	0.0480	0.0480	0.0480
J_M [kgm^2]	0.00003	0.00003	0.00003
N_v [-]	31.17	2616	1570
N_m [-]	31.17	2616	1570
r_R [Ω]	1.6	1.6	1.6
F_v [$\frac{Nm}{rad/s}$]	0.0058	0.0154	0.000923
L_R [H]	0.0023	0.0023	0.0023

Table 3.3. Data on parameters of the actuators for the robot in Fig. 2.5.

LINK	1	2	3
Mass [kg]	10.0	7.	4.15
Link length [m]	0.213	0.026	0.036*
Moment of inertia around the vertical axis [kgm^2]	0.0294	0.055	0.318

Table 3.4. Data on mechanism of the robot in Fig. 2.5.
(*link length is for q^1=0, centre of mass
of the third link is at distance of 0.036[m])

3.3.1. Elements of local servo system

Now we shall consider synthesis of control for the actuator in the
i-th joint under assumption that all the other joints are locked. We
have adopted the model of such system in a form (3.3.4). Thus, the
system to be controlled is described by linear time-invariant model
(system of linear differential equations of the order n_i). The controller has to ensure that for each imposed value of joint angle q^{oi} the
actuator drives the joint to this desired position. This task can be
accomplished by classical position servo system. We shall consider
synthesis of such position servo system around the actuator in the i-th
joint. The synthesis of servo system is well known from classical control theory. Here, we shall briefly repeat some of its basic characteristics, since the servo systems are essential for understanding robot
control.

The servo system can be implemented either by analogue technology, or
by microprocessors. The elements of the servo systems are basically the
same in both cases, and only their implementation differs. First, we
shall consider implementation of servo system by analogue technology,
and afterwards we shall present its microprocessor implementation. The
block-diagram of the servo system is presented in Fig. 3.9. [3].

Let us consider the elements of the servo system. For the sake of simplicity we shall first consider their transfer functions in s-domain,
and afterwards we shall consider the state space model of the servo
system. The elements of servo systems are: detector of the error signal, amplifiers, D.C. electro-motor with its mechanical load (actuator
and the moving part of the robot mechanism), sensor of velocity (e.g.

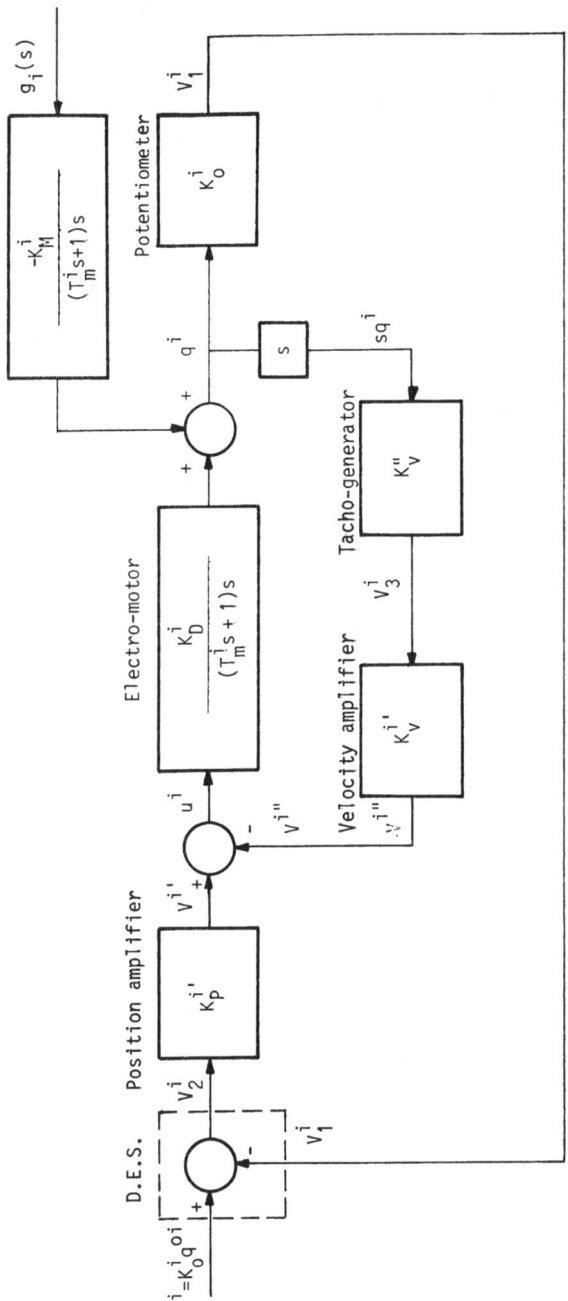

Fig. 3.9. The structural block-diagram of the position servomechanism (electro-mechanic servo system) with D.C. electro-motor (the model of the electro-motor is of the second order – the electrical constant is neglected)

tacho-generator) and sensor of position of the joint (e.g. potentiometer). We assume that the voltage corresponding to desired position of the i-th joint q^{oi} is fed at the input V^i of the servo system. If we implement the servo system by analogue technology and the tactical control level is implemented by microprocessor, then the desired position is fed to the servo system input through D/A converter. The voltage signal from D/A converter is proportional to the desired position, i.e. the voltage at the servo system input is $V^i = K_o^i q^{oi}$, where K_o^i is the proportionality coefficient.

Detector of the error signal. At the output of this element the voltage, which is proportional to the error of the servo system output position, is produced. The error in positioning of servo system is given as a difference between the actual position of the joint q^i and the desired (imposed) position q^{oi}. This element has two inputs: the first input is the input voltage V^i to the servo system which is proportional to desired position q^{oi}, and the second input is the voltage V_1^i which is proportional to actual position and which is obtained from the position sensor placed in the i-th joint axis. Voltage from the position sensor (e.g. potentiometer) V_1^i is proportional to actual position of the joint q^i, i.e. $V_1^i = K_o^i q^i$. (For the sake of simplicity we shall assume that the coefficient of proportionality is the same as for the input voltage V^i and the desired position q^{oi}). The voltage at the output of the detector of the error signal is equal to difference between these two input voltages:

$$V_2^i = V^i - V_1^i = -K_o^i(q^i - q^{oi}) = -K_o^i \Delta q^i \qquad (3.3.8)$$

i.e. the voltage is directly proportional to the error in positioning Δq^i. This element does not introduce any nonlinearity in the servo system. Its transfer function (including the position sensor) can be obtain from (3.3.8) as:

$$\frac{V_2^i(s)}{\Delta q^i(s)} = -K_o^i \qquad (3.3.9)$$

Amplifiers. The role of the amplifier of the position error in the servo system is to amplify the error signal. The amplifier in the velocity feedback loop has to amplify the feedback signal by velocity, which is obtained from the tacho-generator (as a sensor of velocity). In implementation of the servo system by analogue technology these two

amplifiers are implemented by electronic amplifiers whose time constants are considerably lower than the time constants of the electro-
-mechanic components in the servosystem (i.e. of the actuator and the
moving part of the robot mechanism). Due to this, these amplifiers
might be considered as non-dynamic elements. The transfer function of
the amplifier of the position error is given by:

$$\frac{v^{i'}(s)}{v_2^i(s)} = K_P^{i'} \tag{3.3.10}$$

where $K_P^{i'}$ is the gain of the amplifier.

The amplifier in the velocity feedback loop has the following transfer
function:

$$\frac{v^{i''}(s)}{v_3^i(s)} = K_v^{i'} \tag{3.3.11}$$

where $K_v^{i'}$ is the amplifier gain, v_3^i is the output voltage of the tacho-
-generator, and $v^{i''}$ is the output voltage of the amplifier.

D.C. electro-motor and mechanical part of the robot. This element of
the servo system consists of the actuator (which might be either electro-hydraulic actuator, or electro-pneumatic, or D.C. electro-motor
etc.) and the mechanical part of the robot which can rotate around the
axis of the i-th joint of the robot while the other joints are locked.
This element has been already considered in the previous section. The
model of this element in the state space is given by (3.3.4). Its
transfer function in the s-domain is given by (for the actuator model
of the third order $n_i=3$ and matrices $\hat{A}^i$ and $\hat{b}^i$ given by (3.3.6)[*]):

$$\frac{q^i(s)}{u^i(s)} = \frac{K_D^i}{[T_R^i T_m^i s^2 + (T_R^i F_D^i + 1) T_m^i s + 1] s} \tag{3.3.12}$$

where T_R^i is the electric time constant given by:

$$T_R^i = \frac{L_R^i}{r_R^i} \tag{3.3.13}$$

while T_m^i is the electro-mechanic time constant given by:

---

[*] Here, we have neglected the nonlinear amplitude constraint upon the
actuator input. This nonlinearity we shall consider in Sect. 3.3.5.

$$T_m^i = \frac{r_R^i (N_m^i N_v^i J_M^i + \bar{H}_{ii})}{(N_m^i N_v^i C_M^i C_E^i + r_R^i F_v^i)} \qquad (3.3.14)$$

K_D^i and F_D^i are the constants given by:

$$K_D^i = \frac{C_M^i N_m^i}{(N_m^i N_v^i C_M^i C_E^i + r_R^i F_v^i)} \qquad F_D^i = \frac{F_v^i}{N_m^i N_v^i J_M^i + \bar{H}_{ii}} \qquad (3.3.15)$$

We can easily obtain the transfer function (3.3.12) if we remember that the state space model (3.3.4) has been formed on the basis of equations of mechanical and electrical equilibrium in the D.C. electro-motor (3.2.4) and (3.2.5).

However, let us remind that we have neglected the gravity moment around the i-th joint. This gravity moment acts as an external load around the output shaft of the reducer and it might be treated as an external disturbance upon the servo system. The transfer function from the external load moment g_i to the servo system output q^i is given by (if we remember the equation of the moment equilibrium around the output shaft of the reducer (3.2.4)):

$$\frac{q^i(s)}{g_i(s)} = \frac{-K_M^i(1+T_R^i s)}{[T_R^i T_m^i s^2 + (T_R^i F_D^i + 1) T_m^i s + 1]s} \qquad (3.3.16)$$

where

$$K_M^i = \frac{K_D^i r_R^i}{C_M^i N_m^i}$$

In the case when the electrical time constant of the D.C. motor T_R^i is significantly lower than the electro-mechanical time constant T_m^i, then we can neglect T_R^i (which actually means that we adopt the second order model of the actuator $n_i = 2$), the transfer functions (3.3.12) and (3.3.16) get the following forms:

$$\frac{q^i(s)}{u^i(s)} = \frac{K_D^i}{(T_m^i s + 1)s} \qquad (3.3.17)$$

$$\frac{q^i(s)}{g_i(s)} = \frac{-K_M^i}{(T_m^i s + 1)s} \qquad (3.3.18)$$

The input voltage to the actuator u^i is formed combining the output signals of the amplifiers, i.e.:

$$u^i = v^{i'} - v^{i''} \qquad (3.3.19)$$

Tacho-generator. Tacho-generator is the sensor of the velocity. Its output voltage v_3^i is proportional to the velocity of the movement of the joint. As the sensor of the joint velocity usually serves the generator which is placed at the output shaft of the D.C. motor (before the reducer). This generator generates the voltage in its windings which is proportional to the rotational velocity of the output shaft of the motor. Since the rotational velocity of the motor is directly proportional to the angular (or, linear) velocity of the corresponding joint of the robot, the voltage v_3^i is generated at the output of the tacho-generator which is directly proportional to the angular velocity of the joint $\dot{q}^i$. Thus, the transfer function of the tacho-generator can be adopted in the form:

$$\frac{v_3^i(s)}{\dot{q}^i(s)} = \frac{v_3^i(s)}{sq^i(s)} = K_v^{i''} \qquad (3.3.20)$$

Sensor of position. Various sensors can be used to measure the actual position of the joint. One of the most usually applied sensor is simple potentiometer. On its output the potentiometer gives the voltage v_1^i which is proportional to the joint position (angle or linear displacement) q^i. Thus, the transfer function of this element is in the form:

$$\frac{v_1^i(s)}{q^i(s)} = K_o^i \qquad (3.3.21)$$

Obviously, we might assume that all elements in the servo system are linear. The total transfer function of the servo system from the input $q^{oi}(s)$ to its output (joint position) $q^i(s)$ can be obtained by simple combining of the transfer functions of the elements according to their arrangement in the servo system, as presented in Fig. 3.9. So, in the case when the third order transfer function of the actuator and the mechanical part of the robot is adopted (3.3.12), we obtain the transfer functions from the input of the servo $q^{oi}(s)$ to its output $q^i(s)$, and from the load disturbance $g_i(s)$ to the servo system output, in the following forms:

$$\frac{q^i(s)}{q^{oi}(s)} = \frac{K_P^i K_D^i}{K_D^i K_P^i + K_D^i K_V^i s + [T_R^i T_m^i s^2 + (T_R^i F_D^i + 1) T_m^i s + 1] s}$$

$$(3.3.22)$$

$$\frac{q^i(s)}{g_i(s)} = \frac{-K_M^i (1 + T_R^i s)}{K_D^i K_P^i + K_D^i K_V^i s + [T_R^i T_m^i s^2 + (T_R^i F_D^i + 1) T_m^i s + 1] s}$$

where by K_P^i we denote the total gain of the position feedback loop (we shall call this gain position gain), from the joint coordinate - posi-error Δq^i to the input voltage of the actuator $V^{i'}$:

$$K_P^i = \frac{V^{i'}(s)}{\Delta q^i(s)} = K_P^{i'} K_o^i \qquad (3.3.23)$$

while by K_V^i we denote the total gain of the velocity feedback loop (we shall call this gain velocity gain) from the joint velocity $\dot{q}^i$ to the input voltage of the actuator $V^{i''}$.

$$K_V^i = \frac{V^{i''}(s)}{sq^i(s)} = K_V^{i'} K_V^{i''} \qquad (3.3.24)$$

In the case when we adopt the second order transfer function (3.3.17) of the actuator and the mechanical part of the robot (i.e. when $n_i = 2$), we obtain also the second order transfer functions of the complete servo:

$$\frac{q^i(s)}{q^{oi}(s)} = \frac{K_P^i K_D^i}{K_D^i K_P^i + K_D^i K_V^i s + (T_m^i s + 1) s}$$

$$(3.3.25)$$

$$\frac{q^i(s)}{g_i(s)} = \frac{-K_M^i}{K_D^i K_P^i + K_D^i K_V^i s + (T_m^i s + 1) s}$$

In this way we obtain the transfer functions in the s-domain for the classical servo system with feedback loops by position and by velocity.

Let us consider briefly the microprocessor implementation of the position servo system. The direct digital servo contains all the elements as the servo implemented by analogue technology, but some of these elements in digital version are implemented by the microprocessor. Several versions of the microprocessor implementation of servo systems are encountered in practice. One of these possible microprocessor implementations of the servo is presented in Fig. 3.10. The following elements are implemented by microprocessor: the detector of the error

signal, amplifications in the position and in the velocity feedback
loops. The other elements are implemented in analogue technology. As
it can be seen in Fig. 3.10, the signals from the potentiometer and
tacho-generator are feedback to microprocessor through A/D converters[*].
The microprocessor computes (at the tactical control level) the desired
position of the joint q^{oi}. The microprocessor computes control (input
for the actuator u^i) in the following way: it multiplies the error in
positioning $q^i - q^{oi}$ by the position gain K_p^i, and to the computed value
it adds the value of the actual joint velocity $\dot{q}^i$ (obtained from the
tacho generator and through the A/D converter) multiplied by velocity
gain K_v^i. The resulting numerical value of the input signal for the actuator is by D/A converter converted to voltage signal as the actuator
input u^i. The second possible microprocessor implementation of servo
system is presented in Fig. 3.11. In this version the velocity feedback
loop is implemented in analogue technology, so the microprocessor just
has to multiply the error in positioning by the position gain. Here, we
shall not consider in detail the microprocessor implementation of servo
system. The more details on these problems can be found in literature
[4]. We just wanted to point out that the scheme of the digital servo
system is essentially similar to the implementation of the servo system
by analogue technology.

EXAMPLE 3.3.1. For the first joint of the robot in Fig. 3.2. the D.C.
electro-motor is applied, the data on which are given in Table 3.1.
The data on mechanical part of the robot are given in Table 3.2. In
Example 3.3. we have calculated the matrices of the model (3.3.6) for
this joint (if the second and the third joints are locked in positions
$q^{2*} = 0.$ and $q^{3*} = 0.$). Now, we shall determine the transfer function
in the s-domain for this actuator with the mechanism in which just the
first joint is moving. If we adopt the third order model of the actuator $n_1 = 3$, then the transfer function in the s-domain is obtained as:

$$\frac{q^1(s)}{u^1(s)} = \frac{185.33}{(s^2+695.65s+266.17)s} \tag{3.3.26}$$

If we neglect the electrical time constant T_R^1 (i.e. if we adopt the
second order model of actuator $n_1=2$) we obtain the transfer function:

_____

[*] Instead the potentiometer it is often applied *shaft encoder* which
generates directly digital information on the actual position of the
joint, so there is no need to apply A/D converter to convert analogue voltage signal from the sensor to digital number which is used
by microprocessor.

84

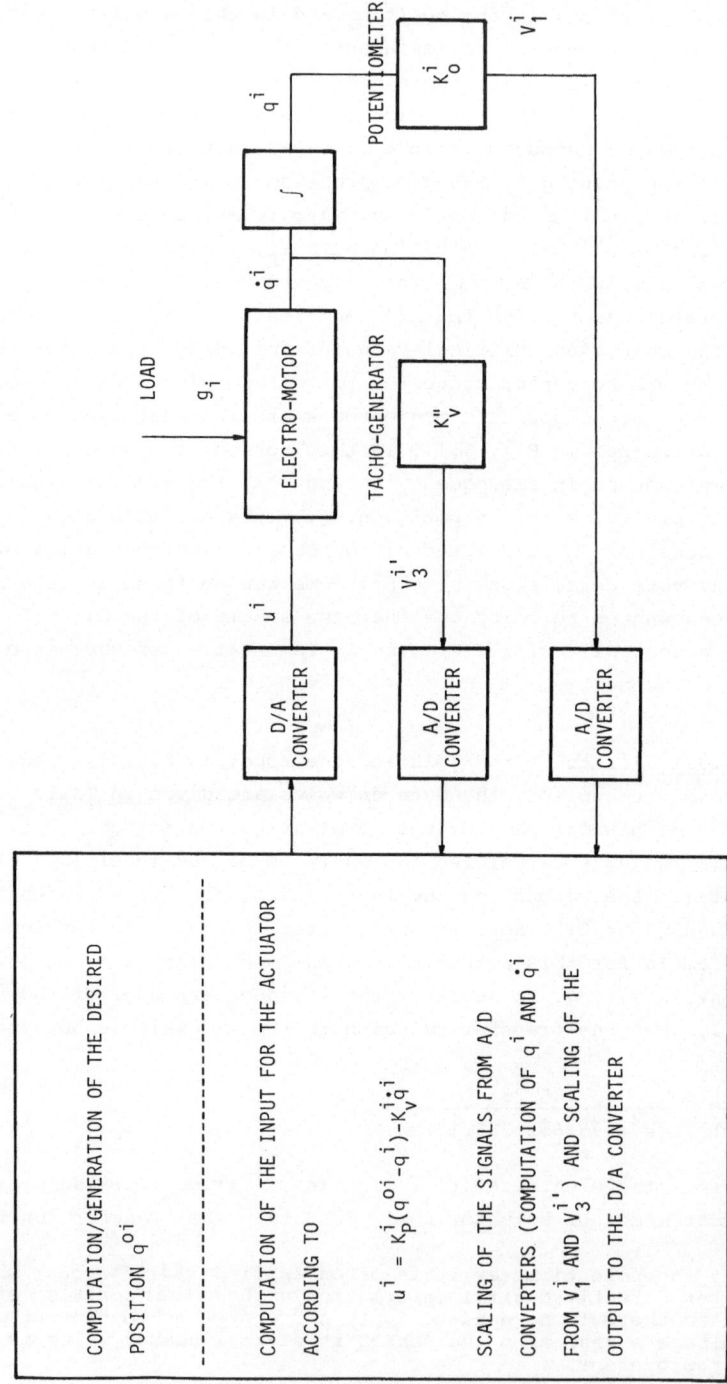

Fig. 3.10. Global scheme of the microprocessor implementation of the position servo system

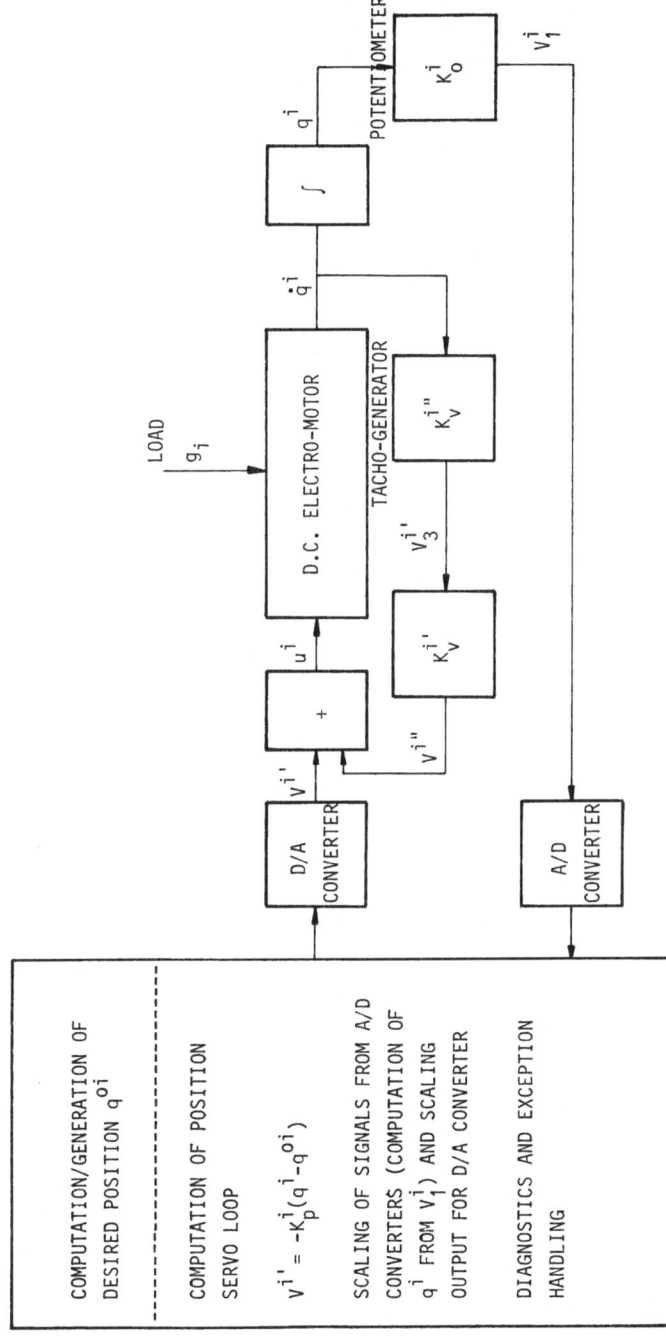

Fig. 3.11. Global scheme of the microprocessor implementation of the position servo system with velocity feedback loop implemented by analogue technology

$$\frac{q^1(s)}{u^1(s)} = \frac{0.26642}{(s+0.382)s} \tag{3.3.27}$$

If we establish servo system around these actuator and joint, then we can obtain the transfer function of the servo as a function of position and velocity gains (this transfer function corresponds to the third order transfer function of the actuator and the moving part of mechanism (3.3.26)):

$$\frac{q^1(s)}{q^{o1}(s)} = \frac{185.33\,K_P^i}{(s^2+695.65s+266.17)s+185.33(K_v^i s+K_P^i)} \tag{3.3.28}$$

Exercises

3.21. Starting from the model of the actuator and the moving part of the robot mechanism in the state space (3.3.4) and the corresponding matrices for D.C. electro-motor, show that the transfer functions in the s-domain for this system are given by (3.3.12) and (3.3.16).

3.22. Repeat the previous exercise but for the second order model of the actuator ($n_i=2$), so that the corresponding transfer functions are given by (3.3.17) and (3.3.18).

3.23. Show that the transfer function of the servo system around the first joint of the robot in Fig. 3.2. as a function of position and velocity gain is given by (3.3.28), if the transfer function of the actuator and the moving part of the robot is given by (3.3.26). Determine also the transfer function of this servo from the external load $g_1(s)$ to the servo system output (the joint angle q^1) as a function of position and velocity gains (for the third order model of the actuator this transfer function generaly is given by (3.3.22)).

3.24. Repeat the previous exercise but if the second joint of the robot in Fig. 3.2. is locked in the position q^{2*} = 1.573 [rad]. (Instruction: the matrices $\hat{A}^i$ and $\hat{b}^i$ have been already determined in the Exercise 3.17, determine the transfer functions for the actuator and the moving part of the robot).

3.25. Determine the transfer functions of the servos as functions of the position and the velocity gains for the actuators and "loads" in Exercise 3.19.

3.26. Determine the transfer functions of the servos (as functions of the position and the velocity gains) around the first three joints of the robot in Fig. 2.5. if we apply D.C. electro-motors, the data of which are given in Table 3.3. (the matrices of the actuators and "loads" models have been already determined in Exercise 3.20).

3.27[*]. Determine the transfer functions in the s-domain for the hydraulic actuator in the i-th joint of robot and for the moving part of the mechanism of robot if we assume that only the i-th joint is moving and all the other joints are locked. Determine also the transfer functions of the servo around this actuator. (Instruction: the matrices of the electro-hydraulic actuator model in the state space have been determined in Exercise 3.9. for two cases $n_i=2$ and $n_i=3$; the procedure is completely identical as for D.C. electro-motors). Calculate the transfer functions of the hydraulic servo system (for the second order model) as functions of the position and velocity gains, if data on the electro-hydraulic actuator are given in Table 3.5. and if this actuator is applied in the first joint of the robot in Fig. 3.2. (assume that $\theta^1=0.12\ q^1$, where θ^1 is obtained in [m], if q^1 is in [rad], and $P_1=0.12\ M_1^*$ where P_1 is in [Nm], if M_1^* is in [N]). Calculate these transfer functions for two positions in which the second joint of the robot is locked: for $q^{2*}=0.$ and for $q^{2*}=1.573$ [rad], (the leakage in hydraulic system is neglected).

Mass of piston [kg]	2.65
Piston area [cm^2]	12.6
Flow/pressure coefficient of servovalve $\left[\dfrac{cm^3/s}{N/m^2}\right]$	0.00075
Coefficient of proportionality of oil flow and the current of servovalve $\left[\dfrac{m^3/s}{mA}\right]$	0.0000833
Coefficient of the viscous friction $\left[\dfrac{N}{m/s}\right]$	30.
Compressibility coefficient [N/m^3]	$1.7\cdot10^7$
Cylinder volume [cm^3]	756.

Table 3.5. Data on electro-hydraulic actuator

3.3.2. Synthesis of servo system in s-domain

In the previous section we have determined the transfer function in the s-domain for the position servo system around the i-th joint of the robot. As we have seen, the transfer function of the servo is obtained by combining the transfer functions of the elements in servo. If the actuator, sensors and detector of the error signal have been already selected, then the transfer function of the servo system depends just on the selection of the gains in the position and velocity feedback loops.

The role of the position servo system is to ensure positioning of the joint q^i into the desired position q^{oi}. As it is well known [3], the transfer function of the system describes the behaviour of the system: it shows accuracy of achieving of the goal position of the joint (in the steady state regime) and it shows the characteristics of the transient process when the output variable is approaching to the desired value (i.e. to the value of the input variable q^{oi}). Obviously, the behaviour of the system can be affected by the selection of the position K_P^i and the velocity K_V^i gains. Thus, our task in synthesis of the servo system for the i-th joint of the robot is practically reduced to the selection of the position and velocity gains. We have to select feedback gains in such a way to satisfy certain requirements regarding the "quality" of the joint positioning.

Let us consider the transfer function of the position servo system. For the sake of simplicity we shall consider the second order transfer function, i.e. we shall neglect electrical time constant T_R^i of the actuator. Let us consider the transfer function (3.3.25) from the servo system input q^{oi} to the output q^i. Let us write this transfer function in the following form:

$$\frac{q^i(s)}{q^{oi}(s)} = \frac{K_D^i K_P^i / T_m^i}{(s-s_1)(s-s_2)} \tag{3.3.29}$$

where s_1 and s_2 are the roots of the characteristic polynome (i.e. the roots of the polynome in the denominator of the transfer function (3.3.25)). These roots might be written in the following form:

$$s_{1,2} = -\xi_i \omega_i \pm j \omega_i \sqrt{1-\xi_i^2} \tag{3.3.30}$$

where ω_i is called *the characteristic frequency of the system* or the undamped natural frequency and ξ_i is *the damping ratio* or, *damping factor* [5]. If we determine the roots s_1 and s_2 from (3.3.25), then we can see that the ω_i and ξ_i are functions of the actuator parameters and the moment of inertia of the mechanism, and of the feedback gains K_p^i and K_v^i. Since the actuator and mechanism parameters are fixed values, we can change ω_i and ξ_i by changing K_p^i and K_v^i.

It is well known that the servo system performance is defined by the characteristic frequency and the damping factor. Let us consider the response of the servo system to the step input function (the amplitude of the step function corresponds to desired angle of the joint q^{oi}). If $\xi_i < 1$, then the response of the servo system to the step function input is given by:

$$q^i - q^{oi} = c_1^i e^{-\xi_i \omega_i t} \sin(\omega_i \sqrt{1-\xi_i^2} t) + c_2^i e^{-\xi_i \omega_i t} \cos(\omega_i \sqrt{1-\xi_i^2} t) \quad (3.3.31)$$

where c_1^i and c_2^i are constants. In this case the response of the servo system is oscillatory around the desired position q^{oi}. The response is presented in Fig. 3.12. If $\xi_i < 1$ the servo system is *underdamped* or *undercritically damped*. In this case the response is fast, it consists of a sinusoidal function superimposed over an exponential function. The output of the servo exponentially approaches desired input position, but it *overshoots* the desired angle.

If $\xi_i > 1$ the servo system is *overdamped* or *overcritically damped*. In this case the response of the servo to step input signal q^{oi} is given by:

$$q^i - q^{oi} = c_3^i e^{-(\xi_i \omega_i + \omega_i \sqrt{\xi_i^2 - 1}) t} + c_4^i e^{-(\xi_i \omega_i - \omega_i \sqrt{\xi_i^2 - 1}) t} \quad (3.3.32)$$

where c_3^i and c_4^i are constants.

The response of the system is also presented in Fig. 3.12. In this case the response of the servo is slower, but it is non oscillatory and there is no overshoot of the final position.

At last, if $\xi_i = 1$ the servo system is *critically damped*. The response of the servo to step function input is given by:

$$q^i - q^{oi} = c_o^i e^{-\omega_i t} \quad (3.3.33)$$

where c_o^i is constant.

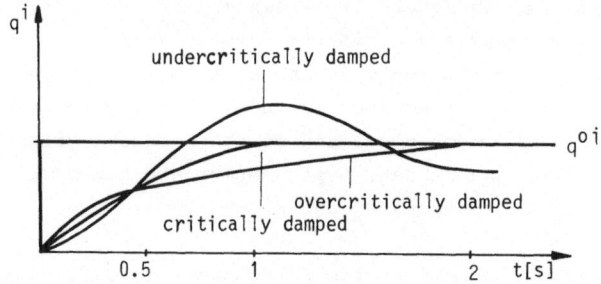

Fig. 3.12. Response of the servo system to step function input

The response of the servo is presented in Fig. 3.12. In this case the
response of the servo is fast but non oscillatory, and without overshoot of the desired position.

Now, we have to establish criteria for selection of servo feedback
gains K_p^i and K_v^i. Actually, we have to set requirements upon the "quality" of the servo system response, which we have to satisfy by appropriate selection of feedback gains.

First requirement concerns appearance of the overshoots in the servo
system response. It is well known, that with the robots the overshoots
of the desired position must not be allowed under any circumstances
[5]. For instance, if the desired position of the robot link is near
some obstacle in the work space and if the joint overshoots the desired position, the link would be driven into the obstacle (i.e. the robot
link would hit the obstacle). This is the reason why for robot it is
always required not to be undercritically damped, i.e. it must not be
allowed to be $\xi_i < 1$. On the other hand, it is required that the servo
response be as fast as possible, in order to achieve fast positioning
of the robot (and ensure fast execution of the set task). So, the most
appropriate response of the robot servo is achieved if $\xi_i = 1$, i.e. if
the servo is critically damped. Since the damping factor ξ_i is direct
function of the feedback gains we have to select the feedback gains to
achieve that $\xi_i = 1$.

The next requirement which has to be met in selection of the feedback
gains concerns the influence of the external load upon the servo output. In the previous section we have noted that the external load is
acting upon servo. In the considered case when all joints are locked

and just the i-th joint is moving the external load upon the servo is
gravity moment g_i. The transfer function (3.3.22), or (3.3.25), shows
the influence of this load upon the servo output (the joint position
q^i). Let us consider the steady state regime when the robot joint
stops, i.e. after the transient process is finished (theoretically it
occurs after infinite time $t \to \infty$, practically, after 3-5 time constants
of the robotic system, i.e. after $3 \div 5 \ (1/\omega_i \xi_i)$). Based on the transfer function (3.3.25), we can determine the *steady state error* between
the actual position of the joint q^i and the desired set position q^{oi}
if the constant gravity moment is loading the servo. Namely, since the
gravity moment is function only of the position of the i-th joint (the
other joints are locked), when the i-th joint stops, we might consider
the gravity moment as constant (i.e. as a step external load). Thus,
the steady state error is given by:

$$|\Delta q^i(\infty)| = |q^i(\infty) - q^{oi}| = \lim_{s \to 0} |s \Delta q_i(s)| = \frac{|g_i(\infty)| K_M^i}{K_D^i K_P^i} \tag{3.3.34}$$

where by $g_i(\infty)$ is denoted steady state gravity moment upon the i-th
joint. It can be seen that the steady state error of the servo is inversely proportional to position gain K_P^i. It is obvious that the steady state error represents the accuracy of the servo positioning.
Thus, we want to keep this error as low as possible. This error also
inform us about the robot *stiffness*. If we apply some external load
upon robot, the robot hand deflects due to finite stiffness of the servo.
This deflection can be also determined by (3.3.34), where the applied
external load (multiplied by the quadratic of the effective lever arm
from the i-th joint to the robot hand) should be substituted for $g_i(\infty)$.
It is required that the stiffness of the robot be as high as possible.
Obviously, if we want to keep steady state error low we must select
the position gain to be as high as possible. Even more, if we calculate the gravity moment $g_i(\infty)$ and if we define allowable steady state
error $\Delta q^i(\infty)$, i.e. if the allowable tolerance is set (or, if a desired
stiffness of the robot is imposed), then using (3.3.34) we can determine position gain K_P^i which satisfies the set requirements.

Based on (3.3.34) it follows that to ensure high precision of the robot
positioning, it is required to select the position gain as high as possible. However, too high position gain involve certain drawbacks. For
example, too high feedback gains amplify also the noise in the system
and, thus, their influence might become too strong.

It is obvious that the position feedback gain directly affects the characteristic frequency of the servo. Based on (3.3.25) and (3.3.30) it is easy to show that the characteristic frequency ω_i and the damping factor ξ_i are given by:

$$\omega_i = \frac{\sqrt{N_m^i C_M^i K_P^i}}{\sqrt{(N_v^i N_m^i J_M^i + \bar{H}_{ii}) r_R^i}} \tag{3.3.35}$$

$$\xi_i = \frac{N_m^i C_M^i \cdot (K_v^i + N_v^i C_E^i) + F_v^i r_R^i}{2\sqrt{N_m^i C_M^i K_P^i} \cdot \sqrt{(N_v^i N_m^i J_M^i + \bar{H}_{ii}) r_R^i}} \tag{3.3.36}$$

The characteristic frequency shows the frequency of oscillations of the servo system output around the final position during the transient process. The robot mechanism has its own resonant frequency with which the structure oscillates due to structural stiffness of the robot links. This frequency is called *structural resonant frequency* or, simply structural frequency. We have shown above that the feedback gains should be selected to ensure the servo system to be (over) critically damped. However, we have to keep in mind the following facts: the damping ratio ξ_i is function of several parameters, and the considered linear model of the actuator is just an approximation of the actual (nonlinear) system behaviour. Thus, although we select feedback gains so that $\xi_i > 1$, the oscillations of the servo system output could appear. The servo system would oscillate with frequency close to characteristic frequency ω_i. If the characteristic frequency of the servo ω_i is close to the structural frequency ω_o, the resonant oscillations of the robot structure will appear, i.e. the oscillations of the servo will excite structural oscillations. It is obvious that such oscillations are not acceptable. Thus, we must require that the characteristic frequency of the servo is under the structural frequency ω_o. However, the structural frequency is difficult to determine theoretically (using some model). Usually it is determined experimentally. Due to unreliability of determination of the structural frequency ω_o, it is often required that the characteristic frequency of the servo ω_i satisfies [5]:

$$\omega_i \leq 0.5\ \omega_o \tag{3.3.37}$$

If the characteristic frequency of the servo satisfies (3.3.37), where ω_o is the guess value of the structural frequency of the robot mecha-

nism, then the servo will not excite the oscillating modes of the robot, i.e. it will not cause resonant structural oscillations.

If the expression for the characteristic frequency (3.3.35) is introduced in (3.3.37), we obtain the upper limit upon the maximum allowable
value of the position gain, i.e.:

$$K_P^i \leq \frac{(0.5\ \omega_o)^2 (\bar{H}_{ii} + N_v^i N_m^i J_M^i) r_R^i}{N_m^i C_M^i} \qquad (3.3.38)$$

In this way we have determined the lower (3.3.34) and the upper limit
(3.3.38) upon the servo position gain. We have to select the feedback
gains to meet all the listed requirements. The following procedure in
selection of the feedback gains can be adopted. First, the position
gain K_P^i is selected according to (3.3.34) to satisfy allowable steady
state error. If the obtained value of position gain satisfy the requirement (3.3.38), then this value of position gain can be accepted.
Otherwise, if the condition (3.3.38) is not satisfied, then we must
compute position gain from (3.3.38) as a maximum allowable value. In
that case the required accuracy around the desired set point will not
be ensured, but we can reduce the steady state error by some other method (see Section 3.4).

Once the position gain K_P^i has been selected, we must determine the velocity gain K_v^i so as to ensure that the damping factor is equal to
one, i.e. to ensure that the servo is critically damped. Based on
(3.3.36) we obtain the velocity gain as:

$$K_v^i = \frac{2\sqrt{K_P^i (\bar{H}_{ii} + J_M^i N_v^i N_m^i) r_R^i}}{\sqrt{N_m^i C_M^i}} - N_v^i C_E^i - \frac{F_v^i r_R^i}{N_m^i C_M^i} \qquad (3.3.39)$$

In this way we select both position and velocity gains for the servo
controlling the actuator in the i-th joint of the robot, assuming that
all the other joints are locked. However, the validity of this solution when all the joints of the robot are moving simultaneously, will be
discussed latter on.

We have considered the second order transfer function of the servo system and we have presented the synthesis of the feedback gains. In the
similar way the synthesis of the feedback gains can be carried on if
the third order transfer function is considered.

EXAMPLE 3.3.2. The transfer function of the servo in the first joint of the robot presented in Fig. 3.2. is given by (3.3.28). The guess value of the structural resonant frequency for this joint is $\omega_o^1 = 12$ [rad/s]. The feedback gains can be calculated based on (3.3.38) and (3.3.39), and their values are:

$$K_P^1 \leq 135.13 \left[\frac{V}{rad} \right]$$

$$K_V^1 = 43.6 \left[\frac{V}{rad/s} \right]$$

(3.3.40)

where in calculating K_V^i we have taken the upper bound for position gain i.e. $K_P^i = 135.13$ [V/rad]. In this case the steady state error is zero since no gravity moment is acting upon the first joint (rotation around the vertical z - axis).

Exercises

3.28. Calculate the position and velocity gains for the servo in the first joint of the robot in Fig. 3.2. if $\omega_o^1 = 12$ [rad/s], using the third order transfer function determined in Exercise 3.23. (for $n_i = 3$), in such a way to satisfy conditions (3.3.38) and (3.3.39).

3.29. Repeat the previous exercise but if the second joint of the robot is locked in the position $q^{2*} = 1.573$ [rad] and $\omega_o^1 \approx 14$ [rad/s]. (Instruction: use the transfer function obtained in Exercise 3.24). How much the feedback gains have changed in respect to the previous case when $q^{2*} = 0.$? In order to prevent the servo to become underdamped in any position of the robot (i.e. to ensure that ξ_i is always equal or greater than one, $\xi_i \geq 1$), which of these two sets of feedback gains should be selected?

3.30. Repeat the previous two exercises for the same joint of the robot in Fig. 3.2, but if we apply as an actuator the D.C. electro-motor which parameters are given in Table 3.6. Compute (using the second order transfer functions of the servo) the servo feedback gains for two different positions of the second joint ($q^{2*} = 0.$ and $q^{2*} = 1.573$ [rad]) and determine the relative variation of

the feedback gains. Explain why in this case the relative variation of the feedback gains is less than for the previous D.C.
electro-motor (in Table 3.1)?

$c_E \left[\frac{V}{rad/s}\right]$	0.62
$c_M \left[\frac{Nm}{A}\right]$	0.61
J_M [kgm^2]	0.035
N_V [-]	190.
N_m [-]	154.
r_R [Ω]	0.3
$F_V \left[\frac{Nm}{rad/s}\right]$	25.13
L_R [H]	$\sim$0.01

Table 3.6. Data on D.C. electro-motor (Exercise 3.30)

3.31. For the servos which transfer functions have been determined in
Exercise 3.25, determine the feedback gains if ω_o^2 = 50 [rad/s],
ω_o^3 = 50 [rad/s], the allowable steady state error in the third
joint is $|\Delta q^3(\infty)|$ = 0.01 [m]. The conditions (3.3.34), (3.3.38)
and (3.3.39) have to be satisfied.

3.32. For the servos which transfer functions have been determined in
Exercise 3.26 (for the robot in Fig. 2.5) calculate the feedback
gains if ω_o^1 = 12 [rad/s], ω_o^2 = 50 [rad/s], ω_o^3 = 50 [rad/s] and
allowable steady state error in the second joint is $|\Delta q^2(\infty)|$ =
0.01 [m]. The conditions (3.3.34), (3.3.38) and (3.3.39) have to
be satisfied.

3.33. For the robot in Fig. 3.13. (so-called Stanford manipulator) compute the servo feedback gains if the data on the moments of
inertia of the mechanism $\bar{H}_{ii}$, data on D.C. electro-motors and data
on structural resonant frequencies of the links are given in Table
3.7. [5]. The conditions (3.3.38) and (3.3.39) have to be satisfied (the steady state errors are not pre-specified).

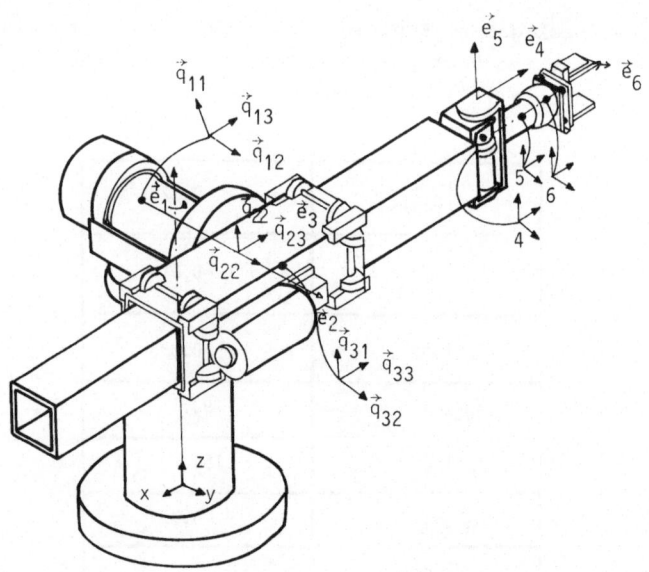

Fig. 3.13. The Stanford manipulator

L I N K	1	2	3
Moment of inertia of mechanism $\bar{H}_{ii}$ [kgm^2]	0.5	2.8	6.2
$J_M N_v N_m$ [kgm^2]	0.953	2.193	0.782
$C_M N_m \left[\dfrac{Nm}{A}\right]$	3.415	8.61	3.41
$C_E N_v \left[\dfrac{V}{rad/s}\right]$	4.043	10.6	4.043
r_R [Ω]	0.9	0.8	0.9
$F_v \left[\dfrac{Nm}{rad/s}\right]$	0.043	0.21	0.043
$f_o = \omega_o/2\pi$ [Hz]	4.	6.	20.

Table 3.7. Data on actuators, moments of inertia and structural
frequencies for Stanford manipulator (for the first
three joints) [5]

3.34.[*] Write the programme (in some high programming language) for computation of servo feedback gains if the input values for the programme are: data on actuator parameters (D.C. electro-motor), data
on the moment of inertia of the mechanism $\bar{H}_{ii}$, data on structural frequency ω_o of the mechanism, data on maximal allowable steady state error $\Delta q^i(\infty)$, and data on maximal gravity moment in
the joint. Check by computer the results obtained in the previous exercise.

3.35.[*] Calculate the servo feedback gains for the hydraulic servo which
transfer function has been determined in Exercise 3.27. (for two
positions of the second joint $q^{2*} = 0.$, and $q^{2*} = 1.573$ [rad])
and which is applied in the first joint of the robot in Fig. 3.2.
The guess value for the structural frequency is $\omega_o^1 = 12.$ [rad/s]
for $q^{2*} = 0.$ and $\omega_o^1 \approx 14.$ [rad/s] for $q^{2*} = 1.573$ [rad]. How
much have the feedback gains varied for various positions of the
second joint? Compare these variations with the results obtained
in Exercise 3.29.

3.3.3. Synthesis of servo system by
       pole-placement method

In the previous section we have presented the synthesis in the s-domain
of the position servo system for the i-th joint of the robot assuming
that all other joints are locked[*]. Now, we shall present the synthesis of feedback gains using the state space model of the system.

The state space model of the actuator and the mechanism (if just the
i-th joint is moving) is given by (3.3.4). As we have explained in the
previous sections, the servo system is realized by introducing feedback loop with respect to the position error of the robot joint (using
the sensor of position, the detector of the position error signal, and
amplifier) and feedback by velocity (using the sensor of the actual
joint velocity and amplifier). The block diagram of the servo in s-domain is given in Fig. 3.9. and it has been explained in Section 3.3.1.
Practically, the servo feedback loops mean that the input voltage

---

[*] Here, we shall not present various classical methods for servo system synthesis (such as for example, Bode's method in the frequency
domain, Niquist method and similar). However, our recommendation is
that the reader should remind of these methods from the literature
[3].

signal u^i is given by (according to (3.3.19)):

$$u^i = -K_p^i(q^i-q^{oi}) - K_v^i\dot{q}^i \tag{3.3.41}$$

Since the position and velocity of the joint are the variables in the system which are measured by sensors they are the *system output* variables. The system output is given, in this case, by:

$$y_i = (q^i, \dot{q}^i)^T \tag{3.3.42}$$

In the general case the output vector is of the order $k_i^y \geq 1$ (here, obviously, $k_2^y = 2$). The output variables are related to the coordinates of the state vector x^i. In general case the relation between the output vector y_i and the state vector x^i is given by[*]:

$$y_i = C_i x^i \tag{3.3.43}$$

where C_i is the output matrix of dimensions $k_i^y \times n_i$. In the particular case, when the output of the system is given by (3.3.42) while the state vector is given by $x^i = (\theta^i, \dot{\theta}^i)^T$ for $n_i = 2$, and if we assume that (3.2.12) holds, then the matrix C_i is the unit matrix I_{ni}. If we adopt the third order model of the system $n_i = 3$, $x = (\theta^i, \dot{\theta}^i, i_R^i)^T$, then the matrix C_i is given by:

$$C_i = \begin{bmatrix} 1 & 0 & 0 \\ 0 & 1 & 0 \\ 0 & 0 & 0 \end{bmatrix} \tag{3.3.44}$$

Now, we can write that the input signal of the actuator (3.3.41) is equal to:

$$u^i = -k_i^T y_i + K_p^i q^{oi} = -k_i^T C_i x^i + K_p^i q^{oi} \tag{3.3.45}$$

where by k_i is denoted the vector of the feedback gains

$$k_i = (K_p^i, K_v^i)^T \tag{3.3.46}$$

---

[*] In general case, the relation between the system output and the system state might be nonlinear, but we shall consider just the linear case.

The model of the complete servo system in the state space (i.e. the model of the actuator and the mechanism (3.3.4) together with the feedback loops (3.3.45)) can be written in the following form:

$$\dot{x}^i = (\hat{A}^i - \hat{b}^i k_i^T C_i) x^i + \hat{b}^i k_p^i q^{oi} \qquad (3.3.47)$$

The model (3.3.47) represents *the closed-loop model* of the system. The task of the control system is to ensure that the system output y_i approaches the desired output $y_i^o = (q^{oi}, 0)^T$. In order to ensure positioning of the joint in the desired set point q^{oi}, we must ensure the stability of the system (3.3.47) around the desired state $x^{oi} = (q^{oi}, 0, 0)^T$, i.e. we must ensure that the eigen-values of the closed-loop matrix of the system $(\hat{A}^i - \hat{b}^i k_i^T C_i)$ are placed in the desired positions in the complex plane. The eigen-values of the closed-loop matrix $(\hat{A}^i - \hat{b}^i k_i^T C_i)$ are the poles of the transfer function of the servo system (3.3.29). It is well known [3] that the system is stable if the eigen-values of the closed-loop matrix are in the left part of the complex plane (i.e. they must be on the left from the imaginary axis). Besides, the poles of the transfer function (3.3.30) must satisfy conditions that the damping factor is $\xi_i > 1$, and that the characteristic frequency is less than $0.5 \; \omega_o$. Thus, in the servo system synthesis (i.e. in the selection of feedback gains), we may specify the desired positions of the system poles in the complex plane. Once the positions of the poles are specified, the vector of the feedback gains k_i can be computed which will ensure desired positions of the eigen-values of the closed-loop matrix of the system. The gain vector is computed from the following condition:

$$\det(\hat{A}^i - \hat{b}^i k_i^T C_i - sI_{ni}) = \bar{K} \prod_{j=1}^{n_i} (s_j^o - s) \qquad (3.3.48)$$

where det( ) denotes the determinant of the corresponding matrix, $\bar{K}$ is the proportionality coefficient, and s_j^o, $j=1,2,\ldots,n_i$ denote the desired positions of eigen-values of the closed-loop matrix in the s-plane. Thus, the synthesis of the servo reduces to determination of the gain vector k_i which has to satisfy the algebraic equation (3.3.48). As it is well known, this procedure for control synthesis is called *pole placement method* [6].

There are numerous problems related to control synthesis by this method if the general case of the multi-input and multi-output system is considered. Here, we shall not treat these problems, since we consider just the synthesis of the feedback gains for local servo systems for individual joints of the robot. Evidently, the local servos are with single input u^i and their output vector is usually given by (3.3.42). In this particular case the application of the pole-placement method for synthesis of servo feedback gains is simple. Nevertheless, we shall briefly mention problem related to the order of the system n_i, the order of the output k_i^y and the number of the eigen-values of the closed--loop system matrix which might be specified.

In the previous section we have considered the second order model of the servo system (for $n_i=2$). If the servo output is given by (3.3.42), then the positions of both eigen-values of the closed-loop matrix can be specified, and based on (3.3.48) we can determine the position and velocity feedback gains to ensure desired places of the eigen-values. This has been practically done in the previous section (but by considering transfer functions), since we have defined poles in order to satisfy certain conditions.

However, if we consider the third order model of the actuator (for $n_i=3$), and if the system output is given by (3.3.42) (i.e. if $k_i^y=2$), then we may specify positions of just two eigen-values, while the position of the third eigen-value is free, i.e. it cannot be defined in advance. Nevertheless, for the considered servo systems it is always possible to ensure that all three eigen-values are in the left part of the complex plane (i.e. it can be ensured that the closed-loop system is stable). However, there is a constraint upon the allowable positions of the eigen-values in the complex plane, if $k_i^y<n_i$. The eigen-values cannot be placed so that the absolute values of their real parts are arbitrary large. Let us show this fact by a particular example. As we have seen in the Section 3.2.2. the matrices of the actuator (3.2.7) have the following form (see Exercise 3.9. for the electro-hydraulic actuators):

$$\hat{A}^i = \begin{bmatrix} 0 & 1 & 0 \\ 0 & a_{22}^i & a_{23}^i \\ 0 & a_{32}^i & a_{33}^i \end{bmatrix}, \qquad \hat{b}^i = \begin{bmatrix} 0 \\ 0 \\ b_3^i \end{bmatrix} \qquad (3.3.49)$$

If we assume that the matrix C_i is in the form (3.3.44) and if we specify two complex conjugate eigen-values $s_{1,2}^o = -\xi_i^o \omega_i^o \pm j\omega_i^o\sqrt{1-\xi_i^{o2}}$, then we can determine the feedback gains based on (3.3.48) as:

$$K_P^i = [\omega_i^{o2}(2\omega_i^o\xi_i^o+a_{22}^i+a_{33}^i)]/a_{23}^i b_3^i$$

$$\text{(3.3.50)}$$

$$K_v^i = [\omega_i^{o2}+2\omega_i^o\xi_i^o(2\omega_i^o\xi_i^o+a_{22}^i+a_{33}^i)-a_{32}^i a_{23}^i+a_{22}^i a_{33}^i]/a_{23}^i b_3^i$$

while the third eigen-value is placed at:

$$s_3^o = a_{22}^i + a_{33}^i + 2\xi_i^o\omega_i^o \qquad \text{(3.3.51)}$$

Based on (3.3.51), we can see that the real part of the third eigen--value depends on the specified eigen-values $s_{1,2}^o$, and if

$$a_{22}^i + a_{33}^i \geq -2\xi_i^o\omega_i^o \qquad \text{(3.3.52)}$$

then the system is not stable, i.e. the third eigen-value is not in the left part of the s-plane. Actually, (3.3.52) determines the upper bound upon the absolute values of the real parts of the specified eigen-values: one must not require that the eigen-values of the clo-sed-loop system have the real parts $\xi_i^o\omega_i^o$ greater (by absolute value) than the limit set by (3.3.52). The real-parts of the eigen-values of the system determine the speed by which the system output y_i approac-hes to the desired output $y_i^o = (q^{oi}, 0)^T$, i.e. the velocity by which the joint approaches to the desired position q^{oi}. If we want to incre-ase the absolute values of the real parts of the eigen-values of the closed-loop system matrix, then we must introduce the feedback by the third state coordinate. In the case of D.C. electro-motors the third state coordinate is the current of the rotor circuit which also might be measured (a resistor might be used as a sensor, but the obtained signal must be filtered using appropriate filters)[*]. In this case the output matrix C_i becomes unit matrix and the vector of the feedback gains is given by:

$$k_i = (K_P^i, K_v^i, K_I^i)^T \qquad \text{(3.3.53)}$$

where K_I^i is the gain in the feedback loop by the robot current (the third state coordinate). Now, all three eigen-values of the closed--loop system matrix might be specified. By pole placement method we can compute the feedback gains (3.3.53) to place the poles of the sys-tem in the desired locations in complex plane.

---

[*] In the case of the electro-hydraulic actuators the third coordinate of the state vector is the difference of pressure in the cylinder which also can be, relatively simply, measured.

The scheme of the servo system is presented in Fig. 3.14. in the case when the feedback loop by the robot current is introduced.

We should note that this method for control synthesis can be also used to determine the feedback gain if only feedback loop by position error is introduced (i.e. if $k_i^y = 1$, $y_i = q^i$). Namely, the servo might be realized only with the position feedback loop, but the performance of such servo system is unsatisfactory regarding the requirements which are encountered in robotics. Actually, such a servo cannot satisfy both requirements regarding the servo damping and the steady state error.

EXAMPLE 3.3.3. The data of the D.C. electro-motor applied in the first joint of the robot in Fig. 3.2, are given in Table 3.1. If the inertia of the mechanism is taken into account, the matrices of the model of the actuator and mechanism (3.3.4) have been calculated in Example 3.3. For this actuator and mechanism we have to determine the servo feedback gains using the pole placement method. Let us assume that the feedback loops are introduced by all three state coordinates, i.e. by joint position, by joint velocity and by rotor current. Let us assume that it is required to place the poles of the closed-loop servo in the following positions:

$$s_{1,2}^o = -6., \qquad s_3^o = -800.$$

By the pole-placement method we can calculate the servo feedback gains as:

$$K_P^1 = 155.43 \left[\frac{V}{rad}\right], \qquad K_V^1 = 50.568 \left[\frac{V}{rad/s}\right], \qquad K_I^1 = 0.267 \left[\frac{V}{A}\right]$$

If the feedback loop by the rotor current is not introduced, we may define just two poles of the closed-loop system. Let us define two real poles:

$$s_1^o = -5., \qquad s_2^o = -6.$$

Based on (3.3.50) we calculate the feedback gains as:

$$K_P^1 = 110.85 \left[\frac{V}{rad}\right], \qquad K_V^1 = 39.372 \left[\frac{V}{rad/s}\right]$$

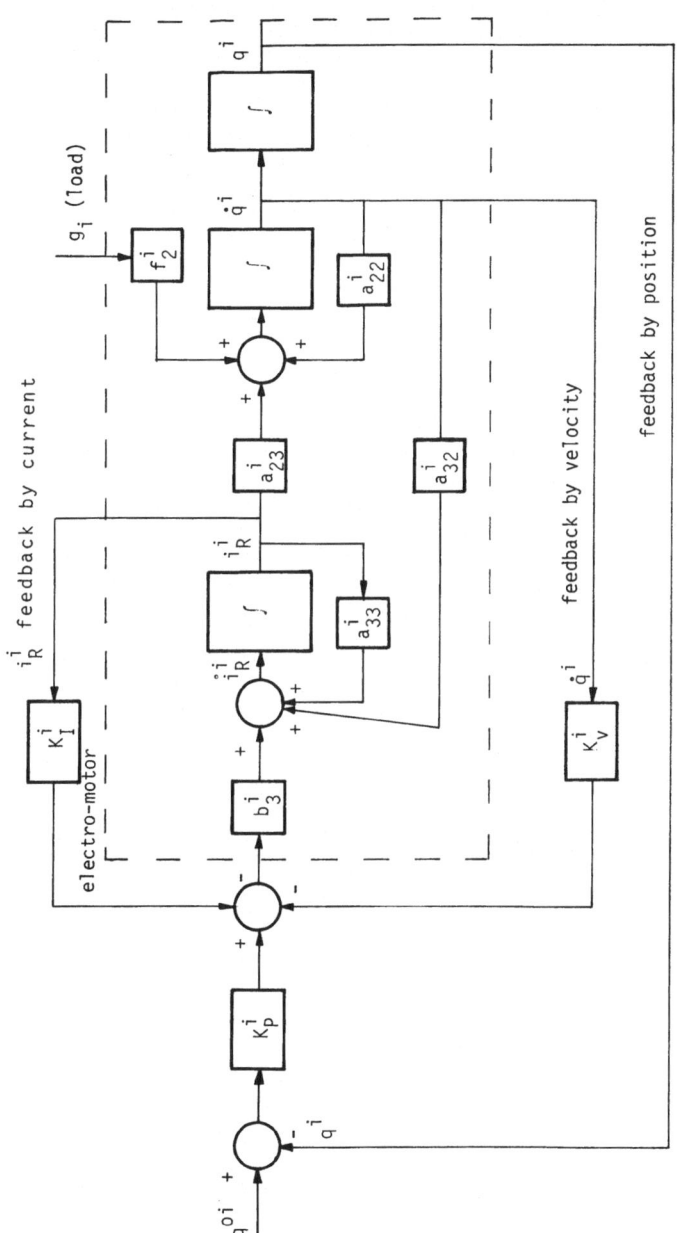

Fig. 3.14. Scheme of servo system with the feedback loop by the rotor current (according to the model of actuator in the state space (3.2.7), (3.3.49))

Exercises

3.36. The second order model of the actuator and the mechanism ($n_i=2$) has the matrices in the following form:

$$\hat{A}^i = \begin{bmatrix} 0 & 1 \\ 0 & a_{22}^i \end{bmatrix}, \quad \hat{b}^i = \begin{bmatrix} 0 \\ b_2^i \end{bmatrix} \tag{3.3.54}$$

Determine the expressions for the feedback gains if the poles of the closed-loop system are prescribed, in two cases:

a) if feedback loops by position and by velocity are introduced $k_i^y = 2$ (two poles are prescribed; examine both cases: when the pair of complex conjugate poles are prescribed and when the pair of real poles is prescribed),

b) if feedback loop by position error is introduced only, $k_i^y = 1$ (one pole is prescribed; examine all three cases: if the prescribed pole is real, if the real part of the pair of complex conjugate poles is prescribed, and if the imaginary part of the pair is prescribed); determine also the second pole which is not prescribed. Is it always possible to stabilize the system by introducing the position feedback only?

3.37. The third order model of actuator and the mechanism ($n_i=3$) has the matrices in the form (3.3.49). If we introduce the feedback by position only, determine the expressions for the position feedback gain so that the closed-loop servo system has:

a) real pole in the prescribed position $s_i^o = -\sigma_i^o$, or

b) real part of the pair of complex poles in the prescribed position $Re(s_{1,2}^o) = -\xi_i^o \omega_i^o$, or

c) imaginary part of the pair of complex poles in the given position $Im(s_{1,2}^o) = \omega_i^o \sqrt{1-\xi_i^{o2}}$ (the complex conjugate pair is $s_{1,2}^o = -\xi_i^o \omega_i^o \pm \omega_i^o \sqrt{1-\xi_i^{o2}}$).

In all three cases determine the expressions for the rest two poles of the closed-loop system. What can you conclude concerning the possibility to stabilize the system by introducing the position feedback loop error only?

3.38.* Write the programme (in some high programming language) for determination of the feedback gains if the input data for the programme are: data on actuator, data on the moment of inertia of the moving part of the mechanism $\bar{H}_{ii}$, prescribed poles of the closed-loop servo system. The programme should include various information feedback structure (if the feedback by position error is introduced only, or if the feedback loops by position and velocity are introduced, or if the feedback loops by all three state coordinates are introduced). (Instruction: use the expressions given in the text and in Exercise 3.37). Using the programme, check the results in Example 3.3.3.

3.39.* Write the programme (in some high programming language) for simulation of the closed-loop servo system if initial state $x^i(0)$ and desired position q^{oi} are given. The input data for the programme are: data on actuator, data on moment of inertia of the moving part of the mechanism $\bar{H}_{ii}$, the vector of feedback gains, the output matrix C_i, the initial state $x^i(0)$ and desired position q^{oi}. (Instruction: use the state space model of the servo system (3.3.47) to compute the first derivative of the system state vector x^i, and then apply some of available numerical methods for integration, e.g. use the simplest Euler method for integration, see Section 5.3.; the integration interval should be selected to be less than 1 [ms]). Simulate the positioning of the first joint of the robot in Fig. 3.2. around desired position $q^{o1} = 0.5$ [rad], if initial state is $x^1(0) = (0, 0, 0)^T$. Data on the model of servo are given in Example 3.3; use two sets of the feedback gains determined in Exercises 3.28. and 3.29. Compare the results of simulation for these two sets of the feedback gains.

3.40.* The data on electro-hydraulic servo system are given in Table 3.5, for $n_i=3$ (see Exercise 3.27). Determine the feedback gains which should place two poles of the closed-loop system in the positions $s^o_{1,2} = -6$. Explain how the feedback loop by the third state coordinate can be implemented in the case of electro-hydraulic servo system?

3.3.4. Effects of inertia variation and gravity moment of the mechanism on the behaviour of servo system

In the previous sections we have considered the synthesis of local static servo systems for the individual joints of the robot mechanism. In doing this we have considered each joint and its actuator independently from the other joints and actuators of the robot. Namely, we have considered the i-th joint of the robot and its actuator, and we have assumed that all the other joints are locked. The model of the i-th joint and actuator is given by (3.3.4). This model includes the moment of inertia of the moving part of the mechanism $\bar{H}_{ii}(q^*)$ around the axis of the i-th joint. This moment of inertia depends on the positions of all the other joints, i.e. it depends on the positions q^* in which the other joints are locked. Up to now, we have considered the behaviour of the system for just one set of positions of all the other joints q^*. For the given values of the vector q^* we have determined the servo feedback gains. Obviously, if the positions q^{j*} (for $j > i$) of the other joints change, the moment of inertia of the mechanism around the i-th joint $\bar{H}_{ii}(q^*)$ will also change. Then the feedback gains calculated for the previous position q^* need not be valid any more[*]. The variation of the positions of the other joints of the robot causes also the change of the gravity moment of the mechanism around the i-th joint. We have neglected the gravity moment in the model (3.3.4), but we have considered its effect upon the servo system positioning (steady state error). The variation of positions of the other joints will reflect the steady state error in the i-th joint.

First, let us consider the effects of the variation of the moment of inertia of the mechanism around the axis of the i-th joint $\bar{H}_{ii}$ upon the performance of the local servo system in the i-th joint. Let us consider what happens if the joints change their positions q with respect to previous positions q^* for which we have calculated the servo feedback gains. The variation of the positions of the joints $q^j \neq q^{j*}$ ($j \neq i$) causes the variation of the moment of inertia of the mechanism around the i-th joint $H_{ii}(q) \neq \bar{H}_{ii}(q^*)$. Let us consider the second order model of the servo system and the synthesis of the servo feedback gains in s-domain (Sect. 3.3.2). We have seen that the damping factor of the servo system is given by (3.3.36). We have also shown that it is required to ensure that the servo is always critically or overcriti-

---

[*] See Exercises 3.28. - 3.30.

cally damped. Let us assume that the feedback gains are selected to ensure that the servo is critically damped for the position of the robot defined by vector q^* (i.e. if the moment of inertia of the mechanism is $\bar{H}_{ii}(q^*)$). In this case the velocity gain K_v^i is given by (3.3.39). If the other joints of the robot are moved to the new positions $q^j \neq q^{j*}$ the damping factor of the servo in the i-th joint becomes:

$$\xi_i = \frac{\sqrt{\bar{H}_{ii}(q^*) + J_M^i N_v^i N_m^i}}{\sqrt{\bar{H}_{ii}(q) + J_M^i N_v^i N_m^i}} \qquad (3.3.55)$$

where $\bar{H}_{ii}(q)$ is the moment of inertia of the mechanism for the new positions of the joints q.

We can distinguish two cases. If $\bar{H}_{ii}(q) < \bar{H}_{ii}(q^*)$, then $\xi_i > 1$, i.e. the servo is overcritically damped in the new position, which means that the selected feedback gains ensure acceptable response of the servo in the new position of the robot. However, if $\bar{H}_{ii}(q) > \bar{H}_{ii}(q^*)$, then $\xi_i < 1$, i.e. the servo is undercritically damped, which cannot be accepted. This means that we must not allow the situation in which $\bar{H}_{ii}(q) > \bar{H}_{ii}(q^*)$ to appear. To prevent this, we must determine $\max_{q'} H_{ii}(q')$, i.e. we must determine the maximum possible value of the moment of inertia of the mechanism around the i-th joint (maximum over all allowable positions of joints). For this position of the robot we must calculate the velocity feedback gain using the expression (3.3.39). By this, we ensure that for all possible positions of the robot joints the servo system in the i-th joint is overcritically, or critically damped.

If we have selected the velocity gain in this way, then we can again distinguish two cases:

a) If $J_M^i N_v^i N_m^i >> \max_{q'}(\bar{H}_{ii}(q^*) - H_{ii}(q'))$, then the variation of the damping factor (3.3.55) for various positions of the robot joints is relatively small. In other words, the response of the servo system is nearly uniform independently on the positions of the other joints of the robot. The servo system is nearly critically damped for all positions of the other joints of the robot, and the satisfactory positioning of the i-th joint might be expected. This case appears if a relatively powerful actuator and large gear reducer with high reduction ratio are applied. Such actuator and reducer have an equivalent moment of inertia of the rotor $J_M^i N_v^i N_m^i$ relatively high with respect to the masses and moments of inertia of the robot links. In practice it is often the case that relatively powerful actuator and large reducer are applied and by this the effects of the variation of the moment of inertia of the mechanism are reduced. Namely, the large equivalent moment of inertia of the actuator rotor tends to mask the variation of the moment of inertia of the mechanism H_{ii}. However, the drawback of such a solution lies in the fact that too large actuators and reducers are applied, which is undesirable from the standpoint of the energy consumption and the prise of the robotic system. On the other hand, large reducers often introduce large backlash and friction in the system which certainly reflects the accuracy of positioning of the servo system. That is why large efforts are directed towards introducing the so-called *direct drive robots*, i.e. the robots with actuators having no reducers. For such robots the problems of backlash and friction are minimized.

b) If $J_M^i N_v^i N_m^i << \max_{q'}(\bar{H}_{ii}(q^*) - H_{ii}(q'))$, then the damping factor might significantly vary depending on the position of the robot (acc. to (3.3.55)). Thus, for the robot configuration q^* for which $\bar{H}_{ii}(q^*) = \max_{q'} H_{ii}(q')$ the critical damping of the servo is obtained (which means fast response of the servo), while for the robot configuration q for which $\bar{H}_{ii}(q) = \min_{q'} H_{ii}(q')$ the damping of the servo becomes very overcritical ($\xi_i >> 1$), which causes slow response of the system. Thus, the servo system has a variable performance depending on the positions of the other joints of the robot (although the response is never oscillatory, nor the overshoots of desired positions appear). To ensure critical damping of the servo for all positions of the other joints, in this case, (and to ensure uniform performance of the robot around all its possible configurations), we must introduce compensation for the effects of the variable moment of inertia of the mechanism (by introducing so-called global control), or we must implement variable gains which change depending on the configuration of the robot (i.e. depending on the actual value of the moment of inertia of the mechanism). These problems will be addressed in Chapters 4. - 6. We must underline that the determination of the variable moment of inertia of the mechanism in general case, is very difficult, since the payload carried by the robot hand might be of variable and unknown parameters (with variable mass, dimensions, shape, moments of inertia and so on). If these parameters are not known in advance, the problem of their identification arises and the adaptation of the servo system to the variation of the payload parameters has to be considered (see Chapter 6).

We have shown in Section 3.3.2. that the position gain can be selected based on Expression (3.3.38), i.e. this expression gives the upper bound of the allowable value of the position gain. As it can be seen from (3.3.38), the variable moment of inertia of the mechanism $H_{ii}(q^*)$ also affects the upper bound of the position gain. However, we have to keep in mind that the resonant structural frequency of the robot also depends on the moment of inertia of the mechanism. This frequency decreases if the moment of inertia of the mechanism increases. We might assume that, in the first approximation, the structural frequency is inversely proportional to the square root of the moment of inertia of mechanism [5].

If $\omega_o(\bar{H}_{ii})$ is the resonant structural frequency for the robot configuration q^* for which the value of moment of inertia of the mechanism around the i-th joint is $\bar{H}_{ii}$, then we can write that the structural frequency $\omega_o(H_{ii})$ for some other configuration of the robot mechanism (for which the value of moment of inertia of the mechanism is H_{ii}) is given by:

$$\omega_o(H_{ii}) = \frac{k}{\sqrt{J_M^i N_v^i N_m^i + H_{ii}}} = \frac{\omega_o(\bar{H}_{ii})\sqrt{J_M^i N_v^i N_m^i + \bar{H}_{ii}}}{\sqrt{J_M^i N_v^i N_m^i + H_{ii}}} \qquad (3.3.56)$$

where k is the proportionality coefficient.

We have assumed that the position servo gain K_p^i is computed based on (3.3.38) for the robot configuration for which the moment of inertia of the mechanism is $\bar{H}_{ii}(q^*)$. If the other joints move in some new (locked) positions q to which corresponds $H_{ii}(q)$, then the characteristic frequency of the servo in the i-th joint becomes, according to (3.3.35) and (3.3.38):

$$\omega_i(H_{ii}) = \frac{\omega_o(\bar{H}_{ii})\sqrt{J_M^i N_v^i N_m^i + \bar{H}_{ii}}}{2\sqrt{J_M^i N_v^i N_m^i + H_{ii}}} \qquad (3.3.57)$$

It is obvious that the characteristic frequency of the servo satisfies the following inequality:

$$\omega_i(H_{ii}) \leq \frac{1}{2}\omega_o(H_{ii}) \qquad (3.3.58)$$

This means that regardless of the variation of the moment of inertia of the mechanism, the condition (3.3.37) is always satisfied, if the position servo gain is selected to satisfy condition (3.3.38). In other words, the constraint upon the maximum value of the position servo gain does not depend on the moment of inertia of the robot mechanism, and we can select the position servo gain just based on (3.3.38). Such position gain will be valid for any configuration of the robot. By so selected position servo gain we ensure that the characteristic frequency of the system is sufficiently below the resonant structural frequency of the system. However, as we have explained above, to ensure that the servo is always (over)critically damped we must select the feedback gains taking into account the maximum possible moment of inertia of the mechanism (or, we might apply variable velocity servo gain).

Based on the above consideration it follows that the effects of the variable moment of inertia of the mechanism upon the servo system performance can be eliminated by selection of powerful actuators and large gear reducers with high reduction ratio, or by applying constant position servo gain and variable velocity servo gain, or by global control which compensates for the effects of the variation of the moment of inertia of the robot mechanism.

The second factor which effects the response of the servo if the configuration of the robot change, is the gravity moment around the axis of the i-th joint $g_i(q^i, q^*)$. In synthesis of local servo system for the i-th joint we have neglected the gravity moment. However, in Section 3.3.2. we have considered the effects of gravity moment as external load upon the performance of the servo system. We have shown that steady state error due to external load is given by (3.3.34). It is obvious that the steady state error is the consequence of the gravity moment (if we assume that all the other joints are locked). When the i-th joint stops, this gravity moment becomes constant which depends on the configuration of the mechanism q^* (but it depends also on the final position of the i-th joint). The error caused by this gravity moment can be calculated by (3.3.34) as the steady state error of the servo.

From (3.3.34) it follows that the error due to gravity moment is inversely proportional to the position servo gain. However, we have seen that the position gain is limited by the condition (3.3.38). Based on (3.3.34) and (3.3.38) we can determine the minimal steady state error which might be achieved:

$$|\Delta q^i_{min}(\infty)| = \frac{|g_i(\infty)|}{(0.5\omega_o)^2(J^i_M N^i_v N^i_m + \bar{H}_{ii})}$$ (3.3.59)

Obviously, this "minimal" steady state error depends on the gravity moment $g_i(q^i, q^*)$, i.e. it depends on the positions of all the joints of the robot. If we determine the maximal possible gravity moment around the i-th joint (for all possible configurations q^*) by (3.3.59) we may determine the steady state error which can appear in the positioning of the i-th joint. Obviously, this error determines the accuracy of the servo in the i-th joint. If this error is not within the allowable tolerances around the desired position of the joint we must compensate for the effects of the gravity moment, i.e. we must eliminate this un-acceptable steady state error.

On the other hand, due to variation of the gravity moment for various configurations of the robot (i.e. for various positions of the locked joints and for various desired positions of the i-th joint), the steady state error of the servo also varies for various configurations of the robot. This means that the performance of the servo in not uniform for all desired set positions, and this is an unacceptable drawback of the servo system (from the standpoint of the robot application and pro-gramming). This is also the reason to consider compensation for the gravity moment.

The compensation for the gravity moment might be achieved in several ways. One of the solutions is to directly compensate for gravity moment, and the other is by application of *the feedback by integral of the position error*. Here, we shall consider the first approach, and the second will be addressed in Section 3.4. It should be noted that in some applications the steady state error due to gravity moment can be eli-minated by brakes in the joints, but this solution is not applicable for elimination of the errors in tracking the desired trajectories of the joints.

Since the gravity moment around the i-th joint is well known function of the joint coordinates (positions), this moment can be computed on--line in the control microprocessor. Based on the computed value of the gravity moment, the microprocessor calculates the voltage signal which corresponds to this moment and generates the control signal for the actuator. This additional control signal produces the driving torque, which compensates for the gravity moment around the joint. The control scheme which includes this direct compensation for the gravity moment is presented in Fig. 3.15. For the second order model of the servo system, the control signal including direct compensation for the gravity moment is computed in the following way (if D.C. electro-motor is applied)[*)]

$$u^i = -K_P^i(q^i-q^{oi})-K_v^i\dot{q}^i+g_i^*(q^{oi},\ q^*)\cdot r_R^i/C_M^iN_m^i \qquad (3.3.60)$$

where by g_i^* is denoted on-line computed gravity moment around the i-th joint in the desired position q^{oi}. Here, it is assumed that the angles or displacements of all the other joints are measured by appropriate sensors and the signals are sent to microprocessor which (based on the obtained values of joint coordinates) computes g_i^*. If the computed gravity moment g_i^* perfectly coincides with the actual gravity moment g_i,

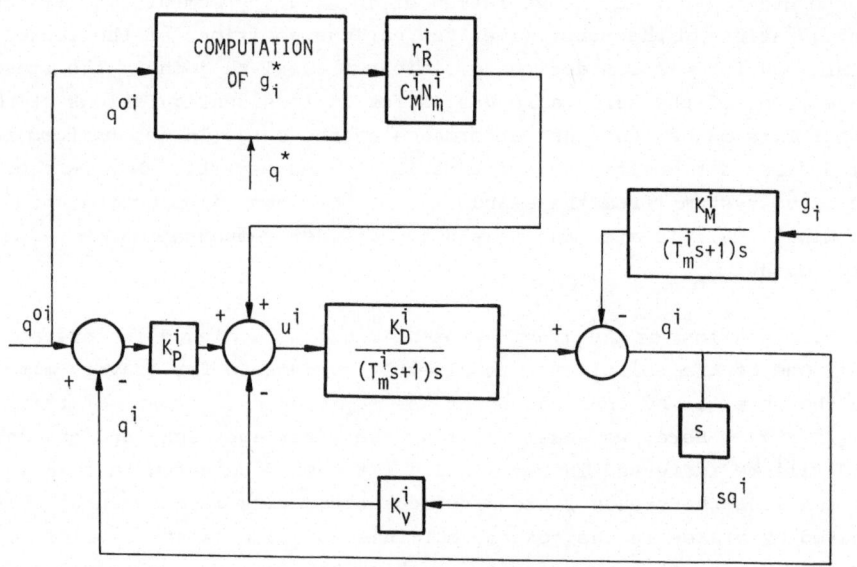

Fig. 3.15. Position servo system with direct compensation for gravity moment

---

[*)] The problem of the direct compensation of the gravity term in the case of the third order model of the actuator will be considered in Chapter 5.

then the control signal (3.3.60) will produce the torque which will competely compensate for the gravity torque and the steady state error due to the gravity moment will be eliminated. The positioning of the servo in the desired position q^{oi} in this case will be very accurate. However, all parameters of the robot mechanism which are required to compute the gravity moment g_i^* (link lengths, positions of the centers of the masses of the links, masses of the links) are usually not known accurately. Thus, in the general case, the gravity moments cannot be ideally accurately calculated, nor they can be ideally compensated for. Nevertheless, in this way we can achieve significant decrease of the positioning error.

In implementing of the presented compensation for the gravity moment, a problem of on-line computation of the gravity moment arises, i.e. a problem of number of additions and multiplications necessary to compute gravity moment. This problem will be addressed in Chapter 5.

EXAMPLE 3.3.4. For the first joint of the robot in Fig. 3.2. the position and velocity servo gains have been computed in Example 3.3.2. for the value of moment of inertia of the robot mechanism which corresponds to the locked position of the second joint $q^{2*}=0$. (the corresponding transfer function of the servo is given by (3.3.28)). If the second joint moves to another locked position, the moment of inertia of the mechanism changes, which causes the variation of the servo damping factor according to (3.3.55). The variation of the damping factor when the angle q^{2*} varies is given in Table 3.8. It can be seen that the damping factor varies significantly if the q^2 varies, but it is always $\xi_1 > 1$ since the velocity servo gain has been calculated for $q^{2*} = 0.$, when the moment of inertia of the mechanism around the first joint is maximal. If we decide to introduce variable velocity feedback gain, then its values for various q^{2*} are also given in Table 3.8. These variable velocity gains are calculated to maintain the damping factor equal to one for all positions of the second joint. Applying such variable feedback gains we maintain the uniform performance of the servo in the first joint independently on the position of the second joint. Obviously, the implementation of the servo system with variable velocity gain is much more complex than if the fixed feedback gains are applied.

The performance of the servo system in the first joint of the robot in Fig. 3.2. with fixed servo gains has been simulated by computer. The responses of the servo for various (fixed) positions of the second

114

joint q^{2*} are presented in Fig. 3.16. under assumption that the selected gains ensure that $\xi_1 = 1$ for $q^{2*} = 0$.

Position of the second joint q^{2*}	ξ_1 for fixed gain	Velocity gain for $\xi_1 = 1$ - [V/rad/s]
0^0	1.	43.6
30^0	1.003	∿43.5
60^0	1.056	41.2
90^0	1.15	37.7
120^0	1.277	33.8
150^0	1.39	∿31.

Table 3.8. Variation of the damping factor and variable velocity servo gain

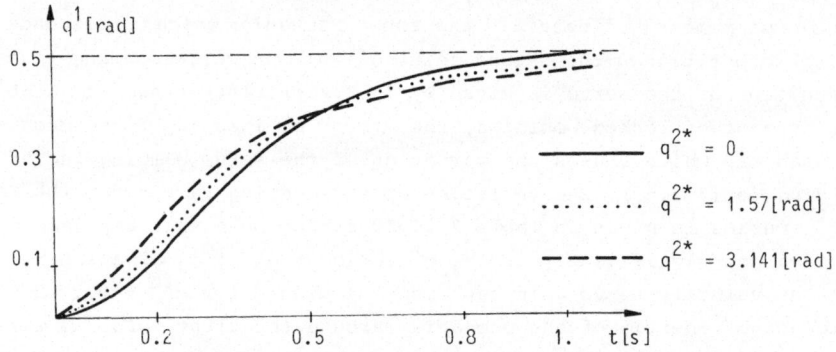

Fig. 3.16. Response of the servo system in the first joint of the robot in Fig. 3.2. for various (locked) positions of the second joint

Let us consider the second joint of the robot in Fig. 2.6. and let us determine the effects of gravity moment upon the servo system in this joint. Let us assume that the D.C. electro-motor is applied, the data of which are given in the second column of the Table 3.3., while the mass of the second link is $m_2 = 9$. [kg], the distance between the second joint and the centre of mass of the second link is $\ell_2^* = 0.2$ [m], the distance between the second and the third joint is $\ell_2 = 0.5$ [m], the mass of the third link is $m_3 = 10$. [kg] and the distance between

the third joint and the centre of mass of the third link is $\ell_3^*=0.2$ [m]
(for $q^3=0.$). If we assume that the structural frequency for the second
joint is $\omega_o^2 = 45$ [rad/s], then we calculate the position servo gain to
satisfy the condition (3.3.38) and obtain the value $K_p^2 = 1414.$ [V/rad].
The steady state error of the servo due to gravity moment can be calculated based on (3.3.59). For various positions of the second and the
third joint various values of the gravity moment are obtained resulting in various steady state errors in the second joint. The variation
of the steady state error in the second joint for various positions of
the second and the third joint is given in Table 3.9.

q^{o2} [rad]	q^{*3} [rad]	$\Delta q^2(\infty)$ [rad]	g_2 [Nm]
0.	0.	0.	0.
1.57	0.	0.000783	86.3
1.57	0.4	0.00110	126.
1.57	0.7	0.0014	154.
1.0	0.7	0.00117	130.

Table 3.9. Steady state errors and gravity moments in the
second joint of the robot in Fig. 2.6.

Exercises

3.41. For the first joint of the robot in Fig. 2.5. determine the servo
feedback gains (data on the robot and actuators are given in Tables 3.3. and 3.4). If the resonant structural frequency is estimated to be $\omega_o^1 = 12$ [rad/s] (for the moment of inertia of the
mechanism around the first joint which corresponds to the position of the third joint of $q^{3*} = 0.6$ [m]) determine the servo
gains to satisfy the following requirement:

a) the critical damping has to be maintained for the following
positions of the third joint $q^{3*}=0.,$ 0.20, 0.40, 0.60 [m]
(Instruction: for each position of the third joint q^{3*} determine the velocity servo gain so that the servo is critically
damped $\xi_1=1$), or

b) the fixed servo gains have to ensure that the system is over-critically damped for all positions of the third joint.

3.42.[*] Let us assume that the electro-hydraulic actuator is applied in the first joint of the robot in Fig. 3.2. Data on this actuator are given in Exercise 3.27. (Table 3.5). The servo gains are selected to satisfy usual requirements, for the position of the second joint q^{2*} = 0. and for the estimated value of the structural frequency of ω_o^1 = 12. [rad/s]. Determine the variation of the damping factor of the servo for various positions of the second joint q^{2*} given in Table 3.8. Also determine the values of the velocity servo gain which should be applied to maintain the critical damping for all positions of the second joint q^{2*} (see Exercise 3.35).

3.43. a) Calculate the errors in positioning of the robot end point due to steady state errors of the servo system in the second joint of the robot in Fig. 2.6. The steady state errors of the servo are given in Table 3.9. (assume that the distance between the centre of mass of the third link and the manipulator end point is 0.3 [m]).

b) Determine the reduction of the end point position error, if the direct compensation for the gravity moment around the second joint is applied, but instead of actual value of the mass of the third link (m_3 = 4. [kg]) an estimated value of $m_3 \approx 3$. [kg] is used for computation of gravity moment.

3.44.[*] Starting from the open-loop model of the servo system (3.3.4) and the closed-loop model (3.3.47) determine the model of sensitivity to variation of the moment of inertia of the mechanism $\bar{H}_{ii}$ and calculate the matrix of the sensitivity model for the servo system given in Exercise 3.41. Determine the eigen-values of the matrix of the sensitivity model for various positions of the third joint q^{3*} of the robot which have been considered in Exercise 3.41.

3.45. Repeat Exercise 3.41 but now assuming that at the end point of the minimal configuration of the robot there is a gripper with a payload of a mass m_p = 1.5 [kg]. If the mass of the payload is not known in advance and if the values of the servo gains are taken as in Exercise 3.41. b), determine the minimal damping factor which can occur when the payload is present.

3.46. Repeat Exercise 3.43 but now assuming that at the end point of the minimal configuration of the robot there is a payload of a mass m_p = 1.5 [kg]. Assume that the direct compensation of the gravity moment is introduced, but in on-line computation of the gravity moment we take an estimated value for $m_p \approx 0$. (since the actual value of the mass of the payload is assumed to be unknown). Determine the steady state errors of the servo in the second joint, and the errors of the positioning of the end point of the manipulator.

3.47. Determine the deflection of the end point of the Stanford manipulator (Fig. 3.13), the data of which are given in Tables 3.7. and 3.10., due to steady state errors of the servos in the robot joints which are caused by gravity moments around the joints. The gravity moments around the robot joints are given in Table 3.10. (the servo gains are computed according to data given in Table 3.7) [5].

JOINT	r [m]	g_i [Nm]
1	0.54	0.
2	0.50	69.3
3	-	81.73
4	0.25	5.54
5	0.25	5.54
6	0.25	0.

Table 3.10. Gravity moments for Stanford manipulator (r is effective lever arm from the joint to the robot end point, g_i is gravity moment around the joint axis; for joints 4 and 5 adopt $K_D^i K_P^i / K_M^i$ = 220 [Nm/rad]) [5]

3.48.* Extend the programme written in Exercise 3.39. to include simulation of the servo system if the constant gravity moment is acting upon the servo system. Apply this programme to simulate response of the robot in Fig. 2.6. (data on servo are given in
Example 3.3.4) if the constant gravity moment is acting around
the second joint and the servo has to drive the joint from initial state $x^2(0) = (0., 0., 0)^T$ into desired position $q^{o2} = 1.573$
[rad]. Simulate the responses of the servo if the constant gravity moment takes various values from Table 3.9.

3.3.5. Nonlinear effects in local servo system

We have applied the linear model of the actuator and moving part of the
mechanism (3.3.4) for synthesis of local servo system. As we have already noted, the linear model of the actuator is an approximation in
which certain nonlinear effects are neglected. Here, we shall briefly
consider the effects of these nonlinearities upon the performance of
the servo system, since we have not taken them into account in the synthesis of the servo feedback gains.

(a) *Effect of the actuator input amplitude constraint*

In the model of actuator and the mechanism (3.3.4), we have denoted
that the amplitude of the input signal for actuator u^i is limited, i.e.
the nonlinearity $N(u)$ given by (3.2.9) has to be taken into account.
However, the synthesis of the local servo system has been performed
neglecting this nonlinearity of the amplitude constraint type.

Let us consider the second order model of the actuator and the mechanism and let us assume that the synthesized servo control is in the
form (3.3.41). If we apply the nonlinearity of the amplitude constraint
type upon this control, we obtain:

$$|-K_p^i(q^i - q^{oi}) - K_v^i \dot{q}^i| \leq u_m^i \tag{3.3.61}$$

Based on (3.3.61) we can define the region in the state space $x^i = (q^i, \dot{q}^i)^T$ for which the control signal does not reach the amplitude
constraint (the upper bound of the control signal). This region in the
state space is presented in Fig. 3.17. For the points within the region the nonlinearities (3.3.61) are satisfied, which means that the

input signal does not reach upper bound of its amplitude. Thus, for all states within this region the amplitude constraint upon the actuator input has no effects, and the servo system behaves as it has been described in the previous sections.

However, for the states out of this region the nonlinearities (3.3.61) are not satisfied, which means that the computed control requires greater signal at the input of the actuator than it is actually allowed for the particular actuator. Thus, the performance of the servo system for

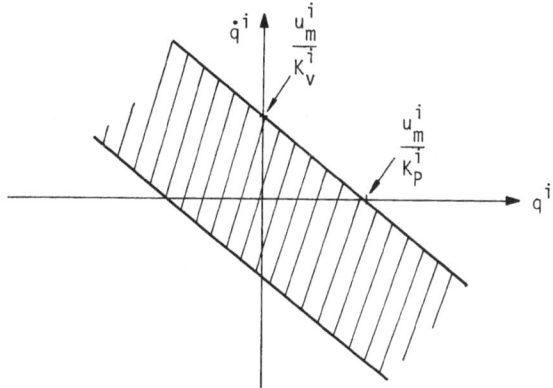

Fig. 3.17. Region of linear behaviour of servo system (for $q^{oi}=0$.)

these states cannot be linear, as we have assumed in the synthesis of the servo gains. It has been shown [7] that for the states out of the region denoted in Fig. 3.17., the closed-loop servo system behaves as its poles were moved to the right in the complex plane. We may say that higher is the required control signal relative to the maximum allowed value u_m^i, the poles of the closed-loop system move more to the right. Theoretically, if the state were as far from the region denoted in Fig. 3.17. that it would require an infinite control signal, then the poles of the closed-loop system would coincide with the poles of the open-loop system. This means, that for the states out of the region presented in Fig. 3.17. the stability of the closed-loop servo system is guaranteed (since the poles of the open-loop system in this particular case of servo-actuator are in the left half of the complex plane, or at the imaginary axis). However, we cannot guarantee the desired speed of the transient process (i.e. the time by which the joint approaches to the desired position q^{oi}). Thus, the region denoted in Fig. 3.17. represents the set of states for which we can expect that the

closed-loop servo system will behave as predicted in the synthesis of
servo gains. If the deviation of the system state around the desired
set position q^{oi} are within this region, then the servo system will
ensure desired positioning of the robot joint. Otherwise, the positioning of the joint will be slower depending on how far is the initial
position of the joint $q^i(0)$ from the imposed desired position q^{oi}.

(b) *Effect of static friction*

In the actuator and on the shaft of the joint there always appears (to
a certain degree) Coulomb friction. This frictional effect has not
been taken into account in the model (3.3.4). The static friction opposes the movement of the joint when it starts to move. Once the joint
is in motion and when it reaches a certain velocity, the static friction drops to zero and dynamic friction appears which opposes the motion. The dynamic friction is usually less than static friction. The
Coulomb friction directly affects *repeatability* of the servo, and,
obviously, repeatability of the robot end point positioning. Due to
this friction the position error of the servo might appear, since the
servo requires an additional signal to overcome the friction moment.
If the robot is well designed and the reducer gears are well fabricated,
then the Coulomb friction might be reduced to minimum and it can be
neglected in the control synthesis. However, if this friction is significant, we can include an additional term in the servo-control. We can
add an impulse signal at the actuator input whenever the joint starts
to move. The servo system with such additional impulse torque for compensation for static friction is presented in Fig. 3.18. The most appropriate way to determine the amplitude of this impulse signal (i.e.
the magnitude of the additional torque) is by experiment.

(c) *Effects of dynamic friction*

Once the joint is in motion the dynamic friction appears instead of
the static friction. This friction also can be reduced to a minimum by
a well design of the robot joint and reducer and by lubrication of
contact surfaces between the moving parts in the robot. However, if
this friction is significant, then it must be compensated for by additional control signal. We may adopt the following approximative expression for dynamic friction moment, i.e. we may write that the moment
due to dynamic friction around the joint is given by:

$$P^i_{t\ dynamic} \approx K^i_d \dot{q}^i + P^i_{o\ dynamic}\ sgn(\dot{q}^i)$$

Thus, we can compensate for dynamic friction moment by additional control signal generated as presented in Fig. 3.18. In this case again the coefficients K^i_d (the dynamic friction coefficient) and $P^i_{o\ dynamic}$ might be determined experimentally (i.e. by identification at the actual robotic system). It should be kept in mind that the actual friction moment is nonlinear function of the joint velocity, but it has been verified experimentally that using the compensation presented in Fig. 3.18. satisfactory results can be achieved [8].

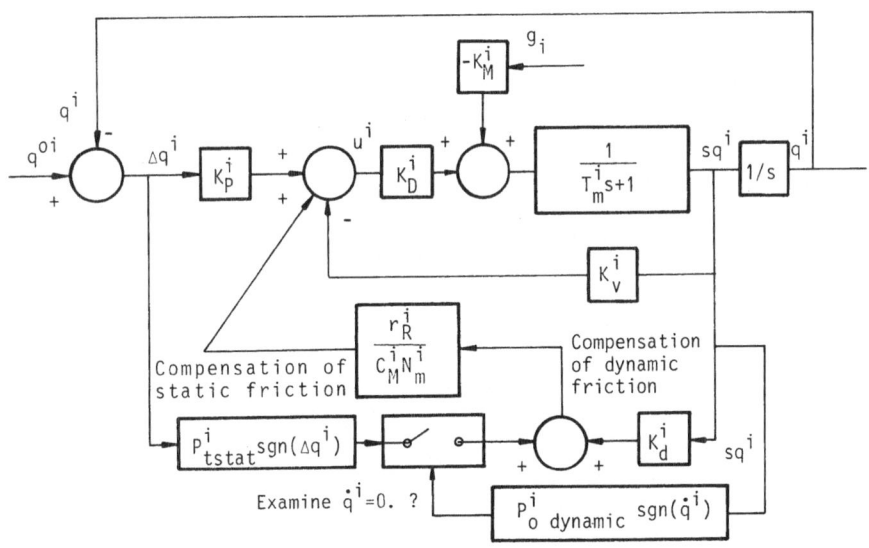

Fig. 3.18. Servo system with friction compensation

(d) *Effect of backlash*

The backlash in the gear reducer and other parts of the robot system might have a significant effects upon the servo repeatability and robot end point positioning. For example, the backlash in the gear reducer in Fig. 3.19. directly reflects the error in the positioning of the robot end point. If the effective lever arm from the joint to the robot end point is relatively long, then very small backlash in the gear reducer causes the significant error in the robot positioning and accuracy of the robot is highly reduced. The repeatability of the robot positioning is also poor if there are backlash in the robot joints

122

since it is hard to keep the joint position always at the same side of
the backlash. The backlash can be hardly compensated for by the control system, since its modelling is very complex and unreliable. Thus,
this effect must be reduced to a minimum by proper design and fabrica-
tion of all "critical parts" in the robotic system. Since the gears
are most critical parts from the standpoint of backlash, their design
and production must be carefully done, if the high accuracy and repeatability of the robot are required. As we have already mention,
these reasons motivated the research towards introducing of so-called
direct-drive robots, i.e. robots which actuators are with no reducers.

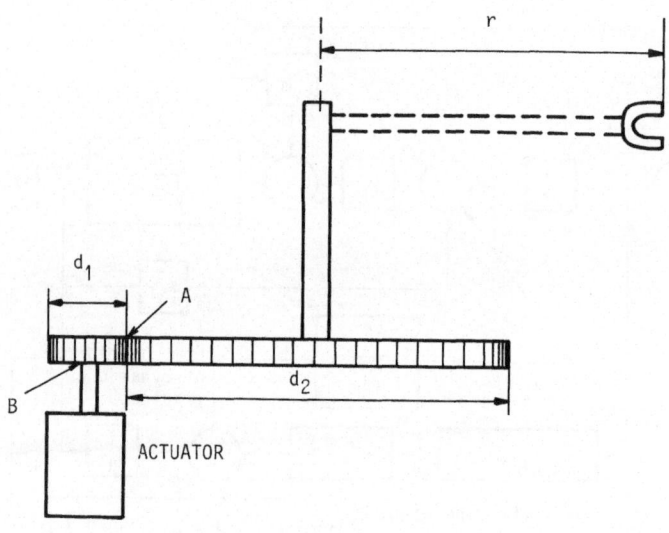

Fig. 3.19. Effect of backlash in gear reducer upon
robot end point positioning

(e) *Other nonlinear effects*

Besides above listed nonlinear effects in the robot servo system there
appear some other nonlinearities which might reflect the servo performance. Their modelling is usually not simple, but they can often be
neglected relative to the effects which have been taken into account
in the linear model of the servo. In other words, we may assume that
the control synthesized on the basis of the linear model is *robust*
enough to overcome these nonlinear effects, so they do not reflect
significantly the servo system performance. Still, in the process of the

robot design and the servo system synthesis and implementation, for each particular robotic system, we must pay attention of these effects [8].

We must underline that the linear model of the servo is rough approximation if electro-hydraulic actuators are considered. Namely, the parameters of the linear models of the hydraulic actuators are functions of the actuator state (or, "the work regime" of the actuator). This problem can be solved if we introduce variable servo gains (depending "on work regime"), or if we select the fixed servo gains but which make the servo system robust enough to ensure desired performance of the servo system regardless of the variation of the actuator parameters.

It should be also noted that if the servo system is to be implemented by microprocessor, then extra nonlinearities due to truncations in the microprocessors and A/D converters have to be taken into account. If the analogue information from the sensors is introduced in microcomputer through A/D converter, then converter obviously truncates the information since it gives a finite number of digits at its output. This truncation also might reflect the accuracy of the servo system and the repeatability of the robot. Thus, the selection of microprocessor and converters must be carefully done.

EXAMPLE 3.3.5. Let us assume that the amplitude constraint upon the input of the actuator u^1 for the servo system synthesized in the Example 3.3.2. is given by $u_m^1 = 27$ [V]. In this case the finite region in the state space for which the inequalities (3.3.61) are satisfied, is given in Fig. 3.20.

Let us consider some state out of this region (e.g. the state $q^1 = 0.3$ [rad], $\dot{q}^1 = 0.$). For this state the control law would require the control signal of $u^1 = 40.539$ [V] which is not allowed. Due to this, the servo behaves as it had the following gains instead the synthesized servo gains (see Example 3.3.2):

$$K_P^{i'} = \frac{u_m^i \cdot K_P^i}{|K_P^i \cdot q^i + K_v^i \dot{q}^i|} = \frac{27}{40.539} \; 135.13 \approx 90 \left[\frac{V}{rad}\right]$$

$$K_v^{i'} = \frac{u_m^i \cdot K_v^i}{|K_P^i q^i + K_v^i \dot{q}^i|} = \frac{27}{40.539} \; 43.6 = 29.04 \left[\frac{V}{rad/s}\right]$$

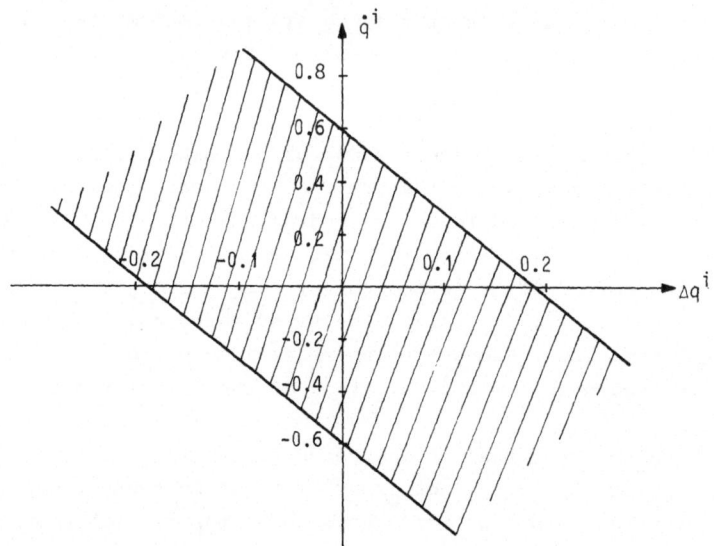

Fig. 3.20. Region of linear performance of
servo system in Example 3.3.5.

Exercises

3.49. Consider the third order model of the servo system (3.3.4), the
state vector of which is given by $x^i = (q^i, \dot{q}^i, i_R^i)^T$, where i_R^i
is the current of the rotor circuit. If the feedback loops by
all three state coordinates are introduced (i.e. if the control
law is given by $u^i = -k_i^T x^i$), if the amplitude of the input signal of the actuator is constraint by (3.3.61), and if the ampli-
tude of the rotor current is constrained by $|i_R^i| < i_{Rmax}^i$, then wri-
te the expressions for the boundaries of the finite region in
the state space for which the servo behaves linearly. Draw this
finite region in the 3-D state space.

3.50. For the servo system in Example 3.3.5, draw in the complex plane
the displacement of the poles of the closed-loop servo, if the
state of the system moves along lines $\dot{q}^i = 0.$, $\Delta q^i < -0.189$, and
$\dot{q}^i = 0.$, $\Delta q^i > -0.189$ [rad]. (Instruction: calculate the values of
the poles of the closed-loop servo for several values of Δq^i satisfying $\Delta q^i < -0.189$, and for several values satisfying $\Delta q^i > 0.198$;
for each state first calculate servo gains as presented in Example
3.3.5., and then calculate corresponding poles). To which positions are approaching these poles if $\Delta q^i \to \infty$?

3.51. Moments due to dynamic friction $P_{o\ dynamic}^{i}$ in the joints of Stanford manipulator (Fig. 3.13) have been measured experimentally. Their values are presented in Table 3.11. In table are also given position gains of individual servos in the robot joints. Determine the errors in the positioning of the manipulator end point (i.e. repeatability) which are caused by the dynamic friction in the joints (in this, assume that the dynamic friction moments act as external loads upon the servos) [5].

JOINT	r [m]	Dynamic friction moment [Nm]	$C_M^i N_m^i K_p^i / r_R^i \left[\frac{Nm}{rad}\right]$
1	0.54	1.91	790.
2	0.50	3.18	1780.
3	-	12.0	27600.
4	0.25	0.565	220.
5	0.25	0.635	220.
6	0.25	0.424	1480.

Table 3.11. Moments due to dynamic friction for Stanfort manipulator [5] (r is effective level arm from the joint to the manipulator end point)

3.52. For electro-hydraulic servo system considered in Exercise 3.35, assume that flow/pressure coefficient of the servovalve varies within the limits K_c^i = (0.0006, 0.0009), while all the other parameters are fixed, as they are given in Table 3.5.

a) Calculate the position and velocity servo gains to satisfy requirements given in Exercise 3.35 (ω_o=12. [rad/s], ξ_i=1), for the following values K_c^i = 0.0006, 0.0007, 0.0008, 0.0009 [$cm^3/s/N/m^2$].

b) If we adopt servo gains calculated for K_c^i = 0.0007 determine the variation of the poles of the closed-loop servo if K_c^i takes values K_c^i = 0.0006, 0.0008, 0.0009. Is it acceptable to adopt fixed servo gains, or is it necessary to implement variable servo gains, as calculated under a)?

3.53. The sketch of the robot is presented in Fig. 3.19. Assume that the backlash appears at the points A and B. If the geometric parameters are as follows: d_1 = 0.05 [m], d_2 = 0.60 [m] and r = 2.5 [m], calculate the repeatability of robot end point positioning due to backlash of 0.0001 [rad] at point A. Next, calculate the robot repeatability due to the backlash at point B of value 0.00005 [rad]. Repeat this calculations, but if the diametar of the larger gear change to d_2 = 0.9 [m], and the backlash at point A increases to 0.0002 [rad] (the backlash at point B is the same as in previous case). Comment the results obtained in these two cases: how the increase of the gear reduction ratio affects the repeatability of the robot?

3.54. For the robot in Fig. 3.19. calculate the repeatability of the robot end point positioning due to truncation of the A/D converter for the servo in the first joint. The A/D converter is connected to the position sensor. The position sensor is located at the output shaft of the actuator, before the gear reducer. The geometric parameters are given in previous exercise. The A/D converter is of 12 bits (i.e. it truncates all angles bellow $360^{\circ}/4096$). Calculate the repeatability for both values of the gear radius d_2.

3.4 Synthesis of Local PID Controller

Up to now, we have considered so-called *static* servo systems for individual joints of the robot. We have considered the movement of the i-th joint while the other joints are kept locked solid. We have shown that the servo system ensures positioning of the i-th joint, but the steady state error appears which is caused by external load - gravity moment of the mechanism around the i-th joint. The gravity moment can be compensated by on-line computation of this moment in control microprocessor (Fig. 3.15). This compensation requires computation of the gravity moment, which might produce problems due to numerical complexity of the expressions for this moment which depends on the robot structure. On the other hand, uncertainty in determining the parameters necessary to compute gravity moment, might cause poor compensation of steady state errors. The second approach to eliminate the steady state error due to gravity moment is by introducing of so-called *dynamic* servo systems [9].

If we allow the other joints to move simultaneously with the i-th joint, then an external moment, due to dynamic coupling, will act upon the i-th joint (see Chapter 4). This dynamic moment will also cause an error in positioning of the i-th joint. The effects of this moment, which is complex function of all joints coordinates, their velocities and accelerations, might also be partially compensated by dynamic servo system.

Here, we shall briefly consider application of dynamic controllers in control of the robot. In the dynamic servo system the feedback loops by the integrals of the state coordinates of the system (or, of the system outputs) are introduced. We shall consider the simplest form of the dynamic controller - so-called PID controller in which the feedback loop by the integral of the position error is introduced.

Namely, it is possible to introduce integrator which integrates the error between the actual position and desired (imposed) position of the joint, and the computed integral of the position error is added to the actuator input signal. In this way PID controller is obtained which is presented in Fig. 3.21. This controller, thus, besides position (P), velocity or, differential (D), includes so-called integral (I) feedback loop [3].

It is obvious that the introduction of the integrator in the controller increases the order of the system model. If the second order model of the actuator and the moving part of the mechanism is considered $(n_i=2)$, then by applying of the integral feedback loop we obtain the third order model of the complete system (controller and actuator). The transfer function of the system (3.3.25) now becomes:

$$q^i(s) = \frac{[K_D^i K_P^i s + K_D^i K_{IN}^i]q^{oi}(s) - K_M^i sg_i(s)}{K_D^i K_{IN}^i + K_D^i K_P^i s + K_D^i K_v^i s^2 + (T_m^i s + 1) s^2} \qquad (3.4.1)$$

where K_{IN}^i denotes gain in the feedback loop by integral of the position error. Based on (3.4.1) we may determine the position error as:

$$\Delta q^i(s) = \frac{-[K_D^i K_v^i s^2 + (T_m^i s + 1) s^2]q^{oi}(s) - K_M^i sg_i(s)}{K_D^i K_{IN}^i + K_D^i K_P^i s + K_D^i K_v^i s^2 + (T_m^i s + 1) s^2} \qquad (3.4.2)$$

128

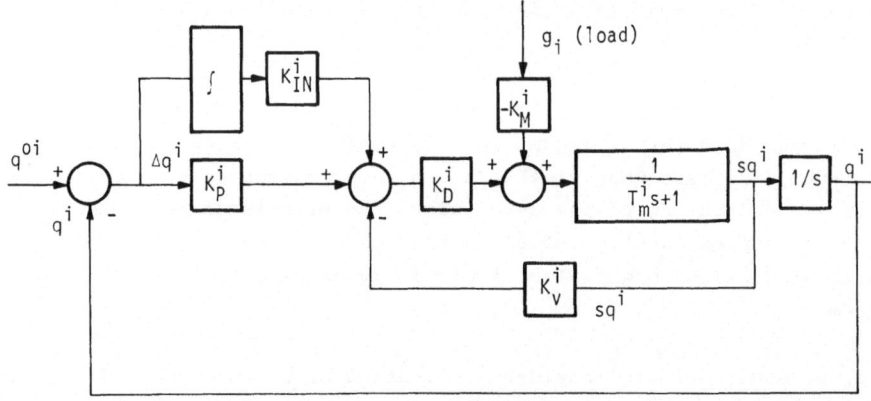

Fig. 3.21. PID (proportional - integral - differential) controller
(for the second order model of actuator, K_{IN}^i denotes
the gain in the integral feedback loop)

If we assume the step external load acting upon the i-th servo (i.e.
we assume constant load moment in the i-th joint), then we obtain steady state error as:

$$\Delta q^i(\infty) = \lim_{s \to 0} s\Delta q^i(s) = 0 \qquad\qquad (3.4.3)$$

Thus, we see that the integral feedback loop eliminates the steady
state error (for step input moment). If all other joints are locked,
then only the gravity moment is acting around the i-th joint when it
stops. This moment in steady state conditions might be assumed to be
constant. Thus, by integral feedback loop we eliminate the positioning
error of the servo caused by gravity moment around the i-th joint. The
PID controller ensures more accurate positioning of the robot joint
than the static controller considered in the previous sections.

Let us consider briefly the problems related to synthesis of PID controller, i.e. let us consider selection of the feedback gains. We may
use various procedures for synthesis of feedback gains: synthesis in
frequency domain (Bode's method), root locus methods, etc. We also may
apply procedures for control synthesis related to state space model of
the system: pole-placement method, optimal regulator approach etc.

Let us consider synthesis of PID controller by pole-placement method,
which we have already considered in Section 3.3.3. The model of the
actuator and the moving part of the mechanism with the introduced

integrator by position error in the state space has the following form
(if we neglect the effects of the external disturbances):

$$\dot{\tilde{x}}^i = \tilde{A}^i \tilde{x}^i + \tilde{b}^i u^i + \tilde{d}^i q^{oi} \tag{3.4.4}$$

where $\tilde{A}^i$ is $(n_i+1) \times (n_i+1)$ matrix, $\tilde{b}^i$, $\tilde{d}^i$, are $(n_i+1) \times 1$ vectors given
by:

$$\tilde{A}^i = \left[\begin{array}{c|c} \hat{A}^i & 0 \\ \hline 1 \quad 0 \quad 0 & 0 \end{array}\right], \quad \tilde{b}^i = \left[\begin{array}{c} \hat{b}^i \\ \hline 0 \end{array}\right], \quad \tilde{d}^i = \left[\begin{array}{c} 0 \\ \hline -1 \end{array}\right] \tag{3.4.5}$$

and $\tilde{x}^i$ is $(n_i+1) \times 1$ *augmented state vector* of the system, which is given
by $\tilde{x}^i = (x^{iT^i}, z^i)^T$. Here, z^i is the new state coordinate which represents the integral of the position error:

$$z^i(t) = \int_o^t (q^i - q^{oi}) dt \tag{3.4.6}$$

In (3.4.4) the control signal (input signal for actuator) is denoted
by u^i. For the PID controller presented in Fig. 3.21 ($n_i=2$) the control
signal is generated as:

$$u^i = -K_P^i(q^i-q^{oi}) - K_v^i \dot{q}^i - K_{IN}^i \int_o^t (q^i-q^{oi}) dt =$$

$$= -K_P^i(q^i-q^{oi}) - K_v^i \dot{q}^i - K_{IN}^i z^i = -\tilde{k}_i^T \tilde{x}^i + K_P^i q^{oi} \tag{3.4.7}$$

where by $\tilde{k}_i$ is denoted $(n_i+1) \times 1$ feedback gain vector which is given by:

$$\tilde{k}_i = (K_P^i, K_v^i, K_{IN}^i)^T \tag{3.4.8}$$

The model of the closed-loop system is obtained by combining (3.4.4)
and (3.4.7):

$$\dot{\tilde{x}}^i = (\tilde{A}^i - \tilde{b}^i \tilde{k}_i^T) \tilde{x}^i + \tilde{b}^i K_P^i q^{oi} + \tilde{d}^i q^{oi} \tag{3.4.9}$$

The closed-loop system matrix is given by $(\tilde{A}^i - \tilde{b}^i \tilde{k}_i^T)$. As described in
Section 3.3.3, the feedback gains $\tilde{k}_i$ have to be selected in such a way
that the eigen-values of the closed-loop system matrix are placed in
the desired locations in the s-plane. In the case of the PID controller, since the order of the system is (n_i+1), we may specify (n_i+1)
eigen-values (if feedback loops by all the state coordinates are implemented). Let us assume that we have specified (n_i+1) eigen-values

so that all s^o_j are in the left half of the s-plane. In that case the feedback gains $\tilde{k}_i$ can be obtained from:

$$\det(\tilde{A}^i - \tilde{b}^i \tilde{k}_i^T - sI_{n_i+1}) = \bar{K} \prod_{j=1}^{n_i+1} (s^o_j - s) \qquad (3.4.10)$$

The feedback gains $\tilde{k}_i$ which satisfy the relation (3.4.10) can be easily determined.

Here, we have assumed that the feedback loops by all state coordinates $\tilde{x}^i$ are implemented. However, it is not always necessary to introduce feedback loops by all state coordinates $\tilde{x}^i$. As we have explained in Section 3.3.3, it is possible to introduce feedback loops just by the system output $\tilde{y}_i$ which is given by (analogously to (3.3.43)):

$$\tilde{y}_i = \tilde{C}^i \tilde{x}^i \qquad (3.4.11)$$

where $\tilde{y}_i$ is $k_i^y \times 1$ output vector, and $\tilde{C}^i$ is $k_i^y \times (n_i+1)$ output matrix. Therefore, the control is introduced as feedback by the system output:

$$u^i = -\tilde{k}_i^T \tilde{C}^i \tilde{x}^i + K_p^i q^{oi} \qquad (3.4.12)$$

where now k_i is a $k_i^y \times 1$ vector of feedback gains.

If this control is introduced in the model (3.4.4), we obtain the closed-loop matrix of the system as $(\tilde{A}^i - \tilde{b}^i \tilde{k}_i^T \tilde{C}^i)$. Now, we may specify k_i^y eigen-values s^o_j of the closed-loop system matrix, while the rest $(n_i+1) - k_i^y$ eigen-values are free (for $k_i^y < n_i+1$):

$$\det(\tilde{A}^i - \tilde{b}^i \tilde{k}_i^T \tilde{C}^i - sI_{n_i+1}) = K \prod_{j=1}^{k_i^y} (s^o_j - s) \cdot \prod_{j=k_i^y+1}^{n_i+1} (s_j - s) \qquad (3.4.13)$$

Based on (3.4.13) we can determine k_i^y gains in feedback loops by system output. We can examine the conditions under which the unspecified eigen-values s_j are also in the left part of the s-plane (i.e. under which they are on the left from the specified eigen-values s^o_j).

Thus, the problem of synthesis of PID controller is reduced to specification of the desired eigen-values s^o_j in the complex plane. The selection of these eigen-values might be performed in various ways. We may require that the eigen-values of the closed-loop system matrix are on the left from the line $Re(s) = -\beta_i$, where β_i is positive number. By this we ensure that the system has sufficiently fast response to step input (i.e. that the position of the joint approaches to desired position q^{oi} faster than $\sim \exp(-\beta_i t)$).

The feedback loop by integral of position error eliminates the steady state error in the controller response to step (gravity) moment. However, if the time variable external moment (disturbance) is acting upon the system, then this feedback loop by integral of position error cannot compensate for the error in the positioning of the joint. Let us assume that the ramp moment disturbance is acting upon the system, i.e. let us assume that the external moment $P_i(t)$ satisfies $dP_i(t)/dt = P_i^B = const$. In this case the Laplace transform of the external disturbance moment is given by $P_i(s) = P_i^B/s^2$. If we replace this moment in the expression for the positioning error in s-domain (3.4.2), we obtain:

$$\Delta q^i(s) = \frac{-[K_D^i K_v^i s^2 + (T_m^i s+1) s^2] q^{oi}(s) - K_M^i P_i^B/s}{K_D^i K_{IN}^i + K_D^i K_P^i s + K_D^i K_v^i s^2 + (T_m^i s+1) s^2} \qquad (3.4.14)$$

Based on this expression we obtain the steady state error as:

$$|\Delta q^i(\infty)| = \frac{K_M^i P_i^B}{K_D^i K_{IN}^i} \qquad (3.4.15)$$

which means that the steady state error of the system response to ramp moment disturbance is inversely proportional to the integral feedback gain.

If we want to reduce the steady state error due to ramp moment disturbance (3.4.15), we should select high integral feedback gain K_{IN}^i. However, to ensure that the pair of the dominant poles of the system is as close as possible to the real axis of the s-coordinate system, and to ensure fast response of the system (i.e. to ensure that the real parts of all the system poles are sufficiently high), we must apply relatively high position and velocity feedback gains. Since the high feedback gains reflect the characteristic frequency ω_i of the system, which is constrained by (3.3.37), and since the high feedback gains are undesirable regarding the effects of the noise upon the system performance, we must make a tradeoff between these opposite requirements. Namely, in specifying of the desired eigen-values of the closed-loop system matrix s_j^o we must take care of the following requirements:

to ensure low steady state error (3.4.15), sufficient damping of the
system, fast response, but also the condition (3.3.37) must be satisfied.

We have explained that if all the joints of the robot are locked, save
for the i-th joint, then just the gravity moment is acting upon the
i-th joint as external disturbance. This moment is constant in steady
state conditions. However, if the other joints are also moving, then
upon the i-th actuator is acting variable moment (load) due to dynamic
interconnections between the joints movements. This dynamic moment is
complex function of actual joints coordinates, velocities and accelerations (see Chapter 4). Therefore, we have to ensure that the error
due to variable external moment disturbance is as small as possible.
Obviously, by PID controller we cannot eliminate this error. It is possible to introduce dynamic controller of higher order, i.e. with feedback loops by higher integrals of the position error. However, such
controller is too complex and its response might be too slow, so they
are not often applied in robotics.

EXAMPLE 3.4. In Example 3.3.4 the PD servo system for the second joint
of the robot in Fig. 2.6. has been synthesized. The data on the D.C.
electro-motor are given in Table 3.3. Let us synthesize the PID controller for the same joint and actuator by pole-placement method. If
the second order model of the actuator is considered (n_i=2), then we
can specify n_i+1=3 eigen-values of the closed-loop system matrix. Let
us select the following places of the eigen-values:

$$s^o_{1,2} = -5.45 \pm j2.8, \qquad s^o_3 = -44.6$$

Based on (3.4.10) we compute the feedback gains as:

$$K^2_P = 1446. \ [\frac{V}{rad}], \ K^2_v = 34.36 \ [\frac{V}{rad/s}], \ K^2_I = 4691.2 \ [\frac{V}{rad/s}]$$

The steady state errors due to constant gravity moment are given in
Table 3.9. for various positions of the second and the third joint of
the robot if PD servo system is applied. It is clear that if the PID
controller is applied these steady state error are reduced to zero. In
Fig. 3.22. are presented the responses of the PD and PID controllers
to step input q^{o2}=0.02+q^2(0)[rad] when the gravity moment is acting around
the joint axis. (We assume that the gravity moment is not compensated
by on-line computation as in (3.3.61)). From Fig. 3.22. we see that

the PID controller eliminates the steady state error, but the overshoot appears.

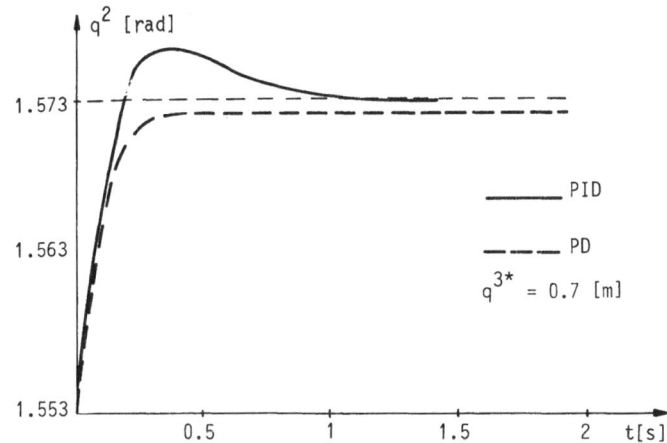

Fig. 3.22. Responses of PD and PID controllers to step input and step moment disturbance

Exercises

3.55. If the order of the model of the actuator and joint is $n_i = 3$, and if the structure of the matrices $\tilde{A}^i$ and $\tilde{b}^i$ is given by (3.3.49) and (3.4.5), then write the expressions for feedback gains of the PID controller which ensure placement of the eigen-values of the closed-loop matrix of the system in desired positions. Assume that the feedback loops are introduced as follows:

a) feedbacks by position, velocity and integral of the position error; determine the expression for unspecified eigen-value,

b) feedbacks by all the state coordinates of the system (by position, velocity, current/pressure, and integral of position error),

c) feedbacks by position and integral of the position error, only; determine the expressions for the rest two unspecified eigen-values and determine conditions to guarantee that these eigen-values are on the left from the specified eigen-values.

3.56. The data on the D.C. electro-motor applied in the second joint of the robot in Fig. 2.5. are given in Table 3.3. Determine the matrices of the model of actuator and joint for $n_i = 3$ (data on the robot mechanism are given in Table 3.4). Determine the feedback gains of the PID controller to satisfy the following conditions: all eigen-values of the closed-loop system matrix should be on the left from the line Re(s) = -3., the dominant pair of eigen-values has to be as close as possible to real axis, the steady state error to ramp moment disturbance (3.4.15) for $P_i^B = 1$ [Nm/s] should satisfy $|\Delta q^2(\infty)| < 0.01$ [m]. Determine the feedback gains under assumption:

a) that the feedback loops by all state coordinates are introduced,

b) that feedback loops by position, velocity and integral of the position error are introduced.

3.57.* How can be PID controller implemented by microprocessor? Determine the expression for numerical integration of the position error by Euler's method for numerical integration. Write the programme (using some higher programming language) for computation of the control signal for PID controller, assuming that the input variables for the programme are: desired position of the joint q^{oi}, actual values of the joint position q^i and velocity $\dot{q}^i$ (which are obtained from the sensors through A/D converters), and the selected feedback gains. Minimize the number of additions and multiplications required for computation of u^i according to (3.4.7).

3.58. Compare the compensations of the steady state error due to gravity moment: by PID controller and by PD servo system with the on-line computation of gravity moment. Compare these compensations from the standpoints of:

a) the number of numerical operations required to compute the control signal (compare for example (3.3.61) and (3.4.7) taking into account the numerical integration in PID controller),

b) the robustness to variations of the payload parameters,

c) the settling time (i.e. the time required for the joint
to settle at the desired position).

3.59.* Extend the programme for computation of the feedback gains on
the basis of the pole-placement method (required in Exercise
3.38) to include the integral feedback loop. (Instruction: use
expressions obtained in Exercise 3.55).

3.5 Synthesis of Local Servo System for Trajectory Tracking

Up to now we have considered local servo system for individual joints
of the robot. The control task of these servo systems is to position
the robot joints in the desired positions. The synthesized servo system for the i-th joint ensures its positioning under assumption that
all the other joints are locked.

As we have shown in Chapter 1. the tasks which are imposed in modern
robotics require not only the accurate positioning of the robot, but
also the tracking of desired trajectories. This means that the input
for the local servo system is not constant (desired) position q^{oi} of
the corresponding joint (the step function), but the desired trajectory $q^{oi}(t)$, i.e. the time variable signal. Since the set input varies
by time, the controller has to ensure that the system output (joint
coordinate q^i) varies with time in the same way, i.e. the controller
output q^i has to track imposed trajectory $q^{oi}(t)$. In other words, the
task of the controller is to minimize the error between the actual
trajectory of the robot joint $q^i(t)$ and the imposed trajectory $q^{oi}(t)$
at each moment of time.

If the input signal for the servo system synthesized in the previous
sections is time variable function $q^{oi}(t)$ (trajectory), the servo system output will not track accurately this trajectory. The error between
the actual trajectory and desired trajectory will appear. Let us assume that the imposed trajectory is ramp function, i.e. that it requires
constant velocity of the joint $q^{oi}(t) = \Omega_i^o t$. The Laplace transform of
this input function is $q^{oi}(s) = \Omega_i^o/s^2$. Let us consider the error of
a static servo system whose transfer function in the s-domain is given
by (3.3.25), assuming that the order of the actuator model is $n_i = 2$. If
we replace the expression for the ramp function input in (3.3.25), then
we obtain the Laplace transform of position error between the actual

joint trajectory and nominal (desired) trajectory as:

$$\Delta q^i(s) = \frac{-[K_D^i K_v^i s + T_m^i s^2 + s] \Omega_i^o / s^2}{K_D^i K_P^i + K_D^i K_v^i s + T_m^i s^2 + s} \tag{3.5.1}$$

The steady state position error in tracking of the desired (input) trajectory is:

$$\Delta q^i(\infty) = \lim_{s \to 0} s \Delta q^i(s) = \frac{-K_D^i K_v^i - 1}{K_D^i K_P^i} \Omega_i^o \tag{3.5.2}$$

Depending on the velocity Ω_i^o of the input trajectory and on the parameters of the servo system, the significant error in tracking of the imposed trajectory might appear.

If we set nominal trajectory which requires more "complex variation" of the joint position by time, the error in servo system tracking might be even higher. For instance, if the nominal trajectory requiring constant acceleration of the joint is imposed ($q^{oi}(t) = a_i^o t^2$), then the Laplace transform of the position error in tracking of this trajectory can be obtained as (taking into account that the Laplace transform of the input trajectory is $q^{oi}(s) = a_i^o / s^3$):

$$\Delta q^i(s) = \frac{-[K_D^i K_v^i s + (T_m^i s + 1) s] a_i^o / s^3}{[K_D^i K_P^i + K_D^i K_v^i s + (T_m^i s + 1) s]} \tag{3.5.3}$$

The steady state error in this case is (under assumption that the delay due to velocity terms are compensated for - see the text to follow):

$$\Delta q^i(\infty) = \frac{-T_m^i a_i^o}{K_D^i K_P^i} \tag{3.5.4}$$

In this case the error might be high depending on the required acceleration a_i^o and on the parameters of the servo. It is obvious that these errors in tracking of desired trajectories are consequence of the delays in the servo system, so that the system output cannot follow the imposed variation of the input signal. However, this delay can be compensated for by introducing of so-called *"pre-compensator"* (or, *feedforward term*).

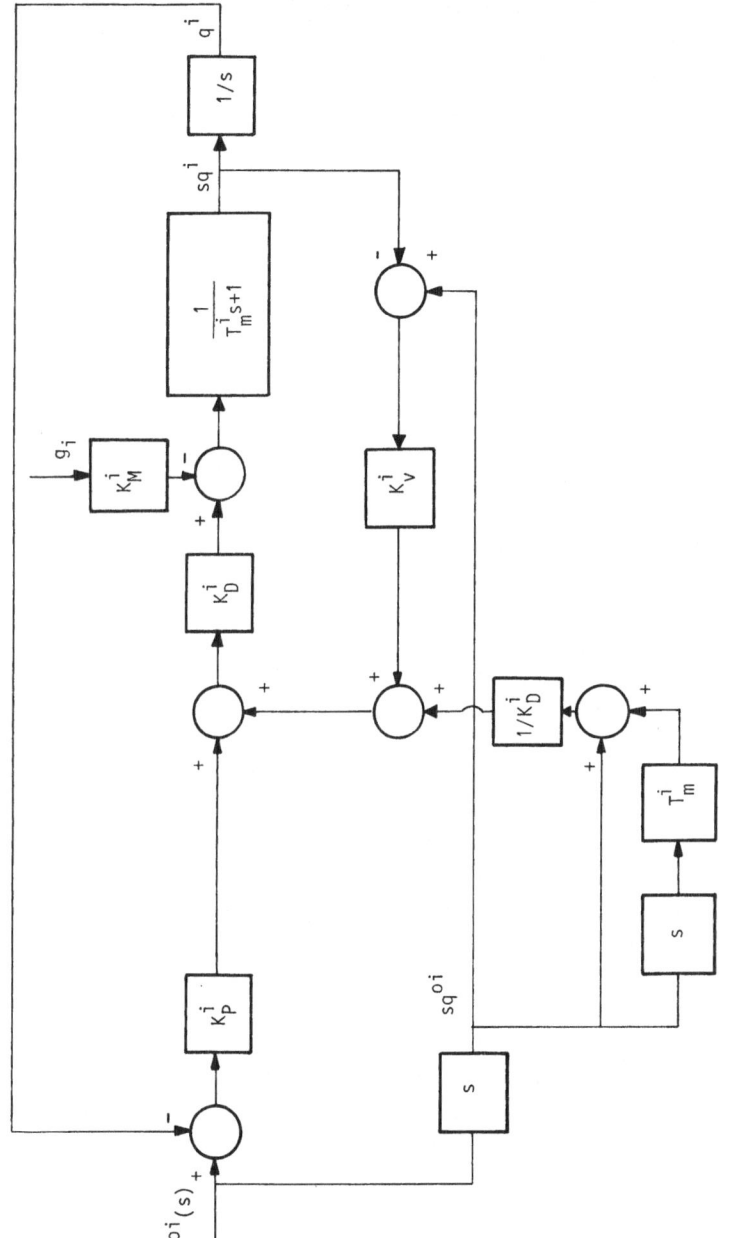

Fig. 3.23. Servo system with feedforward term to compensate for velocity and acceleration variations along nominal trajectory $q^{oi}(t)$

The feedforward term is introduced in the following way: the input signal for servo is not desired trajectory $q^{oi}(t)$, but some modified signal which takes into account the delay in the servo. Namely, the signal which has to "accelerate" (or, "decelerate") the servo proportionally to desired acceleration $\ddot{q}^{oi}(t)$, is added to the original input signal (corresponding to desired trajectory $q^{oi}(t)$). This additional signal has to reduce the error between the imposed input trajectory $q^{oi}(t)$ and actual trajectory $q^i(t)$ of the joint (t.e. the servo output).

First, let us consider the introduction of feedforward term in the s-domain. It can be easily shown that if the feedforward term is introduced as in Fig. 3.23, then the steady state error in tracking the trajectory $q^{oi}(s)=a_i^o/s^3$ is reduced to zero. This means that the delay due to both velocity term and acceleration term is compensated for by the feedforward in Fig. 3.23. Similarly, if the input trajectory is given by $q^{oi}(s) = \Omega_i^o/s^2$, the steady state error is also reduced to zero by the feedforward term added to servo as in Fig. 3.23. Therefore, by the pre-compensator we can ensure that the servo tracks the input time-variable trajectory $q^{oi}(t)$.

Let us consider introduction of the feedforward term using the state-space model of the servo. Let us consider the model of actuator and the moving part of the robot mechanism in the form (3.3.4). We have to ensure tracking of a desired nominal trajectory $q^{oi}(t)$. We consider the second order model of the system $n_i=2$, so that the state vector x^i is given by $x^i = (q^i, \dot{q}^i)^T$. We have to ensure that the actual system state x^i tracks the desired trajectory of the state vector $x^{oi}(t) = (q^{oi}(t), \dot{q}^{oi}(t))^T$. Therefore, we have to ensure that the system state $x^i(t)$ moves along desired *nominal trajectory* $x^{oi}(t)$. To ensure this, we introduce so-called *local nominal programmed control* $u_L^{oi}(t)$ which has to satisfy:

$$\dot{x}^{oi}(t) = \hat{A}^i x^{oi}(t) + \hat{b}^i u_L^{oi}(t) \qquad (3.5.5)$$

The explanation of the physical meaning of the nominal control term is obvious: this is the programmed input signal (so-called open-loop control) which ensures that the system state moves along desired nominal trajectory if the following conditions are met [10]:

(a) if the initial state of the system $x^i(0)$ coincides with the nominal initial state $x^{oi}(0)$, i.e. if $x^i(0) = x^{oi}(0)$, and

(b) if the model of the servo (3.3.4) is ideally accurate and no external disturbance is acting upon it.

Obviously if $x^i(0) = x^{oi}(0)$ and if the model of the system were perfect
(which means that all the parameters of the system were perfectly accurately identified), and if the signal $u_L^{oi}(t)$ computed to satisfy (3.5.5)
is fed to the input of the servo, then the actual state of the system
$x^i(t)$ would be driven along nominal trajectory $x^{oi}(t)$. However, it
is obvious that in the general case none of the above assumptions are
satisfied. Due to this, we have to consider *the model of the deviation*
of the system state $x^i(t)$ *from the imposed nominal trajectory* $x^{oi}(t)$.
Based on the model (3.3.4) and (3.5.5) we can obtain the model of the
state deviation in the following form[*]:

$$\Delta\dot{x}^i(t) = \hat{A}^i\Delta x^i(t) + \hat{b}^i\Delta u^i(t) \qquad (3.5.6)$$

where $\Delta x^i(t) = x^i(t) - x^{oi}(t)$ is $n_i \times 1$ vector of the deviation of the system state from the nominal trajectory $x^{oi}(t)$, and $\Delta u^i(t) = u^i(t) - u^{oi}(t)$
is scalar deviation of the servo input $u^i(t)$ from the nominal programmed control $u^{oi}(t)$. Now, we have to determine the additional control
signal $\Delta u^i(t)$. The role of this signal is to reduce the deviation of
the robot state $\Delta x^i(t)$ from the nominal trajectory, i.e. to ensure that
$\Delta x^i(t) \to 0$. In this way, the problem of synthesis of control for tracking
of the nominal trajectory for (3.3.4) reduces to the problem of synthesis of servo for the model of deviation (3.5.6). Actually, the problem reduces to synthesis of a regulator which has to ensure the stability of the model of deviation around the equilibrium point $\Delta x^i = 0$.
This problem has been already considered in Section 3.3. We have to
synthesize servo which will drive the system state to the desired state of the model of deviation $\Delta x^i = 0$. Since the matrices of the model of
deviation (3.5.6) $\hat{A}^i$ and $\hat{b}^i$ are identical to the matrices of the basic
model of the system (3.3.4), the servo system synthesized in the previous sections can be directly applied for stabilization of the model
of deviation around the nominal trajectory and the nominal control.
Thus, the problem of trajectory tracking is reduced to synthesis of
positional servo system to which local nominal programmed control is
added. This nominal control signal has to compensate for the variations of the velocity and accelerations along the imposed (nominal)

---

[*] Here, we have not taken into account the amplitude constraint upon
the servo input.

trajectory. Obviously, the control signal is calculated as[*]:

$$u^i(t) = u_L^{oi}(t) - k_i^T C_i \Delta x^i(t) \qquad\qquad (3.5.7)$$

where k_i is the feedback gain vector, C_i is the $k_i^y \times n_i$ output matrix (see Section 3.3.3). The feedback gains can be synthesized by one of the previously considered methods: by synthesis in s-domain (Section 3.3.2), or by pole-placement (Section 3.3.3), or by some other method. The scheme of the control is presented in Fig. 3.24. There are two dif- ferences between this scheme and the scheme of simple position servo: the scheme in Fig. 3.24. includes computation of local nominal control as feedforward signal, and velocity feedback gain k_v^i amplifies the er- ror between the actual velocity of the joint and nominal velocity of the joint $\Delta\dot{q}^i(t) = \dot{q}^i(t) - \dot{q}^{oi}(t)$, instead to amplify just the actual velocity signal $\dot{q}^i(t)$ (as in the scheme in Fig. 3.14).

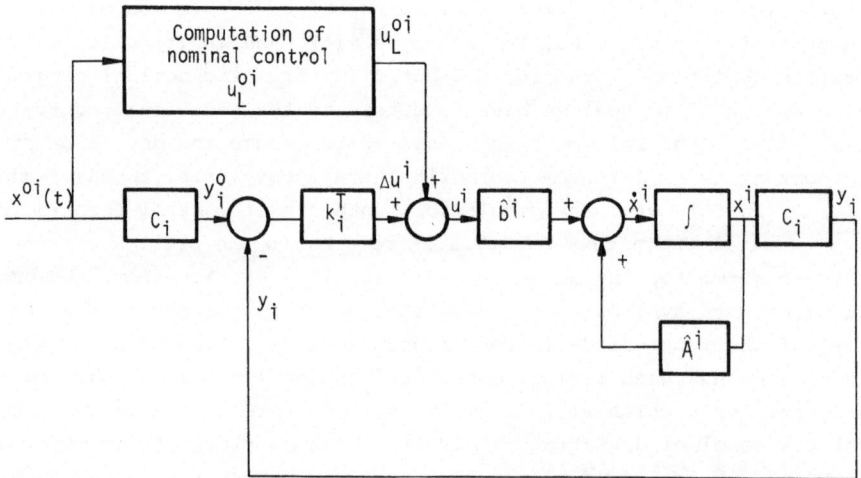

Fig. 3.24. Scheme of servo system with local nominal programmed control (the scheme is given according to the state space model of actuator)

Computation of the local nominal control which satisfies (3.5.6) is very simple. Let us assume that the model of the actuator is of the second order $n_i = 2$. Let us assume that the trajectory $q^{oi}(t)$ is given,

[*] Here, the general case is considered when the feedback loops only by the system output y_i are introduced; if the feedback loops are in- troduced by all state coordinates x^i, then the matrix C_i is the unit matrix.

and that by differentation we get variation of the joint velocity $\dot{q}^{oi}(t)$ and acceleration $\ddot{q}^{oi}(t)$ along the desired trajectory. By this the nominal trajectory of the state $x^{oi}(t)$ and the first derivative of the state $\dot{x}^{oi}(t)$ along the nominal trajectory are defined. Assuming that the matrices $\hat{A}^i$ and $\hat{b}^i$ are in the form (3.3.54), on the basis of (3.5.6) we can compute the local nominal control as:

$$u_L^{oi}(t) = (\dot{x}_2^{oi}(t) - a_{22}^i x_2^{oi}(t))/b_2^i \tag{3.5.8}$$

where by x_2^{oi} is denoted the second coordinate of the state vector (i.e. $x_2^{oi} = \dot{q}^{oi}$ and $\dot{x}_2^{oi} = \ddot{q}^{oi}$). Thus, the nominal control can be easily computed based on (3.5.8) and this is applied in the control scheme in Fig. 3.24. In this, it should be kept in mind that the actuator input amplitude is constrained. Based on (3.2.9), the following inequality has to be satisfied:

$$|u_L^{oi}(t)| \leq u_m^i \tag{3.5.9}$$

If the condition (3.5.9) is satisfied, and if the two above listed conditions were also met, then the nominal control (3.5.8) would drive the the joint along the desired (nominal) trajectory $q^{oi}(t)$. However, if the condition (3.5.9) is not satisfied, the nominal control cannot drive the system along desired trajectory. This means that the actuator cannot realize the desired nominal trajectory. The required velocity or acceleration of the joint are too high, and they cannot be realized by the applied actuator in the robot joint. The desired trajectory should be slow down. In the process of the robot design, the selection of the actuators must be properly done. This means, that the actuators in the robot joints must be selected to ensure realizations of desired velocities and accelerations of all the robot joints (so that the desired speed of the robot is guaranteed). Namely, the robot designer must estimate the velocities and accelerations of the robot joints that will be required in the tasks for which the robot is intended. When he selects actuators, he must check the condition (3.5.9) for each joint and for some representative (test) trajectories. If the condition (3.5.9) is not satisfied, then the actuator in the corresponding robot joint must be re-selected. Otherwise, the robot will not be able to accomplish the desired task[*].

[*] It should be notices that this consideration concerns the case when each joint of the robot is moved independently, i.e. when the robot joints are moved successively, one by one. If the simultaneous motion of all robot joints are required, then the dynamic moments due to coupling between the joints must be taken into account (see Chapter 4).

Analogously to (3.5.9), the nominal programmed control can be determined also in the case when the model of the actuator is of the third order $n_i = 3$ (see Exercise 3.60).

Since this programmed control takes into account only dynamics of the local actuator and single joint, it is called *local nominal control* [10, 11]. It differs from the so-called *centralized nominal control* which takes into account dynamics of the complete robot mechanism and which is computed on the basis of the complete model of the robot dynamics when all the joints are moving simultaneously (see Chapter 4).

The model (3.3.4) takes into account the moment of inertia of the mechanism around the i-th joint (if all the other joints are kept locked). The value of this moment of inertia varies if the other joints change their positions. We have shown (Section 3.3.4) that for the synthesis of the servo feedback gains it is required to consider the maximum value of the moment of inertia of the mechanism around the i-th joint. However, for the synthesis of the local nominal control we must consider the minimum possible value of the moment of inertia of the mechanism around the corresponding joint. If we calculate local nominal control taking into account (in (3.5.8)) some greater moment of inertia, then when the value of the moment of inertia is less, the joint angle would overshoot the desired trajectory. Since the overshoot of the desired input trajectory must not be allowed in any case, this means that we must calculate local nominal control taking into account the minimum possible value of the moment of inertia of the mechanism around the i-th joint (see Exercise 3.62).

EXAMPLE 3.5. For the first joint of the robot in Fig. 3.2. the model of the actuator and the moment of inertia of the mechanism around the joint axis, for $q^{2*} = 0.$, is given by (3.3.4) and the elements of the matrices $\hat{A}^i$ and $\hat{b}^i$ are given by (see Exercise 3.18):

$$\hat{A}^i = \begin{bmatrix} 0 & 1 \\ 0 & -0.38 \end{bmatrix} \qquad \hat{b}^i = \begin{bmatrix} 0 \\ 0.27 \end{bmatrix}$$

The corresponding transfer function in s-domain is given by (3.3.27). We want to synthesize the servo which has to ensure tracking of nominal trajectory $q^{o1}(t)$ given by:

$$q^{o1}(t) = \begin{cases} \dfrac{a^o t^2}{2} , & 0 < t \le \dfrac{\tau}{2} \\[3mm] \dfrac{a^o \tau^2}{8} - \dfrac{a^o}{2}(t - \dfrac{\tau}{2})^2 + \dfrac{a^o \tau}{2}(t - \dfrac{\tau}{2}), & \dfrac{\tau}{2} < t \le \tau \end{cases} \qquad (3.5.10)$$

This trajectory is presented in Fig. 3.25 for $a^o = 2$ [rad/s^2].

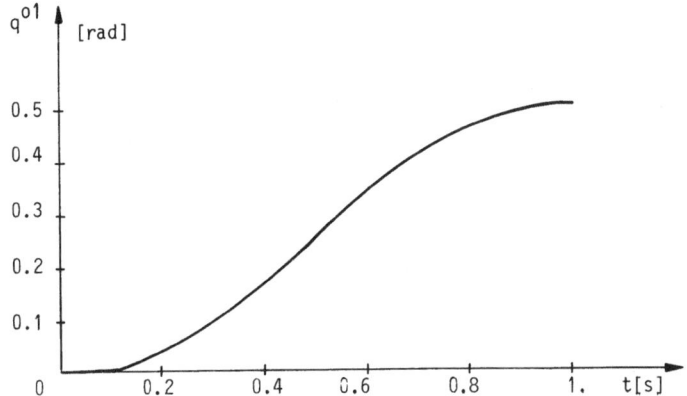

Fig. 3.25. Nominal trajectory for the first joint of robot in Fig. 3.2.

Compensation of the error in tracking of the trajectory due to variation of the velocity and acceleration can be realized by control scheme given in Fig. 3.23, or by scheme in Fig. 3.24. In this case the local nominal control is given by:

$$u^{o1}(t) = \begin{cases} (a^o + 0.38 a^o t)/0.27, & 0 < t \le \dfrac{\tau}{2} \\[3mm] [-a^o + 0.38 a^o (\tau - t)]/0.27, & \dfrac{\tau}{2} < t \le \tau \end{cases} \qquad (3.5.11)$$

while the servo gains can be selected as in Example 3.3.2. (i.e. by (3.3.40)). This programmed control compensates for the variation of the velocity and acceleration along the nominal trajectory, and so the accurate tracking of this trajectory is ensured. In Fig. 3.26. the results of the simulation by digital computer of the tracking of nominal trajectory (3.5.10) are given. Two cases are presented: if the nominal programmed control is not introduced and if the local nominal control (3.5.11) is included in the control scheme. It can be seen that the tracking in the latter case is much better, since in the former case the servo does not compensate for the variation of velocity and acceleration along the nominal trajectory. We assume that the initial error between the actual position of the joint $q^1(0)$ and the nominal position $q^{o1}(0)$ is $\Delta q^1(0) = q^1(0) - q^{o1}(0) = 0.2$ [rad]. The Fig. 3.26. shows how this error $\Delta q^1(t)$ decreases along the trajectory in the two above mentioned cases. The digital simulation of the dynamics of the single joint and complete robot will be addressed in Section 5.3.

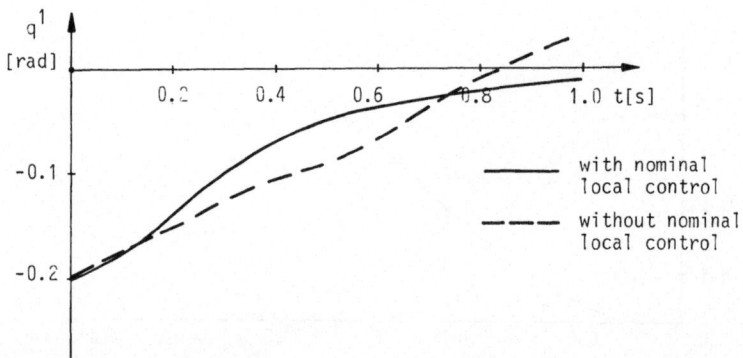

Fig. 3.26. Tracking of the nominal trajectory with and without nominal programmed control for the first joint of robot in Fig. 3.2.

Exercises

3.60. For the third order model of the servo $(n_i=3)$ in the form (3.3.4) where the matrices of system are in the form (3.3.49), show that the local nominal control is calculated according to the following equations (assuming that $q^{oi}(t)$, $\dot{q}^{oi}(t)$, $\ddot{q}^{oi}(t)$ are given and that $x^i = (q^i, \dot{q}^i, i_R^i)^T = (x_1^i, x_2^i, x_3^i)^T)$:

$$x_3^{oi}(t) = (\dot{x}_2^{oi}(t) - a_{22}^i x_2^{oi}(t))/a_{23}^i$$

$$\dot{x}_3^{oi}(t) = (\ddot{x}_2^{oi}(t) - a_{22}^i \dot{x}_2^{oi}(t))/a_{23}^i, \quad \ddot{x}_2^{oi}(t) \approx (\dot{x}_2^{oi}(t) - \dot{x}_2^{oi}(t-\Delta t))/\Delta t$$

$$u^{oi}(t) = (\dot{x}_3^{oi}(t) - a_{23}^i x_2^{oi}(t) - a_{33}^i x_3^{oi}(t))/b_3^i$$

Draw the corresponding control scheme (analogous to the scheme in Fig. 3.23. - Δt is short sampling interval).

3.61. For the Stanford manipulator shown in Fig. 3.13. the data on
which are given in Table 3.7, the servo feedback gains (for the
static servo systems) for all joints have been computed in Exercise 3.33. Determine on the basis of (3.5.2) and (3.5.4), the
steady state errors for:

a) the input trajectory given by $q^{oi}(t) = 0.2\ t$ (in the s-domain
$q^{oi}(s) = 0.2/s^2$),

b) the input trajectory given by $q^{oi}(t) = 4.\ t^2$ (in the s-domain
$q^{oi}(s) = 4./s^3$), and determine the errors in positioning of
the manipulator tip point due to these steady state errors,
if the effective lever arms r between the joints and the robot tip are given in Table 3.11. [5].

3.62. The nominal programmed control for the first joint of the robot
in Fig. 3.2. is given by (3.5.11) for the nominal trajectory
(3.5.10). The matrices of the servo model have been computed for
the position of the second joint $q^{2*}=0$. Calculate these matrices
for the position of the second joint $q^{2*} = 1.573$ [rad]. Show that
if the nominal control (3.5.11) is applied in the first joint
when the second joint is in the position $q^{2*} = 1.573$ [rad], than
there can occur overshoot (assume that initial error is zero
$\Delta q^1(0) = 0$.). Determine the nominal programmed control using the
value of the moment of inertia of the mechanism which corresponds
to the position of the second joint $q^{2*} = 1.573$ [rad]. Which nominal control has to be applied if the position of the second
joint q^{2*} is not known?

3.63.[*] Write the programme (in some high programming language) for computation of nominal local programmed control for the case when the
model of the servo is of the second order (3.5.8) and when the
third order model of the servo is considered (see Exercise 3.60).
Minimize the number of additions and multiplications for these
computations. Next, introduce the check of the condition (3.5.9).
By combining this programme with the programme written in Exercise 3.57 write the programme for microprocessor implementation
of the servo which ensures tracking of the desired nominal trajectory (the inputs for this programme are nominal trajectory
$q^{oi}(t)$, actual joint position $q^i(t)$, velocity $\dot{q}^i(t)$, and rotor current
$i_R^i(t)$, and the parameters of the actuator, the moment of inertia

of the mechanism around the joint axis $\bar{H}_{ii}$, the servo gains, and the output of the programme is the computed input signal for the actuator $u^i(t)$).

3.64.[*] Extend the programme written in Exercise 3.39. for simulation of the servo system, to include the simulation of tracking of the input nominal trajectory with and without local nominal control. Assume that the desired nominal trajectory is given by (3.5.10). Check the results of simulations in Example 3.5. for the first joint of the robot in Fig. 3.2.

References

[1] Vukobratović M., Applied Dynamics of Manipulation Robots: Modelling, Analysis and Examples, Springer-Verlag, 1989.

[2] Vukobratović M., Potkonjak V., Dynamics of Manipulation Robots: Theory and Application, Monograph, Series: Scientific Fundamentals of Robotics 1, Springer-Verlag, 1982.

[3] Chestnut H., Mayer R.W., Servomechanisms and Regulation System Design, John Wiley and Sons, Inc., New York, 1963.

[4] Cadzow A.J., Martens R.H., Discrete-Time and Computer Control Systems, Prentice-Hall, Englewood Cliffs, New Jersey, 1970.

[5] Paul P., Robot Manipulators: Mathematics, Programming, and Control, The MIT Press, 1981.

[6] Chen C.T., Introduction to Linear System Theory, Hort, Rinehard, and Winston, New York, 1970.

[7] Stojić M.R., Lazarević S., "Stability Properties of a Regulator System with Control Upper Magnitude Constraints", Proc. Third International Symposium on Network Theory, Split, 1975.

[8] Kubo T., Anwar G., Tomizuka N., "Application of Nonlinear Friction Compensation to Robot Arm Control", Proc. of IEEE Conference on Robotics and Automation, 722-277, 1986.

[9] Johnson D.C., "Accomodation of External Disturbances in Linear Regulator and Servomechanism Problems", IEEE Trans. AC-16, 552-554, 1971.

[10] Vukobratović M., Stokić D., Kirćanski N., Non-Adaptive and Adaptive Control of Manipulation Robots, Monograph Series: Scientific Fundamentals of Robotics 5, Springer-Verlag, 1985.

[11] Vukobratović M., (Ed.), Introduction to Robotics, Springer-Verlag, 1988.

Appendix 3.A
Local Optimal Regulator

In Chapter 3. we have considered synthesis of local servo systems for individual robot joints (and actuators) using the transfer function of the system and by pole-placement method (using the state space model of the system). Other methods for synthesis of simple linear local servo system for control of robot joints will not be considered in this book. Here, we shall briefly present synthesis of local controller by minimization of standard quadratic criterion, i.e. we shall consider synthesis of local optimal controller for one single joint and actuator of the robot (when all the other joints are locked). The model of the considered system is again given by (3.3.4). We shall consider synthesis of optimal regulator for positioning of the individual joint in the desired position q^{oi}.

In the servo system synthesis, considered in Chapter 3, the designer of the control system has to specify the desired locations of the closed-loop system poles in the s-plane, and by this he speciffies desired response of the system to some typical input signals (e.g. to step input signal and so on). In doing this, the designer does not take into account the magnitudes of the input signals for actuators that will be generated by the controller, nor the energy required by actuators to realize desired positioning of the joint (or, tracking of desired nominal trajectory). To satisfy requirements regarding the system response and, at the same time, to minimize the energy consumption, it is demanded to synthesize control by minimization of some criterion. This criterion should include both desired response of the system and requirements regarding magnitudes of the actuator input signals. Various criteria for control synthesis have been established. One of these criteria is so-called standard quadratic criterion which includes both above mentioned requirements. The standard quadratic criterion is defined in the following form (assuming that the model of the system is in the form (3.3.4)) [1, 2, 3]:

$$J_i(x^i(0)) = \frac{1}{2} \int_0^\infty (\Delta x^{iT} Q_i \Delta x^i + u^i r_i u^i) dt \qquad (3.A.1)$$

where Δx^i is $n_i \times 1$ vector of the state deviation from the desired nomi-
nal state x^{oi} (x^{oi} is defined as $x^{oi} = (q^{oi}, 0, 0)^T$ for $n_i = 3$, and $x^{oi} = (q^{oi}, 0)^T$, for $n_i = 2$, since the control problem is to ensure positioning
of the joint in the desired position q^{oi}), $\Delta x^i = x^i - x^{oi}$, Q_i is $n_i \times n_i$ po-
sitive definite matrix, $r_i > 0$ is positive number. For the system descri-
bed by the state space model (3.3.4), we have to synthesize control law
which will minimize the criterion (3.A.1). The minimization of the first
member in the criterion $\Delta x^i Q_i \Delta x^i$ represents requirement to minimize the
error between the actual state of the system x^i and the desired (nominal) state x^{oi}. By this, it is required that the control law ensures
desired performance of the system, i.e. to ensure that the joint is
driven to desired position q^{oi}. The second member in the criterion
$u^i r_i u^i$ represents minimization of the control signal amplitudes, by
which it is (indirectly) taken into account energy consumption of the
actuator during the positioning of the joint.

Therefore, the following problem is set: to synthesize the control of
the linear system (actuator+joint), the model of which is given by
(3.3.4), in such a way as to minimize criterion (3.A.1). This optimization problem can be solved analytically. Here, we shall not present
the proof of the solution of the standard quadratic regulator, since
it can be found in literature [1, 2, 3]. It has been shown that the
control which minimizes criterion (3.A.1) is obtained in the form which
is linear by deviation of state Δx^i:

$$u^i = -r_i^{-1} \hat{b}^{iT} K_i \Delta x^i = -k_i^T \Delta x^i \qquad (3.A.2)$$

Here K_i denotes $n_i \times n_i$ positive definite matrix which is the solution
of the matrix algebraic equation of Ricatti type:

$$K_i \hat{A}^i + \hat{A}^{iT} K_i - K_i \hat{b}^i r_i^{-1} \hat{b}^{iT} K_i + Q_i = 0 \qquad (3.A.3)$$

Therefore, the control law is obtained as a linear feedback with respect to deviation of the system state from the nominal (imposed) state
x^{oi}. The feedback gains are given by:

$$k_i^T = r_i^{-1} \hat{b}^{iT} K_i \qquad (3.A.4)$$

where k_i is $n_i \times 1$ vector of feedback gains.

The control scheme practically represents servo system with feedback loops by all state coordinates. The scheme is identical to the one presented in Fig. 3.14. (or, to the one in Fig. 3.9), the only difference being the algorithm by which the feedback gains are determined. In the schemes in Figs. 3.9. and 3.14, the feedback gains are obtained by selection of the positions of the poles of the closed-loop system in the s-plane. In the case of the optimal quadratic regulator the feedback gains are computed by minimization of the quadratic criterion (3.A.1), which reduces to solving of the equation (3.A.3).

However, in the synthesis of the optimal regulator, the problem of selection of the weighting matrices Q_i and r_i arises. It is obvious that the selection of these matrices represents the tradeoff between the requirements concerning the system response (i.e. the speed by which the desired nominal state is reached), and the requirements which concern the energy consumption, i.e. the amplitudes of the actuator input signals. If r_i is low relative to the members of the matrix Q_{ij} then this means that the fast positioning of the joint is required regardless to energy consumption, i.e. regardless how high amplitudes of the actuator input signals are demanded. Otherwise, if r_i is high relative to the members of Q_i, then too high actuator input signals u^i are not allowed, but this might reflect the time required for joint positioning in the desired position q^{oi}. The relation between the selection of the weighting matrices and the system performance might be complex, in general case.

It can be shown, that the poles of the closed-loop system (with the optimal quadratic regulator) satisfy the following inequality [2, 3]:

$$\lambda(\hat{A}^i - \hat{b}^i r_i^{-1} \hat{b}^{iT} K_i) \leq - \frac{\lambda_m(W_i)}{2\lambda_M(K_i)} \qquad (3.A.5)$$

where W_i is $n_i \times n_i$ matrix given by $W_i = Q_i + K_i \hat{b}^i r_i^{-1} \hat{b}^{iT} K_i$, λ_m denotes the minimal eigen-value of the corresponding matrix, and λ_M denotes the maximum eigen-value of the corresponding matrix, $\lambda(...)$ denotes eigen--values of the matrix in the brackets. Based on (3.A.5), we can get an estimate of positions of the poles of the system if the optimal quadratic regulator is applied.

To ensure desired performance of the system, when the optimal regulator is applied (i.e. to ensure sufficiently fast response of the system) instead criterion (3.A.1) we may consider criterion

$$J_i(x^i(0)) = \frac{1}{2} \int_0^\infty (\Delta x^{iT} Q_i \Delta x^i + u^i r_i u^i) \exp(2\beta_i t) dt \qquad (3.A.6)$$

where β_i is desired stability degree of the closed-loop system ($\beta_i > 0$ is positive real number). By minimization of criterion (3.A.6) we obtain the standard quadratic regulator with prescribed stability degree. It can be shown that the control of the system (3.3.4) which minimizes (3.A.6) is again obtained in the form (3.A.2), but, now, K_i is $n_i \times n_i$ positive definite matrix which is the solution of the matrix equation [2, 3]:

$$K_i(\hat{A}^i + \beta_i I_{ni}) + (\hat{A}^i + \beta_i I_{ni})^T K_i - K_i \hat{b}^i r_i^{-1} \hat{b}^{iT} K_i + Q_i = 0 \qquad (3.A.7)$$

where I_{ni} is the $n_i \times n_i$ unit matrix. It can be also shown that, in this case, the poles of the closed-loop system satisfy the following inequality:

$$\lambda(\hat{A}^i - \hat{b}^i r_i^{-1} \hat{b}^{iT} K_i) \leq -\beta_i - \frac{\lambda_m(W_i)}{2\lambda_M(K_i)} \qquad (3.A.8)$$

i.e. the poles are on the left side from the line $Re(s) = -\beta_i$ (see Fig. 3.A.1) in the s-plane. Therefore, by the selection of β_i (the prescribed stability degree) we can specify positions of the poles of closed-loop system, i.e. we can ensure that the joint approaches the desired position q^{oi} faster than $\sim\exp(-\beta_i t)$. By this regulator we ensure the stability degree of the closed-loop system, and the speed of approaching to the desired system state is guaranteed.

Obviously, the optimal regulator requires feedback loops by all the state coordinates x^i. If all the state coordinates are not measurable (i.e. if we do not want to introduce feedback loop by current (pressure), or by velocity), then we may apply so-called output regulator [3], or we may apply observer to reconstruct the system state on the basis of measured outputs [4]. In the above discussion, we have restricted ourselves to synthesis of optimal regulator for positioning of the single joint of the robot (when all the other joints are locked). However, the optimal regulator can be synthesized to ensure tracking of the desired trajectory [1-3]. These problems will not be considered in this book.

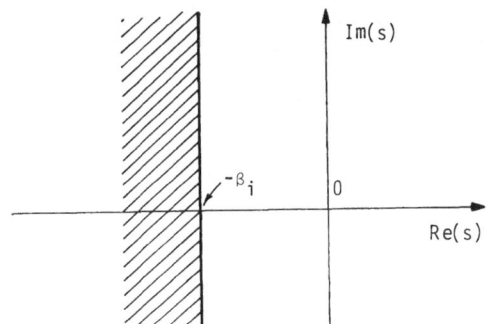

Fig. 3.A.1. Region in the s-plane which is on the left
from the line $Re(s) = -\beta_i$

EXAMPLE 3.A.1. For the first joint of the robot in Fig. 3.2, a D.C. electro-motor is applied, data on which are given in Table 3.1. If we take
into account the moment of inertia of the robot mechanism, the matrices of the model of the actuator and the joint (3.3.7) have been calculated in Example 3.3. For this system we have to synthesize the optimal linear regulator. If we select the following weighting matrices:

$$Q_i = \begin{bmatrix} 1.0 & 0. & 0. \\ 0. & 0.1 & 0. \\ 0. & 0. & 0.01 \end{bmatrix} \qquad r_i = 0.1, \qquad \beta_i = 3.$$

then the feedback gains are obtained according to (3.A.4):

$$k_i = (129.8, \ 43.314, \ 0.057)^T$$

The poles of the closed-loop system are

$$s = -5.92, \qquad s_2 = -5.72, \qquad s_3 = -708.8$$

and obviously, they satisfy condition (3.A.8).

References

[1] Kalman R.E., "On the General Theory of Control Systems", Proc. of
    IFAC, Butterworth Scientific Publications, London, 1960.

[2] Athans M., Falb L.D., Optimal Control, McGraw-Hill, New York, 1966.

[3] Anderson O.D.B., Moore J.B., Linear Optimal Control, Englewood
    Cliffs, New Jersey, Prentice-Hall, 1971.

[4] Luenberger D.G., "An Introduction to Observers", IEEE Transactions
    of Automatic Control, Vol. AC-16, No. 6, December 1971

Chapter 4
Control of Simultaneous Motions of Robot Joints

4.1 Introduction

In the previous chapter we have considered the synthesis of local servo system for one single joint of the robot. We have assumed that only the considered joint can move, while all the other joints of the robot are kept locked. The robots of the first generation often have implemented such a solution of the control system. In this case the joints of the robot are moving successively one by one until the goal position of the robot hand is reached. During the movement of one joint (until it reaches its desired position) the rest of the joints are kept locked (i.e. they are fixed). The local servos, synthesized in Chapter 3, satisfy such robotic tasks which require the successive movements of the robot joints.

However, such solution of the control system is not satisfactory for the tasks which are imposed to modern robotic systems. The reasons for this are numerous. Consider the task requiring only positioning of robot hand in various positions in work space (assume that tracking of the nominal trajectory in space is not required). The successive movements of the robot joints are obviously slow. The time required by the robot hand to reach its goal position is equal to sum of the time periods required for each single joint to reach its goal position. If we realize simultaneous motion of all joints towards their goal positions (corresponding to the desired position of the robot hand), then the time required to reach desired position of the hand is equal to the time period necessary to position the joint for which duration of movement, from the initial to the goal position, is the longest one. It is obvious that in the latter case the positioning of the robot hand is considerably faster. This fact is very important regarding the application of robots in industry, where it is often required to execute each task as fast as possible. On the other hand, many robotic tasks require that the robot hand tracks some desired specified path in the working space. Also it is often required that the robot hand keeps certain orientation in space (or, changes the orientation in some desired manner) during the robot motion. Such tasks cannot be realized by successive movements of the robot joints. These tasks requires simultaneous movements of all joints of the robot.

If the joints of the robot are moving simultaneously, their movements are dynamically coupled: the movement of each joint affects the movements of the other joints. The load moment acting upon each joint is variable and depends on the movements of all joints of the robot. The following problem arises: whether the local servo systems synthesized independently for each "isolated"joint, are capable to ensure simultaneous movements of all joints, or not? If we assign desired positions at the inputs of the local servos and allow the joints to move simultaneously, would they reach the goal positions in desired way (without overshoots, with minimal steady-state errors, etc.)? Similarly, if we set desired trajectories of the joints at the inputs of local servos, would the servo systems ensure satisfactory tracking of these trajectories assuming that the joints are moving simultaneously? In other words, the question arises whether the coupling between the joints can "destabilize" the local servos and disturbs the desired movements of the robot joints to such extend that the performance of the robot becomes unacceptable? To answer these questions it is necessary to analyze dynamic interconnections between the joints and to analyze the stability of the robotic system. In this chapter we shall consider the performances of the robotic system if all the joints are moving simultaneously and if only the local servo systems are applied to control the robot joints. On the basis of such analysis we shall determine whether we can accept the local servo systems (which control each joint independently), or we must introduce additional feedback loops to compensate for the effects of coupling between the joints, in order to ensure simultaneous movements of all robot joints.

4.2 Coupling Between Joints

Let us consider simultaneous positioning of several joints of robotic system. Performance of the robot is described by the mathematical model of the dynamics of the mechanism (3.2.2) and the mathematical models of the actuators (3.2.8), or by mathematical model of the entire system (3.2.27). The movement of a single joint (if all the other joints of the robot are locked) is described by the model (3.3.4), which includes moment of inertia of the mechanism around the axis of the considered joint. Let us assume that joints are controlled by local servo systems. These local servos have been synthesized assuming that only the corresponding joint is moving while the other joints are locked. The scheme of the control system of the entire robotic system (if only local servo systems are applied) is presented in Fig. 4.1. When the other joints start to move the dynamic forces and moments, described by (3.2.2), appear. These dynamic forces act as external load upon the servo systems around the robot joints. Let us consider how these forces affect the performances of local servo systems.

(a) *Variable moment of inertia.* We have already shown (Sect. 3.3.4) that the moment of inertia of the robot mechanism around the axis of the i-th joint varies if the other joints change their angles (or linear displacements). This means that H_{ii} in the model (3.2.2) is function of all joint coordinates q^j for $j>i$. We have already shown that in order to prevent overshoots in the servo system, we must determine the maximal value of the moment of inertia of the mechanism $H_{ii}(q)$. Then, for this maximum value of the moment of inertia we have to determine the feedback gains so as to ensure the servo system to be critically damped. If this procedure is followed, then when the joints coordinates change their values and the moment of inertia becomes $H_{ii} < \max_q H_{ii}(q)$, it is ensured that the servo system around the i-th joint is overcritically damped (and, thus, the overshoots cannot appear). However, if all the joints of the mechanism are moving simultaneously, during the movement of the i-th joint, the moment of inertia of the mechanism H_{ii} around this joint is varying, which might cause an uneven behaviour of the servo system. The local servo is always critically or overcritically damped (i.e. the overshoots of the set goal position cannot appear), but still the performance of the servo might be unsatisfactory. The compensation for the variable moment of inertia might be achieved by variable feedback gains, or by global control which will be considered in Chapter 5.

(b) *Cross-inertia members.* If we consider the matrix of inertia $H(q)$, we can see that besides the diagonal elements $H_{ii}(q)$ (which variation we have considered above) there are also off-diagonal elements $H_{ij}(q)$ which represent so-called cross-inertia members. These members represent the effects of accelerations of the other joints upon the load moment around the i-th joint $H_{ij}(q)\ddot{q}^j$. With some robot structures the values of these members might be close to the values of the eigen-inertia terms H_{ii} (i.e. $H_{ij} \approx H_{ii}$), and therefore their effects might be significant. This inertia coupling

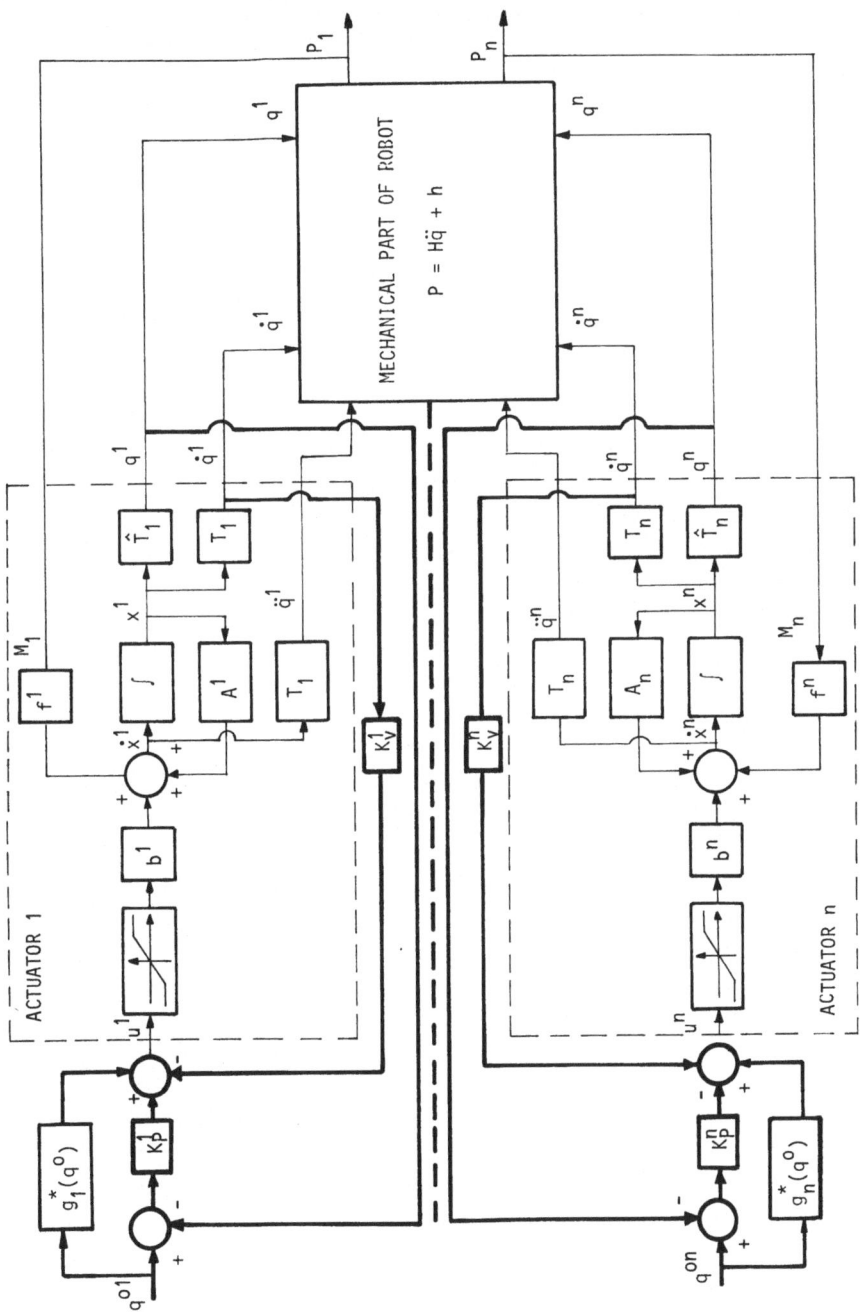

Fig. 4.1. Control scheme consisting of local servo systems

between the robot joints represents external load upon the i-th joint. When the j-th joint is accelerated or deceleration the inertia coupling causes the external load upon the servo in the i-th joint (the moment is given by $H_{ij}(q)\ddot{q}^j$). As we have already explained, the external load upon the shaft of the actuator (D.C. motor) causes errors in the positioning of the joint, or in the tracking of trajectories. However, these moments are significant if the accelerations are relatively high. When the accelerations drop to zero (when the robot stops), these moments also vanish and they do not affect the positioning of the joints, i.e. they do not cause steady state errors in robot positioning. These moments can cause errors in tracking trajectories, if high accelerations $\ddot{q}^j$ are demanded. If accurate tracking of fact trajectories is essential for implementation of a given robotic task, then these moments have to be compensated for. As we have shown in previous chapter the constant external load can be compensated in various ways. However, these moments due to cross-inertia interconnections are variable and they depend on the positions of the robot joints q^j and on the accelerations $\ddot{q}^j$ of the joints. Therefore these moments cannot be efficiently compensated by integral feedback loop. The compensation of these dynamic moments by global control will be considered in Chapter 5.

(c) *Gravity moments*. The effects of gravity moments upon the local servos have already been considered in Section 3.3.4. Here, we shall only note that in simultaneous movements of several joints gravity moments become variable, causing errors both in positioning and in tracking of trajectories. Thus, the gravity moments are not any more constant load upon the servo. Therefore, they may produce, if not compensated by global control, uneven tracking of desired trajectories, and uneven positioning of the robot joints.

(d) *Centrifugal and Coriolis forces (torques)*. The vector $h(q, \dot{q})$ in the model of the mechanism dynamics (3.2.2) includes centrifugal and Coriolis forces which depend on the joint velocities $\dot{q}$. Since these forces are directly proportional to squares of the joint velocities, they are significant only if the joints are moving at high velocities. When the robot is starting to move or is stopping these forces are negligible, which means they do not affect the positioning of the robot in any desired positions and do not cause

steady state errors. However, these forces cause errors in the
tracking of fast trajectories. If the robot joints are moving at
high velocities, these forces act as external loads upon the servo
systems around the joints. If the precise tracking of fast trajectories is not required the effects of these forces can be ignored.
If the tracking of fast trajectories is essential, we must take
into account centrifugal and Coriolis forces in the synthesis of
control. Since the requirements upon the modern robots in industry
are growing (regarding the accuracy and speed of the desired trajectories and so on), it is often required to compensate for these
dynamic forces.

Example 4.2. Let us consider the robot in Fig. 3.2. For the servo
around the first joint of this robot, we have synthesized the servo
gains in Example 3.3.2. assuming that all the other joints of the manipulator are locked (we have assumed that the joints are locked in
the positions in which moment of inertia of the mechanism around
the first joint H_{11} gets its maximal value). Let us consider the behaviour of this servo if all the other joints are moving. Let us consider the positioning of the first joint. Desired position of the joint is $q^{o1}=0.5$ [rad] and the initial position of the joint is $q^1(0)=0.$
[rad]. If all the other joints are locked, then the positioning of the
servo systems is as shown in Fig. 4.2. (bold line). The results are
obtained by digital simulation using the complete robot dynamic model.
If the second joint is moving in such a way that it is accelerated at
$\ddot{q}^2=1$ [rad/s^2] (while the third joint is fixed), then upon the first
joint is acting the moment due to cross-inertia term $H_{12}(q)\ddot{q}^2$. The positioning differs from the previous case.

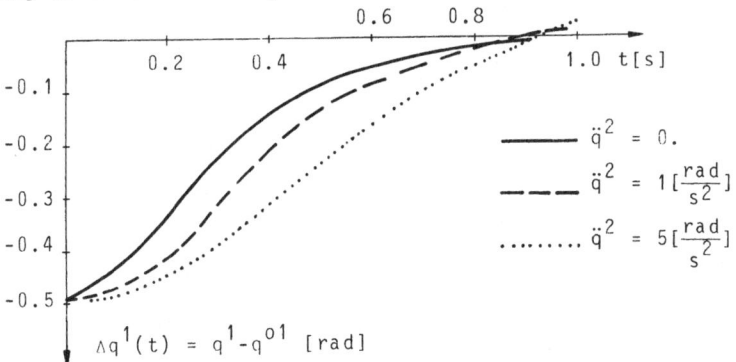

Fig. 4.2. Positioning of the first joint of the robot in Fig. 3.2.
for various accelerations of the second joint

The results of digital simulation are presented in Fig. 4.2, too (dashed line). It can be seen that the effects of cross-inertia term is relatively weak. However, if the acceleration of the second is extremely high $\ddot{q}^2$=5. [rad/s], the effects of these forces will be considerably stronger, which can be verified by the simulation of the positioning of the first joint (also presented in Fig. 4.2. - dotted line).

Exercises

4.1. For the robot of cylindrical structure, presented in Fig. 2.5, (the data on which are given in Table 3.4), the servo feedback gains for the third joint have been synthesized in Exercise 3.32. Determine the "steady state error" due to Coriolis forces when the first joint is moving at velocity $\dot{q}^1$=2 [rad/s], and the third joint is moving at velocity $\dot{q}^3$=1 [m/s].

4.2. For the manipulator in Fig. 3.2. and for the servo system in its first joint, servo gains have been synthesized in Example 3.3.2. Determine the steady state error in the first joint (when the second joint is in q^2=0.) due to inertial coupling between the first and the second joint if the second joint is moving at acceleration of $\ddot{q}^2$=5. [rad/s^2]. Check whether this result complies with the results of simulation in Fig. 4.2. Explain the differences in these results.

4.3.[*] If the actuator for the first joint of the robot in Fig. 3.2. is hydraulic actuator, for which the servo has been synthesized in Exercise 3.35., determine the steady state error due to inertial coupling between the first and the second joint, if the second joint is moving at acceleration of $\ddot{q}^2$=5 [rad/s^2]. Compare the obtained results with the results in Exercise 4.2.

4.4.[*] Write the programme (in one of high-level programming language) for simulation of robot performance. The simulation should include the complete dynamic model of the robot. Write this programme for particular robot in Fig. 3.2. assuming that the robot joints are driven by D.C. actuators (data on actuators are given in Table 3.1). Each joint is controlled by its servo system, and the feedback gains are synthesized as presented in the previous chapter. The input data for the programme are: data on actuators (for

computing the matrices of the models of actuators (3.2.7)), data
on robot (for computing matrix H and vector h acc. to (3.2.3)),
data on servo gains (for computation of actuators inputs acc. to
(3.3.41)), data on desired positions of the joints q^{oi}, data on
initial state of the robot $x(0) = (x^1(0), x^2(0), x^3(0))^T$, $x^i(0) = (q^i(0), \dot{q}^i(0))$, and data on integration interval. Simulate the simultaneous positioning of all three joints from the initial state
$x(0) = (0, 0, 0, 0, 0, 0)^T$ towards desired positions $q^{o1} = 0.2$ [rad],
$q^{o2} = 0.1$ [rad] and $q^{o3} = 0.05$ [m]. (Instruction: At each sampling interval for given $x(t)$ compute matrix H and vector h (using programme written in Exercise 3.5), compute inputs for actuators acc.
to (3.3.41), then compute the first derivative of the state $x(t)$
acc. to (3.2.27) and apply some method for numerical integration –
for example simple Euler's method, see Chapter 5. and Exercise 3.39).

4.3 Analysis of Linearized Model of Robot

In the previous section we have considered the effects of various components of the dynamic moments (3.2.2) on the performance of the local
servo systems. It is obvious that the effects of dynamic coupling between the joints can be significant. Therefore, it is necessary to examine these effects more precisely. Actually, we have to examine whether the local servo systems can be successfully applied for control of
simultaneous movements of several robot joints (see Fig. 4.1). The coupling between robot joints are dynamic moments around joints which
are described by the dynamic model of the mechanism (3.2.2). These dynamic forces are in general case, complex functions of angles, velocities and accelerations of joints. Therefore, the analysis of the effects of these forces upon the servo system performances is not simple,
specially if the trajectory tracking problem is addressed. To do this
we have to apply various methods for analysis of complex systems which
have been developed within the control system theory.

One of the ways to analyze the performances of complex nonlinear systems is by their linearization. The nonlinear model is linearized,
and then the linear model is analyzed. The basic motive for such an
approach lies in the fact that many methods for analysis of complex
linear systems have been developed. However, the results of analysis
of linear model cannot guarantee the equal performance of the original
nonlinear system, since the linear model is just an approximation of

the actual system[*]. Namely, if the stability of the nonlinear model
of the system is analyzed by testing the stability of the linear model
of the system, then we obtain just necessary, and not sufficient, conditions for stability of the nonlinear model of the system. The meaning
of this is as follows: if linear model of the system is not stable, then
we can be sure that the nonlinear model also is not stable; however,
if the linear model is stable, then the nonlinear model is not neces-
serily stable. The stability of the linear system guarantees the stability of the nonlinear model only in some region around the point, in
the state space, around which the linearization has been performed.
However, the linear analysis is often applied for analysis of the nonlinear systems, due to its simplicity and since it offers an insight
in the system performance in the close region around the point in the
state space around which the nonlinear model has been linearized. Latter on we shall explain some other reasons which justifies the application of the linear analysis for robotic systems (see Chapter 5).

Here, we shall briefly describe how we can obtain the linear model of
the robot, and we shall consider application of some methods for stability analysis of the linear systems to the robotic systems. We assume
that just the local servos are applied for control of robot joints [1,
2, 3].

4.3.1. Linear model of robot

Let us consider the entire model of the robot in a so-called centralized form (3.2.27)

$$\dot{x} = \hat{a}(x) + \hat{B}(x) N(u) \qquad (4.3.1)$$

where the vector function $\hat{a}(x)$ and matrix $\hat{B}(x)$ are given in (3.2.27).
The vector function $\hat{a}(x)$ and matrix $\hat{B}(x)$ are complex nonlinear functions of the system state x. Let us consider the system in the surroundings of the point x^o in the state space. Let us assume that the "nominal" state x^o corresponds to desired positions of the robot joints
q^{oi}, i.e. $x^o = (x^{o1}, x^{o2},...,x^{on})^T$, $x^{oi} = (q^{oi}, 0)^T$ for $n_i=2$. Let us
assume that we can determine the input signals $u^o=(u^{o1}, u^{o2},...,u^{on})^T$

---

[*] Nonlinear model of the system is also an approximation of the actual system but, obviously, much better approximation, i.e. it is
much closer to the actual system performance.

such that the system is in equilibrium in the point x^o, i.e. that it is satisfied:

$$0 = \hat{a}(x^o) + \hat{B}(x^o)u^o \tag{4.3.2}$$

The model of state deviation around the point x^o, u^o can be written in the following form, based on (4.3.1) and (4.3.2):

$$\Delta\dot{x} = \bar{a}(x^o, \Delta x) + \hat{B}(x^o, \Delta x)N(u^o, \Delta u) \tag{4.3.3}$$

where $\bar{a}$ is N×1 vector function given by:

$$\bar{a} = \hat{a}(x^o+\Delta x)-\hat{a}(x^o) + (\hat{B}(x^o+\Delta x)-\hat{B}(x^o))u^o$$

Here, Δx denotes N×1 vector of state deviation from the equilibrium point x^o, i.e. $x(t) = x(t)-x^o$, Δu is n×1 vector of deviation of the input signals from the "nominal" signals u^o, i.e. $\Delta u(t) = u(t) - u^o$, $N(u^o, \Delta u)$ denotes the nonlinearity of the amplitude saturation type according to (3.2.9) [*)]

$$N(u^{oi}, \Delta u^i) = \begin{cases} -u^{oi}-u_m^i & \text{for} & \Delta u^i \leq -u_m^i-u^{oi} \\ \Delta u^i & \text{for} & -u_m^i-u^{oi} < \Delta u^i < u_m^i-u^{oi} \\ -u^{oi}+u_m^i & \text{for} & \Delta u^i \geq u_m^i-u^{oi} \end{cases} \tag{4.3.4}$$

The model of deviation of the state coordinates from "nominal" point x^o (4.3.3) has the same form as the original model (4.3.1). The model of deviation is the system of differential nonlinear equations.

We want to linearize the model (4.3.3) in order to obtain linear approximation of the robotic system which will serve for analysis of the stability of the entire robotic system. The linear model is obtained in the form:

$$\Delta\dot{x} = A_L(x^o)\Delta x + B_L(x^o)N(u^o, \Delta u) \tag{4.3.5}$$

where A_L is N×N matrix and B_L is N×n matrix. Matrices A_L and B_L are obtained according to:

[*)] Here it is assumed that u^o satisfies $|u^o| < u_m^i$.

$$A_L = \frac{\partial \bar{a}(x^o, \Delta x)}{\partial \Delta x} \bigg|_{\text{for } \Delta x=0} , \qquad B_L = \hat{B}(x^o, 0) \qquad (4.3.6)$$

Obviously, the matrices A_L and B_L depend on x^o, i.e. on the point in the state space around which the model is linearized, and this we shall consider latter on. Based on (4.3.3) and (4.3.6) we obtain:

$$A_L = \frac{\partial \hat{a}}{\partial \Delta x}\bigg|_{\Delta x=0} + \frac{\partial \hat{B}}{\partial \Delta x}\bigg|_{\Delta x=0} \cdot u^o \qquad (4.3.7)$$

If we introduce $\hat{a}$ and $\hat{B}$ as defined in (3.2.27) we obtain:

$$A_L = A + F(I_n - HTF)^{-1} \frac{\partial H}{\partial \Delta x} TF(I_n - HTF)^{-1}[HTAx^o + HTBu^o + h(q^o, \dot{q}^o)] +$$

$$+ F(I_n - HTF)^{-1}[\frac{\partial H}{\partial \Delta x}(TAx^o + TBu^o) + HTA + \frac{\partial h}{\partial \Delta x}] \qquad (4.3.8)$$

where $\partial H/\partial \Delta x$ and $\partial h/\partial \Delta x$ are matrices of adequate dimensions.

To obtain matrices A_L and B_L of the linear model of the system it is necessary to determine the linear model of the mechanical part of the robotic system, i.e. it is necessary to determine the following matrices:

$$\frac{\partial H(q^o)}{\partial q} - \text{matrix of dimensions } n \times n \times n$$

$$(4.3.9)$$

$$\frac{\partial h(q^o, \dot{q}^o)}{\partial q} , \quad \frac{\partial h(q^o, \dot{q}^o)}{\partial \dot{q}} - \text{both of dimensions } n \times n$$

These matrices represent the first derivatives of the matrices $H(q)$ and $h(q, \dot{q})$ by q and $\dot{q}$, for $q = q^o$ and $\dot{q} = \dot{q}^o$.

Thus, the problem how to determine the linear model of the robot dynamics reduces to determination of matrices (4.3.9), i.e. to linearization of the dynamic model of the mechanical part of the robotic system. The matrices (4.3.9) can be determined in several ways:

(a) The matrices of the linear model (4.3.9) can be obtained analytically. This means that we can write the original model of dynamics of the mechanical part of the robot in analytical form (3.2.2), and we can analytically determine the first derivatives of the elements of the matrix $H(q)$ and of the vector $h(q, \dot{q})$ by the joint angles (displacements) q and their (rotational or linear) velocities $\dot{q}$. By this we obtain the matrices (4.3.9). However, since the elements of matrices $H(q)$ and $h(q, \dot{q})$ are complex nonlinear functions on q and $\dot{q}$, determination of their derivatives analytically might be very tedious and complex task (which might be subject to many errors). This is especially the case with robots with many rotational joints which models are often very complex functions of q and $\dot{q}$. These were the reasons which motivated development of methods for automatic computer-aided determination of the linear model of mechanical part of the robot by analytical differentiation of the matrix H and of the vector h [1].

(b) The matrices (4.3.9) might be also determined applying various numerical procedures, which might be set at a digital computer. These procedures are completely analogous to numerical procedures for computation of nonlinear dynamic models of robots (for computation of matrix $H(q)$ and vector $h(q, q)$). As we have already explained, several algorithms have been developed for numerical computation of the matrices H and h of the mathematical model of dynamics of the robot mechanism, and these procedures are based upon recursive relations between the velocities of the robot links, between their accelerations, between their forces and moments. Parallel to these procedures for determination of the matrices H and h, the algorithms for determination of matrices (4.3.9) have been developed. These algorithms are based upon the same principles as the algorithms for determination of the basic matrices H and h. The algorithm based on D'alembert's principle for numerical determination of the matrices (4.3.9) of the linear model of the dynamics of open kinematic chains, is presented in detail in [1]. This algorithm set at the digital computer enables computation of numerical values of the matrices (4.3.9) for given q^o, and $\dot{q}^o$, for arbitrary type and structure of robot (i.e. for robots with arbitrary number and types of joints, etc.). In this way, using such program for digital computer, we can simply determine the matrices of the linear model of the robot, but just numerically (i.e. for the given data on q^o and $\dot{q}^o$ we determine the numerical values of the elements of the matrices (4.3.9)).

(c) The matrices of the linear model of the robot (4.3.9) can be determined by various procedures for identification. We assume that the nonlinear dynamic model of the robot is available and based on this model we can numerically identify the linear model which

describes the performance of the actual system. The aim is, obviously, to determine the linear model whose response to various inputs
is as close as possible to the response of the actual system.

We shall briefly describe a simple procedure for numerical identification of the linear model of the robotic system. Let us consider the
"exact" nonlinear model of the state deviation of the robot (4.3.3)
around the point x^o. Let us assume that certain signals $\delta u(t)$ are fed
at the system inputs. For such control signals we may integrate numerically the system of differential equations (4.3.3) and obtain the
trajectories of the state vector $\delta x(t)$ which represent the response of
the robotic system to the given inputs $\delta u(t)$ (around the nominal point
x^o, u^o), for some selected initial state $\delta x(0)$:

$$\delta \dot{x}(t) = \bar{a}(x^o, \delta x) + \hat{B}(x^o, \delta x) N(u^o, \delta u(t)) \qquad (4.3.10)$$

Let us assume that we have selected the input signal to satisfy the
amplitude constraint (4.3.4), i.e. such that $N(u^o, \delta u(t)) = \delta u(t)$. The
input signal might belong to various classes of functions (of time).
Let us observe the response of the nonlinear model (4.3.10) in m_1 dis-
crete time instants t_i, $i=1,2,\ldots,m_1$. The matrices of the linear model
(4.3.5) have to be determined in such a way that they satisfy the following equations:

$$[\delta \dot{x}] = [A_L(x^o), B_L(x^o)] [-\tfrac{\delta x}{\delta u}-] \qquad (4.3.11)$$

where $[\delta \dot{x}]$ is $N \times m_1$ matrix given by $[\delta \dot{x}] = [\delta \dot{x}(\bar{t}_1), \delta \dot{x}(\bar{t}_2),\ldots,\delta \dot{x}(\bar{t}_{m1})]$
and $[\delta x/\delta u]$ is $(N+n) \times m_1$ matrix given by:

$$[-\tfrac{\delta x}{\delta u}-] = [\dfrac{\delta x(\bar{t}_1)}{\delta u(\bar{t}_1)}, \dfrac{\delta x(\bar{t}_2)}{\delta u(\bar{t}_2)},\ldots, \dfrac{\delta x(\bar{t}_{m1})}{\delta u(\bar{t}_{m1})}]$$

Based on (4.3.11) it is possible to determine the pair of matrices
$A_L(x^o)$, $B_L(x^o)$ via generalized inversion of matrix $[\delta x/\delta u]$, i.e. using
the following equations:

$$[A_L(x^o), B_L(x^o)] = [\delta \dot{x}][-\tfrac{\delta x}{\delta u}-]^T \{[-\tfrac{\delta x}{\delta u}-][-\tfrac{\delta x}{\delta u}-]^T\}^{-1} \qquad (4.3.12)$$

Here, it is assumed that the matrix in { } is non-singular. In this
way we can determine the matrices $A_L(x^o)$, $B_L(x^o)$ of the linear model
of the robotic system around the point x^o, u^o.

The procedure for numerical identification of the linear model (4.3.5) on the basis of nonlinear model (4.3.3) is simple. We have to simulate behaviour of the nonlinear model of the state deviation around the point x^o, u^o (4.3.5), if the selected input signals $\delta u(t)$ are applied. The values of the state coordinates $\delta x(\bar{t}_i)$ and of the first derivatives of the state coordinates $\delta \dot{x}(\bar{t}_i)$, $i=1,2,\ldots,m_1$ (i.e. the response of the nonlinear model) are memorized for m_1 time instants during the simulation. In this, we have to select the sufficiently large number m_1. Next, we have to compute the matrices $A_L(x^o)$, $B_L(x^o)$ according to (4.3.12), using the memorized values of $\delta x(\bar{t}_i)$ and $\delta \dot{x}(\bar{t}_i)$.

In computing the matrices $A_L(x^o)$, $B_L(x^o)$ we have to take large number of data $\delta x(\bar{t}_i)$ (i.e. to simulate the system performance sufficiently long and to memorize many "instants" m_1), in order to cover the surroundings of the point x^o and take into account as much information on the system behaviour as possible. By this we ensure that the response of the linear model "sufficiently" well approximates the response of the original nonlinear model in the surroundings of the point x^o, u^o. Actually, the response of the linear model of the system has to "fit" the response of the original nonlinear model of the system. However, if the number of data to be memorized increases, then numerical problems in solving of the equation (4.3.11) also increase. Namely, numerical errors in matrix inversion and matrix product which are required in (4.3.12) increase if the dimensions of the matrices increase. Generally speaking, this procedure for identification of the linear model of the robotic system is subject to numerous possible numerical errors, which might even lead to unreliable and inadequate linear model. It is necessary to ensure that the input signal $\delta u(t)$ is "sufficiently complex" to excite all significant modes of the robotic system (i.e. its frequency bandwidth must be sufficiently wide). The above presented procedure is efficient for identification of linear model if the original nonlinear model (4.3.3) is stable. Otherwise, if we apply some arbitrary input signal $\delta u(t)$, then it might happen that $\delta x(t) \to \infty$, and then the identification becomes invalid due to large numerical errors. Therefore, first the system (4.3.3) is stabilized introducing some feedback loops (for example, simple local servos around the robot joints might be applied to "approximately" stabilize entire robot). Then, the linear model of the stabilized robot is identified. At last, we omit from the linear model the introduced feedback loops and by this we obtain open-loop linear model of the robotic system. However, often it is not easy to "approximately" stabilize robotic system.

Due to this, it must be underlined that the numerical aspects highly
limit the applicability of this procedure for identification of the
linear model of the robot. Therefore, the more complex and more sophisticated procedures for identification of the linear model of the system might be applied. In these procedures the numerical problems due
to operations with large amount of data are overcome in various ways.

In the above described identification procedure we have used the nonlinear model of the robot to determine the linear model. However, we
have to notice that the linear model of the robot might be identified
also by recording the responses of the actual robotic system to various input signals $\delta u(t)$, i.e. the linear model might be identified experimentally. Based on records of the responses of the actual robot
$\delta x(t)$ to various input signals δu, it is possible to identify linear
model which approximatively describes the performance of the robot dynamics around the nominal point x^o, u^o.

Here, we have presented how we can directly identify the linear model
of the entire robotic system which includes the models of robot actuators. Since the models of actuators are linear (or, we may say that
the linear models are good approximation of their actual performances),
we may identify only the matrices of the linear model of the mechanical part of the system (4.3.9). If these matrices are identified, it is
easy to determine the matrices of the linear model of the entire robotic system based on (4.3.8). The identification of the linear model of
the mechanical part of the system can be carried out applying the procedure described above.

Example 4.3.1. The model of dynamics of the mechanical part of the robotic system presented in Fig. 3.2. is given by differential equations
(3.2.3). Since this model is relatively simple, it is easy to determine the corresponding matrices (4.3.9):

$$\frac{\partial H}{\partial q^1} = \begin{bmatrix} 0 & 0 & 0 \\ 0 & 0 & 0 \\ 0 & 0 & 0 \end{bmatrix}$$

$$\frac{\partial H}{\partial q^2} = \begin{bmatrix} -2m_3 \ell_1 \ell_3^* \sin q^2 & -m_3 \ell_1 \ell_3^* \sin q^2 & 0 \\ -m_3 \ell_1 \ell_3^* \sin q^2 & 0 & 0 \\ 0 & 0 & 0 \end{bmatrix}$$

$$\frac{\partial H}{\partial q^3} = \begin{bmatrix} 0 & 0 & 0 \\ 0 & 0 & 0 \\ 0 & 0 & 0 \end{bmatrix} \qquad (4.3.13)$$

$$\frac{\partial h}{\partial q} = \begin{bmatrix} 0 & -m_3 \ell_1 \ell_3^* \dot{q}^2 (\dot{q}^2 + 2\dot{q}^1)\cos q^2 & 0 \\ 0 & m_3 \ell_1 \ell_3^* (\dot{q}^1)^2 \cos q^2 & 0 \\ 0 & 0 & 0 \end{bmatrix}$$

$$\frac{\partial h}{\partial \dot{q}} = \begin{bmatrix} -2m_3 \ell_1 \ell_3^* \dot{q}^2 \sin q^2 & -2m_3 \ell_1 \ell_3^* (\dot{q}^2 + \dot{q}^1)\sin q^2 & 0 \\ 2m_3 \ell_1 \ell_3^* \dot{q}^1 \sin q^2 & 0 & 0 \\ 0 & 0 & 0 \end{bmatrix}$$

If the matrices of the actuators models (D.C. electro-motors) are given by (3.2.11), then, using the data on the robot and the applied actuators given in Tables 3.1. and 3.2., we compute the matrices of the linear model A_L, B_L around the point in the state space $q^{o2} = 0.5$ [rad], $\dot{q}^{o1} = 0.$, $\dot{q}^{o2} = 0$. (the values of other state coordinates might be selected arbitrary, they have no effects upon these matrices). We compute these matrices according to (4.3.8):

$$A_L = \begin{bmatrix} 0 & 1 & 0 & 0 & 0 & 0 \\ 0 & -0.832 & 0 & 2.56 & 0 & 0 \\ 0 & 0. & 0 & 1 & 0 & 0 \\ 0 & 2.56 & 0 & -15. & 0 & 0 \\ 0 & 0 & 0 & 0 & 0 & 1 \\ 0 & 0 & 0 & 0 & 0 & -45. \end{bmatrix}$$

$$\qquad (4.3.14)$$

$$B_L = \begin{bmatrix} 0 & 0 & 0 \\ 0.58 & -1.75 & 0 \\ 0 & 0 & 0 \\ -1.78 & 10.43 & 0 \\ 0 & 0 & 0 \\ 0 & 0 & 0.361 \end{bmatrix}$$

This way of determination of the linear model of the robot might be easily applied in the cases of simple robot structures. However,

168

in the cases of complex robot structures with many degrees of freedom, this analytic differentiation is very tedious job, so in these cases it is much more convenient to apply numerical algorithms set at the digital computers to obtain linear model of such robots. As already mentioned, the algorithms for automatic computer-aided generation of linear models of the robot with arbitrary structure in analytical forms have been developed. Therefore, the matrices (4.3.13) can be generated automatically by such software packages [4].

Exercises

4.5. Show that the matrices of the linear model of the robot are obtained in the form (4.3.8) if we start from the nonlinear model of robot (3.2.27), (4.3.6) and (4.3.7).

4.6. For the robot presented in Fig. 2.5., the dynamic model of which has been determined in Exercise 3.1, determine analytically the matrices (4.3.9) and the matrices of the linear model (4.3.8), if D.C. electro-motors are applied (and the models of actuators are linear and of the second order).

4.7. If data on the manipulator in Fig. 2.5. are given in Table 3.4., determine the values of the matrices (4.3.9) (computed in Exercise 4.6) for the following nominal state coordinates: a) $x^o = (0, 0, 0, 0, 0, 0)^T$, b) $x^o = (0, 0, 0, 0, 0.5, 0)^T$, c) $x^o = (0, 1., 0.2, 0., 0.5, 1.)^T$. If data on actuators are given in Table 3.3. compute the values of the elements of the matrices (4.3.6) for these three cases.

4.3.2. Analysis of stability of linear model of robot with position control

The linear model of the robot, obtained by one of the procedures described in the previous section, represents the time-invariant system of N linear differential equations. To examine the performance of this system we may apply various well-known methods for analysis of linear multi-input multi-output systems. In this particular case, we shall address the problem of stability of such system. We want to examine whether the robotic system is stable around the nominal point x^o (i.e. around the desired goal positions of the robot joints) if we apply only

local controllers and the joints are moving simultaneously. Here,
for the sake of simplicity, under the notion of stability we shall assume asymptotic stability [5].

To examine stability of a complex linear system we may apply various
procedures for stability analysis. One of the most often applied is to
determine eigen-values of the matrix of the linear model. If the open-
-loop system is considered, we may analyze the stability of the system
by examining the eigen-values of the matrix A_L:

$$\det(A_L - sI_N) = K_a \prod_{j=1}^{N} (s_j^o - s) \qquad (4.3.15)$$

where I_N is the $N \times N$ unit matrix, K_a is the coefficient of proportionality, s_j^o $j=1,2,\ldots,N$ are the eigen-values of the matrix A_L. By analysis of the eigen-values s_j^o we can easily conclude on stability of the
open-loop linear model of the system.

Now, let us consider the stability of the closed-loop robotic system
if the local servo systems are applied (Fig. 4.1). In this case the
control law is given by:

$$\Delta u = -KC\Delta x = -KC(x(t) - x^o) \qquad (4.3.16)$$

where K is $n \times K_c$ matrix of feedback gains, C is the output matrix of
dimensions $K_c \times N$. The matrix of feedback gains K includes the position
gains K_P^i, velocity feedback gains K_v^i and the gains in the feedback loops by the robots currents K_I^i (if such feedback loops are applied),
and it is given by:

$$K = \begin{bmatrix} K_P^1 \ K_v^1 \ K_I^1 & & & & \\ & K_P^2 \ K_v^2 \ K_I^2 & & & 0 \\ & & \cdot & & \\ & & & \cdot & \\ & & & & \cdot \\ & 0 & & & K_P^n \ K_v^n \ K_I^n \end{bmatrix}$$

It is obvious that the matrix of feedback gains is diagonal and given by $K = \text{diag}(k_i^T)$. The servo feedback gains k_i are synthesized as described in the previous chapter. The output matrix C is also diagonal matrix, given by $C = \text{diag}(c^i)$. In the case when local feedback loops are applied by all state coordinates x^i, then the output matrix C is the N×N unit matrix. Here, by K_c is denoted the order of the system output which is given by

$$K_c = \sum_{i=1}^{n} k_i^y$$

If the control (4.3.16) is applied, i.e., if the local servo feedback loops are closed, the matrix of the closed-loop linear model of the system is given by $A_L - KC$. To analyze the stability of the robot with local servo systems around the desired positions of joints q^{oi}, we have to determine eigen-values of the closed-loop system matrix:

$$\det(A_L - B_L KC - sI_N) = K_a \prod_{j=1}^{N} (s_j^o - s) \tag{4.3.17}$$

where K_a is the proportionality coefficient and s_j^o, $j=1,2,\ldots,n$ are the eigen-values of the matrix $A_L - B_L KC$.

Next, we have to analyze positions of the eigen-values s_j^o of the closed-loop system matrix in the complex plane. As it is well known, if all eigen-values of s_j^o of the system matrix lie in the left part of the complex plane, then the linear model of the robot is stable. If any of the eigen-values lies in the right part of the complex plane (i.e. if any eigen-value has positive real part), then the system is unstable. Thus, we have to test if

$$\text{Re}(s_j^o) \le 0 \quad \text{for} \quad j=1,2,\ldots,N \tag{4.3.18}$$

However, to ensure satisfactory performance of the robot, i.e. to ensure sufficiently fast response of the system, it is required that all the eigen-values have to lie at the left-hand side of the line $\text{Re}(s) = -\alpha$, where α is prescribed stability degree of the particular robot. As it has been already explained in Sect. 3.3.3. this requirement means that the linear model of the system has all time constants less than $1./\alpha$, i.e. the response of the linear model of the system to the step inputs (positions) has to approach to the desired positions faster than $\sim\exp(-\alpha t)$. Therefore, instead to examine the condition (4.3.18) we have to test condition:

$$Re(s_j^o) \leq -\alpha, \quad \text{for} \quad j=1,2,\ldots,N \qquad (4.3.19)$$

If the condition (4.3.19) is satisfied we may be sure that the linear model of the robotic system approaches the desired state x^o faster than $\exp(-\alpha t)$ even when all joints of the robot are moving simultaneously towards their goal positions q^{oi} (which correspond to x^o).

However, to ensure satisfactory performance of the entire robotic system, it is not only required that the system has sufficiently fast response (and short settling time), but it is also necessary to prevent overshoots of the goal positions (i.e. in the position control it is required that all joints do not overshoot their set goal positions). In Sect. 3.3. we have explained the reasons why the overshoots of desired positions of the robot joints are not acceptable. The local servo systems have been synthesized to prevent the appearance of the overshoots of the goal positions, if each joint is moving independently. However, if simultaneous motions of several joints are permitted the overshoots in some joints might appear due to dynamic coupling between the joints motions. Actually, it is quite obvious that the eigen-values of the linear model of the entire robotic system differs from the eigen-values of the models of the decoupled subsystems (The decoupled subsystems are independent joints with their local servo systems). We have to examine whether the eigen-values of the closed-loop linear model of the robot (with local servo systems applied) satisfy the condition:

$$Im(s_j^o) = 0 \quad \text{for} \quad j=1,2,\ldots,N \qquad (4.3.20)$$

Actually, we require that all eigen-values of the linear model of the robotic system are real. However, this condition is too restrictive. The system performance is mainly affected by the so-called dominant eigen-values which are the closest to the imaginary axis (i.e. which are with the greatest real parts), while the eigen-values which are far from the imaginary axis in the left half of the complex plane have considerably less effects (since they are related to shorter time constants of the system). Therefore, we can relax the requirement (4.3.20), so that we require that the dominant eigen-values of the linear model matrix have to be real (or very close to the real axis), while the other eigen-values (far in the left half of the complex plane) might be complex. However, for all eigen-values it is generally required that they should be as close as possible to the real axis.

This means that the analysis of the stability of the linear model of
the robot with the closed local servo feedback loops is relatively
simple: we have to determine the eigen-values of the matrix of the
closed-loop linear model, and then, we have to examine whether these
eigen-values satisfy the conditions (4.3.19) and (4.3.20) (this latter
condition might be partially fulfilled, as explained above). If these
conditions are met we may state that the linear model of the robotic
system is exponentially stable around the point x^o, u^o [3] with exponential stability degree α. We can also state that the linear model
has no overshoots. However, by this the exponential stability of the
entire nonlinear model of the system is not proved, which means that
we cannot guarantee that the actual system will have the satisfactory
performance if several joints are moving simultaneously[*]. If the condition (4.3.19) is not met (even for $\alpha=0$.) we may state that the robot
system is not stable. Namely, the stability of linear model of the robot gives necessary (but not sufficient) conditions for stability of
the nonlinear model of the robotic system. Therefore, if condition
(4.3.19) is not met we may be sure that the synthesized local servo
systems do not ensure satisfactory performance of the robot. In that
case there is no need for further analysis of the stability of the nonlinear model of the robot, but we have to re-select the local controllers until we stabilize the linear model of the system. When we determine the local servos which stabilize the linear model of the system,
we have to examine the stability of the nonlinear model of the system
(see Appendix 4.A).

In the analysis of stability of the linear model of the robot in the
above described way, the main problem lies in determination of eigen-
-values of the system matrix since this matrix might be of relatively
high order. If we consider a robot with n=6 degrees of freedom (6 joints) and if all its actuators are modelled by simple linear models of
orders $n_i=3$, the entire system is of order $n \times n_i=18$. Computation of the
eigen-values of such high order matrix is related to numerical problems. However, there are numerous algorithms (and corresponding com-

---

[*] It has been shown [6] that the robotic system in general can be
asymptotically stabilized around the desired positions of the
joints if we close local servo feedback loops by position error and
by velocity. However, the exponential stability of the nonlinear
model of the robot is not guaranteed if the linear model is exponentially stable. Also we cannot guarantee that nonlinear model
will have no overshoots. The tracking of input trajectories by local servos if several joints are moving simultaneously must be examined for each particular robot separately (see Appendix 4.A).

mercially available computer programmes) for computation of the eigen-
-values of the high order matrices. Therefore, the problem of computation of eigen-values of linear model of robot will not be addressed
here.

We have presented above how we can form the linear model matrices of
the closed-loop robotic system if only local static servo systems are
applied. It is easy to extend this procedure to the case when local
dynamic controllers are applied. It is quite obvious that in that case the order of the entire model of the system becomes higher (for
example if in n=6 joints we apply PID controllers, and if we consider
the models of actuators of the orders $n_i=3$, the matrix of the linear
model of the system is of the order $n \times (n_i+1)=24$ - see Exercise 4.9).

Up to now we have considered the linear model of the robot around one
point x^o in the state space (and around the corresponding nominal control signals u^o which satisfiy (4.3.2)). As we can see in equations
(4.3.8), (4.3.9) the values of the elements of the matrices of the
linear model of robot depend on the point x^o around which the nonlinear model is linearized. (As we already explained the inertia matrix H
depends on the joints positions q^o and the vector h depends on the
joints positions q^o and velocities $\dot{q}^o$). Therefore, in general case,
for various points in the state space x^o we obtain various linear
models of the system. If we consider eigen-values of the open-loop
linear model matrix, we see that they move in the complex plane as it
is moving the point x^o in the state space around which the system is
linearized. If we select the local feedback gains, the matrix of the
closed-loop linear model of the entire robot varies depending on the
point x^o, and therefore its eigen-values also vary. According to this
considerations, if we want to examine the performance of the robot
when the local controllers are applied and when the several joints are
moving simultaneously, we have to examine the linear model of the robot around several points in the state space. Theoretically, we should
examine the system in the infinite number of points x^o, but practically this can be reduced to examination several "characteristic" points
in the state space. It cannot be precisely defined around which and
around how many points in the state space we have to determine the
linear models of the robot and to analyze the system stability. For
each particular robotic system we have to analyze variations of the
eigen-values of the system depending on the point in the state space around
which the system is linearized. In this way, we have to examine whether

the conditions (4.3.19) and (4.3.20) are met for "all" points in the
state space (i.e. for all possible positions and velocities of the robot joints), and to examine whether the local controllers stabilize
the robotic system in the "entire" state space[*]. Namely, the goal is
to determine such local controllers which will ensure stability of the
robotic system around "all" points in the state space. Since the linear model of the robot varies as varies the point around which the system is considered, to keep the eigen-values of the linear model in
approximately fixed positions in the complex plane (i.e. to ensure
uniform performance of the robot independently on the specified goal
positions of the robot) we should have to apply variable local gains.
Actually, we should have to apply gains which vary in dependence on
the actual positions (and velocities) of the robot joints. However,
the implementation of the variable feedback gains is not simple, so
we prefer to apply such unique local feedback gains, which could ensure that the robot performance is (approximatively) uniform in all points in the state space. Therefore, the aim is to determine such feedback gains which ensure that the positions of the dominant eigen-
-values in the complex plane of the linear model of the robot move mi-
nimally if the point x^o around which the system is linearized is "moving" in the state space. Selection of such uniform local controllers,
might be realized by various iterative procedures set at a digital
computer (see Chapter 5).

Example 4.3.2. For the robot in Fig. 3.2. the matrices of the open-loop
linear model have been determined in Example 4.3.1. For all three joints the local static controllers have been synthesized (see Example
3.3.2). The feedback loops by position and velocity of the joint have
been introduced in local servo. Data on actuators have been given in
Table 3.1. and data on the robot mechanism have been given in Table
3.2. In Table 4.1. the local feedback gains are given and eigen-values
of the local decoupled subsystems (i.e. if each joint with its actuator is considered independently on the rest of the system). Based on
(4.3.17) the eigen-values of the matrix of the linear model of the entire closed-loop robotic system are determined. These eigen-values are
given in Table 4.1, too. The eigen-values of the open-loop linear

---

[*] It is obvious that the state space of each particular robotic system is limited by the kinematic constraints (allowable rotations or
displacements of the robot joints) and by the actuators capabilities (in regard to the allowable velocities and accelerations of the
joints).

Subsystem joint	Local servo gains	Eigen-values of decoupled subsystem	Eigen-values of open-loop linear model	Eigen-values of closed-loop linear model
1	$K_P^1 = 135.13$	-6.	0.	-3.45
	$K_V^1 = 43.6$	-6.	0.	-14.7
2	$K_P^2 = 126$	-25.	0.	-22.3
	$K_V^2 = 8.65$	-25.	-0.38	-25.
3	$K_P^3 = 1674.4$	-25.	-15.44	-25.
	$K_V^3 = 13.44$	-25.	-44.98	-90.

Table 4.1. Local servo gains, eigen-values of decoupled subsystems, eigen-values of the open-loop and closed-loop linear model of the robot in Fig. 3.2.

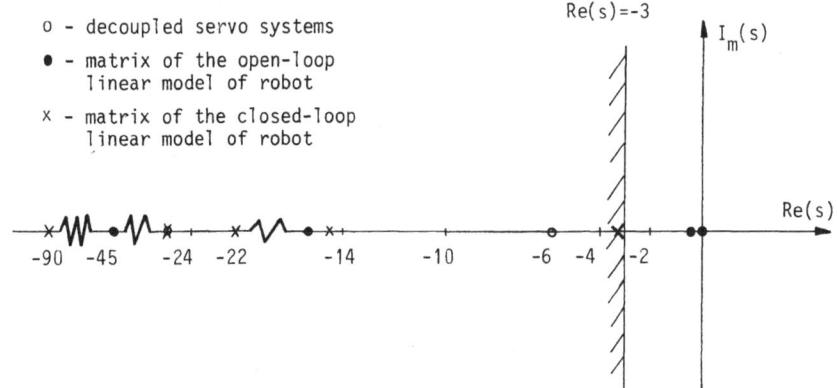

o - decoupled servo systems

● - matrix of the open-loop linear model of robot

x - matrix of the closed-loop linear model of robot

Fig. 4.3. Eigen-values of decoupled subsystems and of matrices of open-loop and closed-loop linear model of the robot in the complex plane

model of the entire robot are also given in Table 4.1. The eigen-values in all three cases (eigen-values of the local decoupled servos, eigen-values of the open-loop and closed-loop linear model of the robot) are presented in Fig. 4.3. in the complex plane. We see that the

eigen-values of the closed-loop system matrix are at the left - hand
side of the line Re(s) = -3. which means that the linear model of the
robot is exponentially stable with the stability degree α = 3. Since
all the eigen-values of the closed-loop system matrix are real, this
means that the condition (4.3.20) is also satisfied. However, this
does not guarantee that the nonlinear model of the robot (which is,
obviously, much closer to the actual robotic system) is exponentially
stable and that it has satisfactory performance when several joints
are moving simultaneously.

Exercises

4.8. For the robot in Fig. 3.2. (which has been considered in Example
3.2, local servo gains are given in Table 4.1, data on the robot
are given in Tables 3.1. and 3.2.) write the matrix of the linear
model with the closed local servo loops and check the eigen-values given in Table 4.1.

4.9. Starting from the matrices of the open-loop linear model (4.3.8),
write the expressions for the matrices of system with the closed
local servo loops, if in all n joints PID controllers are applied
(i.e. if local feedback loops by position error, velocity and integral of position error are introduced around each joint). For
the robot in Fig. 3.2. (for which the matrices of the open-loop
linear model are given by (4.3.14)) write the expression for the
matrix of the linear model if in all three joints are applied PID
controllers (write this matrix as a function of local servo gains).

4.10. For the robot in Fig. 2.5. the matrices of the open-loop linear
model have been determined in Exercise 4.6. Determine the matrix
of the closed-loop linear model in the following cases: a) if in
all joints just local position feedback loops are introduced,
b) if in all joints local position and velocity feedback loops
are introduced, c) if in all joints local PID controllers are applied.

4.11. In Exercise 4.7. the matrices of linear model of the robot in Fig.
2.5. are determined around three different "nominal" states. Local servo systems have been synthesized and the obtained servo
feedback loops arc given in Table 4.2. Determine the matrices of
the linear model of the robot if these local controllers are ap-

plied. Determine these matrices for the same three "nominal" states as in Exercise 4.7.

Gain Joint	Position feedback gain [Nm/rad] [N/m]	Velocity feedback gain [Nm/rad/s] [N/m/s]
1	40.2	11.9
2	1723.8	17.
3	1036.3	10.4

Table 4.2. Local servo gains for robot in Fig. 2.5. (data on robot are given in Tables 3.3. and 3.4)

4.4 Synthesis of Decentralized Control for Simultaneous Motions of Robot Joints

In previous sections we have considered the stability of the robotic system with position control if all joints are moving simultaneously. We have analyzed the stability of the robot using linear model of the system, and this is obviously an approximative procedure. In doing this we have restricted ourselves to consider just the case when local servo systems are applied around robot joints. Certainly, the stability analysis using linear model of robot can be applied in the cases when any other control law is applied for control of the robot (see Chapter 5). However, decentralized control of robots (by local servo systems synthesized for each joint independently) is still most frequently encountered in practice, and therefore we shall pay special attention to this control law.

As we have seen in the previous section, in position control of robot in local servos we may apply nominal control signal u^o which has to compensate for gravity moments in the desired goal position (see Fig. 4.1). This additional control signal u^o need not to be introduced, since we can compensate for the effects of gravity moments applying some other methods. We have synthesized local servo systems in Ch. 3. We have seen that in simultaneous positioning of several joints of robot, dynamic coupling between the joints appears and this coupling is not compensated by local servos. If the effects of dynamic coupling during positioning of the joints are "strong", the local servos might become inefficient for simultaneous position control. By analysis of stability of robot around the set goal positions of the joints we may check whether the local servos, synthesized for each joint independently, are sufficient to ensure satisfactory simultaneous positioning of the robot. Instead approximative linear analysis presented in previous section, we have to analyze the entire nonlinear model of the robot (see Appendix 4.A). If the analysis shows that the robot positioning is not satisfactory (regarding the speed of transient process etc.) when selected local servos are applied, we have to re-select local servos. Local servos can usually ensure efficient simultaneous positioning of robot joints.

The problem is much more complex if the tracking of desired path of a robot endeffector in workspace is required. As we have already explained, in many robot tasks it is required that the robot hand tracks some desired path in space $s^o(t)$, and, even more, the velocity and accelerations of the robot hand have to vary along this path in some prescribed way. To achieve this, the control system has to determine, at the tactical control level (see Chapter 2), trajectories of all robot joints $q^{oi}(t)$ which correspond to the desired path of the hand $s^o(t)$. The robot joints must simultaneously realize these trajectories in order to ensure that the robot hand moves along desired path (and change its velocity and acceleration in prescribed manner). The requirements regarding the dynamic performance of the robot might be very strong in such control tasks. Often it is required to ensure not only accurate tracking of the joints angles trajectories $q^{oi}(t)$ but to ensure tracking of desired variations of the joints velocities and accelerations (in order to synchronize the joint motions and realize desired variations of the robot hand velocities and accelerations). If desired nominal trajectories of joints angles are with high velocities and accelerations, then the dynamic moments (coupling between the joints) might be very high and they might have strong effects upon the system behaviour. Here, we shall consider implementation of such robot tasks by simple local servo systems.

Let assume that desired path of the robot hand is assigned by specification of external coordinates of the robot as a function of time $s^o(t)$ (by this also the desired velocities and accelerations of the robot hand are given). At the tactical control level trajectories of the joints are computed $q^{oi}(t)$ to correspond to the set desired path of the endeffector $s^o(t)$. The computed trajectories $q^{oi}(t)$ of the joints are forwarded to local servos which have to ensure their tracking, i.e. they have to ensure that the actual joint angles are as close as possible to $q^{oi}(t)$ at each moment t during the task execution (and by this to achieve that actual joint velocities $\dot{q}(t)$ are as close to desired trajectories $\dot{q}^o(t)$, and the same holds for accelerations). The problem is whether the local servos systems can adequately track input trajectories $q^{oi}(t)$ if all joints of the robot are moving simultaneously. Actually, our problem is to examine whether the local servo systems can ensure tracking of joints trajectories and realization of the desired gripper movement.

The problem of trajectory tracking by local controller has been already addressed in Sect. 3.5. We have shown that if we introduce feedforward term in local servo we can ensure satisfactory tracking of desired input trajectory $q^{oi}(t)$, assuming that all the other joints are locked.

The feedforward term is synthesized in the form of local nominal control which has to satisfy (3.5.5). Actually, if only the i-th joint is moving and all the other joints are kept locked, the dynamics of the i-th joint and its actuator is described by

$$\dot{x}^i = \hat{A}^i x^i + \hat{b}^i N(u^i) \qquad i=1,2,\ldots,n \tag{4.4.1}$$

where notations have been explained in Sect. 3.3. Local nominal control is synthesized to satisfy the following equations[*]:

$$\dot{x}^{oi} = \hat{A}^i x^{oi} + \hat{b}^i u_L^{oi} \tag{4.4.2}$$

where $x^{oi}(t)$ is the nominal trajectory of the i-th subsystem. (Under *subsystem* we assume actuator and joint if all the other joints are locked, and the subsystem model is given by (4.4.1)). Calculation of the *local nominal control* $u_L^{oi}(t)$ has been presented in Sect. 3.5. This local nominal control ensures implementation of nominal trajectories for decoupled subsystems (4.4.1) assuming that no perturbation is acting upon the system, that the model is perfect and that $x^i(0)=x^{oi}(0)$.

---

[*] For the sake of simplicity we shall assume that the local nominal control is synthesized using the same $\bar{H}_{ii}$ as for the synthesis of the local servo gains, although this is not appropriate approach (see Sect. 3.5).

If these conditions are not fulfilled, the state of the subsystem
(4.4.1) deviates from the nominal trajectory $x^{oi}(t)$. The model of
deviation of the subsystem from its nominal trajectory is given by:

$$\Delta \dot{x}^i = \hat{A} \Delta x^i + \hat{b}^i \Delta u^i \qquad (4.4.3)$$

where we have neglected the amplitude constraint upon the actuator
input u^i. As we have shown in Sect. 3.5. the control Δu^i which stabilizes the model of deviation of the decoupled subsystem (4.4.3) reduces to local servo system (which has been synthesized in Sect. 3.3).
Therefore, the control which stabilizes the individual joint around
the nominal trajectory $q^{oi}(t)$ (i.e. which stabilizes the decoupled
subsystem (4.4.1) around $x^{oi}(t)$) is in the form:

$$u^i(t) = u_L^{oi}(t) - K_p^i(q^i - q^{oi}(t)) - K_v^i(\dot{q} - \dot{q}^{oi}(t)) \qquad (4.4.4)$$

assuming that only the position and velocity feedback loops are introduced. Obviously, such control is very simple: each joint is
controlled independently from the rest of the system. Each local controller has only information on the actual state of the corresponding
joint (position and velocity). Therefore, such local controllers represent *decentralized control* of robot.

We may apply this control (4.4.4) for tracking of nominal trajectories
of the joints $q^{oi}(t)$ also in the case when all joints are moving simultaneously, i.e. we may try to ensure simultaneous tracking of nominal
trajectories of all joints of the robot by such decentralized control
law. However, this control has been synthesized neglecting the actual
coupling between the joints which appears when all joints are simultaneously moving. The simultaneous motion of all joints is described by
the model:

$$\dot{x}^i = A^i x^i + b^i u^i + f^i P_i \qquad (4.4.5)$$

where P_i is driving torque around the i-th joint given by model
(3.2.2).

If we compare the model of entire robot when all joints are moving
simultaneously (4.4.5), with the model of decoupled subsystems (4.4.1),
we can see that the coupling between the joints has not been taken
into account in control law (4.4.4). The coupling between the actuators

is given by $f^i P_i$. Actually, upon the actuators are acting dynamic moments P_i which are complex functions of the joints positions, velocities and accelerations (3.2.2). In synthesis of (4.4.4) we have taken into accounts just "decoupled dynamics" of the robot mechanism, i.e. we have assumed that upon the actuator is acting driving torque $\tilde{P}_i$ given by:

$$\tilde{P}_i = \bar{H}_{ii}\ddot{q}^i \qquad (4.4.6)$$

where $\bar{H}_{ii}$ is an estimate of the moment of inertia of the mechanism around the i-th joint (see Sect. 3.3). Therefore, in synthesis of control (4.4.4) difference between actual coupling $f^i P_i$ and coupling $f^i \tilde{P}_i$:

$$f^i(P_i - \tilde{P}_i) = f^i[H_i(q)\ddot{q} + h_i(q, \dot{q}) - \bar{H}_{ii}\ddot{q}^i] \qquad (4.4.7)$$

has not been taken into account, which means that effects of dynamic interconnections between the actuators are not compensated by the local controllers (4.4.4). We have to examine the influence of these factors (4.4.7) upon the stability of the entire robot around the nominal trajectories. In other words, we have to examine whether the dynamics of the robotic system (4.4.7) which has not been taken in the synthesis of the control (4.4.4) can "spoil" tracking of the desired nominal trajectories. To answer these questions we have to analyze so-called *practical stability* of the system around the nominal trajectory $x^o(t) = (x^{o1}(t), x^{o2}(t),...,x^{on}(t))$. However, this analysis is out of the scope of this book and it can be found elsewhere [7, 8]. By analysis of the practical stability of the robot around the nominal trajectory we get an answer to the question whether the control law (4.4.4) is able to overcome the effects of the factor (4.4.7) upon the stability of the robot. If the influence of coupling is relatively weak, the decentralized control (4.4.4) might overcome it, and the implementation of the imposed nominal trajectories of the joints $q^{oi}(t)$ can be ensured, regardless the fact that in the control synthesis we have not taken into account the actual coupling between the joints. In such a case, the control (4.4.4) would be sufficient to realize simultaneous tracking of all nominal trajectories of the joints, and by this to realize desired nominal path of the robot hand.

The scheme of the control (4.4.4) is presented in Fig. 4.4. The control represents n local controllers around the robot joints. The decentralized structure of this control is quite obvious. The main

182

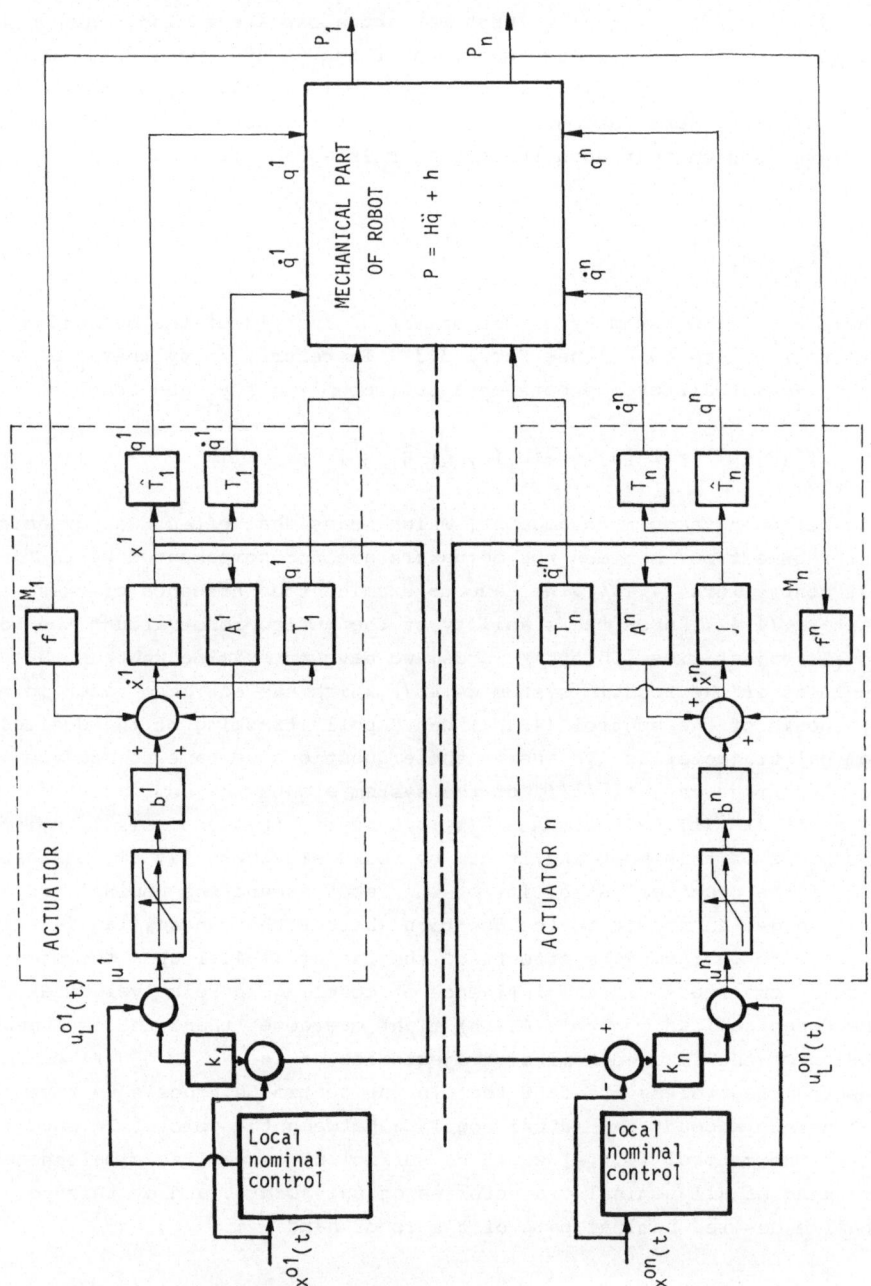

Fig. 4.4. Control scheme: local servo systems and
local nominal control

advantage of this control law relative to other control laws (which will be considered in the next chapters) lies in its simplicity. Namely, the control (4.4.4) includes only local servo loops and the local nominal control. The local nominal control does not include computation of nominal driving torques P_i and, thus, it does not require computation of the dynamic model of the mechanical part of the robot. Therefore, the local nominal control which is computed via (3.5.5) can be easily computed on-line using cheap microprocessors.

However, we must underline once again that the control (4.4.4) does not compensate for the dynamic coupling between joints, and, therefore, the tracking of fast trajectories by such control might be poor. If relatively slow trajectories are to be realized (so that the dynamic coupling between joints is relatively weak) and if high accuracy in trajectory tracking is not required, then the local controllers are usually sufficient. This is specially the case if powerful actuators and reducers with high reduction ratio are applied. In that case the constant value of the equivalent moment of inertia of the actuator is high relative to the variable values of the moment of inertia of the mechanism, and, as we have seen in Section 3.4.4, the equivalent moment of inertia of the actuator masks the effects of the variable moment of inertia of the mechanism. Powerful actuators and reducers might reduce to a high extend relative effects of coupling upon the behaviour of local servos. If the powerful actuators are applied, then relatively high local gains might be applied and local servos are less sensitive to effects of an external load (dynamic coupling) upon them.

However, in modern robotics it is required to ensure accurate tracking of fast trajectories. On the other hand, it is quite obvious a tendency to apply cheaper actuators requiring less power and to apply smaller reducer gears in order to reduce their weight and to reduce the backlash and friction. As we have already mentioned in Chapter 3, the number of installed robots with direct-drive actuators is increasing. The equivalent moment of inertia of such actuators is relatively small. In these cases the application of only local controllers (4.4.4) usually is not acceptable, but we have to apply a control which takes into account dynamic coupling between joints [9]. Such dynamic control which compensates for dynamic coupling between joints will be considered in the next sections.

It should be noticed that the decentralized control (4.4.4) is still applied to control many current robots on the market.

The reasons for these are extreme simplicity of the implementation of this control, and the fact that majority of robots in industrial practice are not indented for precise tracking of fast trajectories, but they just ensure position control and tracking of relatively slow nominal motions.

It should be also noticed that the linear analysis, considered in the previous section, can be used for approximative analysis of the robot performance if the decentralized control (4.4.4) is applied for simultaneous tracking of the nominal trajectories of all joints. Around "each" point on the nominal trajectory $x^o(t)$ in the state space the linear model of the entire robotic system is varying. Generally speaking, we should determine the linear models of the robot around an infinite number of points at the nominal trajectory $x^o(t)$, and examine their stability when local controllers (4.4.4) are applied. In this way we can get an answer whether the robot is stable in the surroundings of the nominal trajectory $x^o(t)$. Actually, as we have already explained, we get only necessary condition for nonlinear system stability, but not sufficient conditions. However, in practice, instead an infinite number of points it is sufficient to examine the system stability around several "characteristic" points at the nominal trajectory $x^o(t)$. For these "characteristic" points, the linear models have to be determined and their stability has to be checked. Since the linear models vary along the nominal trajectory, the eigen-values of the linear model matrix also vary. To keep uniform performance of the robot along the desired nominal trajectory, we should introduce variable local feedback gains. As we have already explained, implementation of variable feedback gains is not simple. Therefore, we should try to select constant local servo gains which ensure approximately uniform behaviour of the robot along the nominal trajectory. However, for each particular robot we should examine whether we can determine such unique local servo gains so that the robot system behaves uniformly in the entire working space and for all allowable working regimes (velocities and accelerations).

Example 4.4. For the robot in Fig. 3.2. we have synthesized local controllers (4.4.4) around its joints in Examples 3.3.2, 3.5. and 4.3.2. (local nominal control and local servo feedback gains). The nominal trajectories of the robot joints are selected in such a way that the joints move from the point A defined by coordinates (0.0 [rad], 0.4 [rad], 0.0 [m]) towards the point B with coordinates (0.5 [rad],

0.6 [rad], 0.08 [m]). The joints velocities have triangular profiles. The movement should be accomplished in $\tau = 1$ [s]. The desired nominal trajectories of the joints are given in Fig. 4.5. The digital simulation of simultaneous tracking of these nominal trajectories by the synthesized local controllers are presented in Fig. 4.6. (for all three joints).

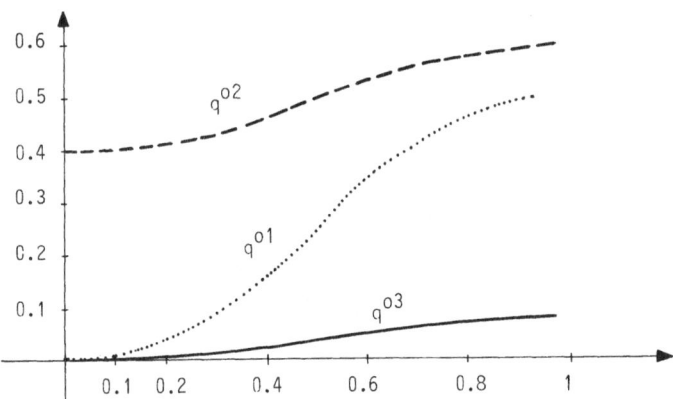

Fig. 4.5. Nominal trajectories of joints of the robot in Fig. 3.2.

In the figure also is presented the tracking of the nominal trajectories if local nominal control (feedforward) is not introduced, but only the local servo feedback loops are applied (dashed line). It can be seen that in this latter case the tracking is relatively poor, since the controllers do not compensate for the variation of the accelerations along the nominal trajectories even at the subsystem level. The simulation in both cases assumes that the initial errors of the joints coordinates in respect the nominal trajectories $q^{oi}(0)$ are $\Delta q^1(0) = -0.1$ [rad], $\Delta q^2(0) = -0.05$ [rad], $\Delta q^3(0) = -0.01$ [m], $\Delta\dot{q}^i(0) = 0$ for i = 1, 2, 3.

Exercises

4.12. Draw detail scheme of control considered in Example 4.4 for the robot in Fig. 3.2. The control includes local nominal control and local (static) servo systems in all three joints (feedback loops by position and by velocity as in (4.4.4)). Which is the total number of feedback loops in this control scheme?

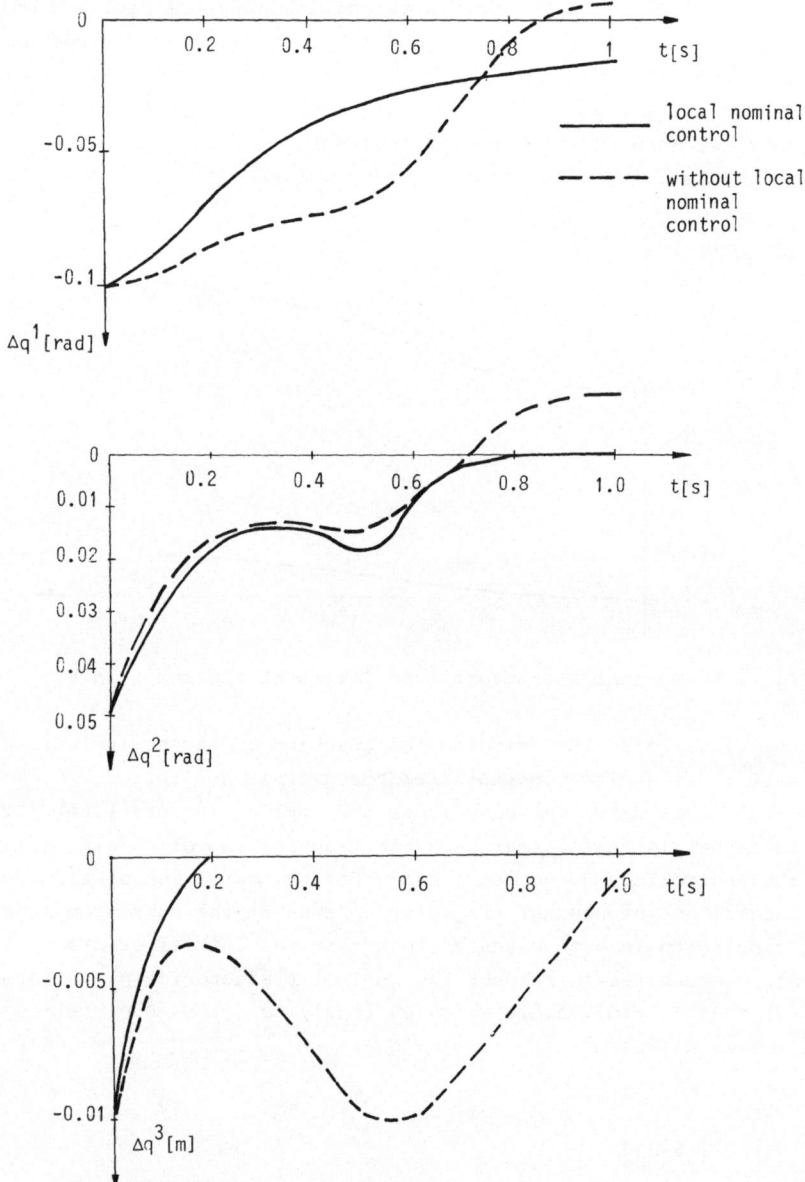

Fig. 4.6. Simulation of tracking of nominal trajectories for
robot presented in Fig. 3.2. with local servo systems with or without local nominal control

4.13. Compare the simulation of nominal trajectory tracking for the first joint of the robot in Fig. 3.2. in two cases: when all the other joints are locked (Example 3.5. - the results of simulation are given in Fig. 3.26) and when all the joints simultaneously track their nominal trajectories (Example 4.4. - the results of simulation are given in Fig. 4.6). Explain the reasons which cause differences in trajectory tracking in these two cases. Would these differences appear in the tracking of trajectory of the third joint of this robot?

4.14. Determine the minimal number of real operations (multiples and adds) that has to be performed to compute nominal local control and local servo control according to (4.4.4) in all three joints of the robot in Fig. 3.2. (the control scheme has been considered in Exercise 4.12). Assume that the second order models of actuators are used to compute local nominal control and assume that local servo loops are introduced by position and velocity (assume that nominal q^{oi}, $\dot{q}^{oi}$ and actual q^i, $\dot{q}^i$ are given). Determine the minimal number of microprocessors which have to be implemented in parallel in order to compute input signals for all three actuators acc. to (4.4.4) at each 3 [ms] (sampling period within which the control computer has to compute new values of control signals for actual q^i, $\dot{q}^i$ and nominal q^{oi}, $\dot{q}^{oi}$), if we apply microprocessors:

a) INTEL-80-80 (assume that one floating point add takes about 0.8 [ms], and one floating-point multiply takes about 1.5 [ms]), or

b) INTEL-80-87 (addition takes about 35 [µs], and multiply takes 65 [µs]).

Assume that computer time is consumed only for additions and multiplications.

4.15. Explain in detail the statement presented in Section 4.4. that with the direct-drive actuators application of only local servos for control of simultaneous motion of joints might be poor (take into account equivalent moment of inertia of such actuator and allowable local feedback gains).

4.16.* Extend the programme written in Exercise 4.4. for simulation of
simultaneous motion of all joints to include local controllers
with local nominal control (use also the program written in
Exercise 3.64). Check the results of simulation in Example 4.4.
(Fig. 4.6).

4.4.1. Synthesis of nominal programmed control

In the previous text we have explained that the simple local controllers cannot always ensure simultaneous motions of the robot joints
along desire nominal trajectories. It is often required to apply some
control law which takes into account dynamic coupling between the motions of the joints. A possible extension of decentralized control is
to introduce so-called centralized nominal programed control.

In Section 4.3. we have considered simultaneous position control of
robot joints and we have considered analysis of linear model of the
robot around the desired positions of the robot joints. In that case
the control system has to ensure that the robot state will be driven
toward the desired nominal state x^o (which corresponds to the desired positions q^{oi} of the joints). We have assumed that the nominal control signals u^{oi}, determined based on (4.3.2), are introduced at the actuators
inputs. These nominal signals ensure that the robot is in equilibrium
in the point x^o, i.e. that $\dot{x} = 0$. for $x = x^o$. However, if the robot
hand has to move along some defined path $s^o(t)$ in workspace (with desired variations of velocities and accelerations), then the system
state $x(t)$ has to move along nominal trajectory $x^o(t)$ which corresponds to nominal trajectories of the joints $q^{oi}(t)$. It is obvious that
the local nominal control $u_L^{oi}(t)$ synthesized in Section 3.5. ensures
tracking of the trajectory $q^{oi}(t)$ of the i-th joint, only if all the
other joints are kept locked. If all joints are moving simultaneously,
$u_L^{oi}(t)$ does not ensure moving of the actual subsystem state $x^i(t)$ along
desired trajectory $x^{oi}(t)$, since the influence of the other joints
(dynamic moment P_i) has not been accounted in computation of $u_L^{oi}(t)$.
Therefore, in order to compute feedforward which will take into account the coupling between the joints along the nominal trajectories
$q^{oi}(t)$, we have to consider the entire model of the system (4.3.1).

We have to determine programmed control $u^{oi}(t)$ which satisfies (analogously to (4.3.2)) [2, 6]:

$$\dot{x}^o(t) = \hat{a}(x^o(t)) + \hat{B}(x^o(t))u^o(t), \qquad t \in (0, \tau) \qquad (4.4.8)$$

Control $u^{oi}(t)$ represents a set of time-variable signals $u^o(t) = (u^{o1}(t), u^{o2}(t), \ldots, u^{on}(t))^T$ which have to satisfy (4.4.8) along the desired nominal trajectories. If these signals are fed at the inputs of actuators of the robot, they will realize the movement of the robot state along the nominal trajectory $x^{oi}(t)$ assuming that the following conditions are fulfilled (analogously as in Section 3.5):

(a) the actual initial state of the system $x(0)$ has to coincide with the nominal initial state $x(0) = x^o(0)$,

(b) the model of the entire system (4.4.8) must be perfect (all parameters of the system must be perfectly identified, etc.),

(c) no disturbance is acting upon the system.

If all these conditions are satisfied, the signals $u^{oi}(t)$ would drive the robot joints so that their angles vary in time as prescribed $q^{oi}(t)$. Thus, if the control satisfying (4.4.8) is applied, the system is in equilibrium along the trajectory $x^{oi}(t)$.

The model of deviation of the system state around the nominal trajectory $x^{oi}(t)$ and the nominal control $u^{oi}(t)$ can be observed. The model of deviation is obtained in the form which is equivalent to the model of deviation around the point x^o, u^o (4.3.3):

$$\Delta\dot{x}(t) = \bar{a}(x^o(t), \Delta x(t)) + \hat{B}(x^o(t), \Delta x(t))N(u^o(t), \Delta u) \qquad (4.4.9)$$

We have to keep in mind that the $N \times 1$ vector $\bar{a}$ and the $N \times n$ matrix $\hat{B}$ are time variable functions of $\Delta x(t)$ since they depend on nominal trajectory $x^o(t)$. Similarly, the constraint upon the amplitude of the input signals u^i becomes time dependent $N(u^{oi}(t), \Delta u^i)$ since it depends on $u^{oi}(t)$. The model of deviation is in equilibrium for $\Delta x = 0$. (since if $\Delta u = 0$, then it follows $\Delta\dot{x} = 0$.).

Analogously to the linearization of the model of the state deviation around the point x^o, u^o which is obtained in the form (4.3.5), we can obtain linear model of deviation of the robot state around the nominal trajectory $x^o(t)$ and control $u^o(t)$:

$$\Delta \dot{x} = \tilde{A}_L(x^o(t))\Delta x + \tilde{B}_L(x^o(t))N(u^o(t), \Delta u) \qquad (4.4.10)$$

where the matrices $\tilde{A}_L$ (N×N) and $\tilde{B}_L$ (N×n) are, now, time dependent, since they depend on nominal trajectory. Now, we should analyze stability of time-variable linear model. However, it is sufficient to consider linear model of the robot in a several "characteristic" points at the nominal trajectory and we have to examine eigen-values of the matrices of these linear models.

The nominal control $u^{oi}(t)$ which satisfies (4.4.8) represents *programmed* control, since it is a function just on time and it does not depend on the actual state of the system. Since it is computed using the total centralized model of the robot (4.4.8) it is called *centralized nominal programmed control* (to differ it from the local nominal control which is computed using the models of local (decoupled) subsystems (3.5.5) without taking into account the coupling between the subsystems). The centralized nominal control represents feedforward along the trajectory which compensates not only for the delays in the system caused by the variations of the joints velocities and accelerations along the nominal trajectory, but also compensates for the dynamic coupling between the joints of the robot caused by their simultaneous motions. This centralized nominal control compensates for the effects of nominal coupling between the joints and therefore it reduces the effects of the interconnections between the joints.

Let us briefly explain the above statement that the nominal centralized control $u^o(t)$ reduces the effects of coupling between the joints. Let us consider the model of the robot in the form (4.4.5). The centralized nominal control satisfies (analogously to (4.4.8)):

$$\dot{x}^{oi}(t) = A^i x^{oi}(t) + b^i u^{oi}(t) + f^i P_i^o(t), \quad i=1,2,\ldots,n \qquad (4.4.11)$$

where $P_i^o(t)$ is the so-called nominal driving torque in the i-th joints which must satisfy:

$$P_i^o(t) = H_i(q^o(t))\ddot{q}^o(t) + h_i(q^o(t), \dot{q}^o(t)) \qquad (4.4.12)$$

The model of the robot actual performance is given by (4.4.5). From (4.4.5) and (4.4.11) we can obtain the model of deviation of the system state from the nominal trajectory $x^o(t)$ and the nominal control $u^o(t)$ (analogously to (4.4.9)):

$$\Delta \dot{x}^i = A^i \Delta x^i + b^i \Delta u^i + f^i \Delta P_i, \quad i=1,2,\ldots,n \tag{4.4.13}$$

where $\Delta x^i = x^i - x^{oi}(t)$ is the vector of deviation of the state of the i-th actuator (subsystem) from its nominal trajectory, $\Delta P_i = P_i - P_i^o(t)$ is the deviation of the moment in the i-th joint from the nominal moment $P_i^o(t)$, $\Delta u^i = u^i - u^{oi}(t)$ is the deviation of the input signal of the i-th actuator from the nominal control $u^{oi}(t)$. For the sake of simplicity, in (4.4.13) we have neglected the constraint upon the amplitude of the actuator input signal. The control Δu^i has to be synthesized to stabilize the model of deviation of the state (4.4.13) around the nominal trajectory and control. The synthesis of the control Δu^i is performed in the following way. Instead the actual model of the state deviation (4.4.13) we consider approximative model in the following form:

$$\Delta \dot{x}^i = A^i \Delta x^i + b^i \Delta u^i + f^i \Delta \tilde{P}_i, \quad i=1,2,\ldots,n \tag{4.4.14}$$

where $\Delta \tilde{P}_i$ is the deviation of the moment from the nominal moment, but for approximative model of the robot dynamics (4.4.6). Therefore, $\Delta \tilde{P}_i$ is given by

$$\Delta \tilde{P}_i = \bar{H}_{ii} \Delta \ddot{q}^i \tag{4.4.15}$$

where $\bar{H}_{ii}$ is the estimated constant value of the moment of inertia of the mechanism around the i-th joint. It is obvious that the model (4.4.14) represents the set of models of the individual joints and their actuators considered independently one from another. Now, the control Δu^i, which stabilizes the model (4.4.14), can be obtained as a set of local servo systems around the robot joints. The local servos can be synthesized for each joint independently from the other joints. The synthesis of such independent local servos has been considered in Chapter 3. The local servo systems stabilize the approximative model of deviation (4.4.14). The local servo systems are synthesized to stabilize each joint decoupled from the other joints. If just position and velocity servo loops are introduced in each local servo, then the control signal for each actuator is obtained as

$$u^i(t) = u^{oi}(t) - K_p^i(q^i - q^{oi}(t)) - K_v^i(\dot{q} - \dot{q}^{oi}(t)) \tag{4.4.16}$$

The scheme of the robot control, if the centralized nominal control and local servo systems around robot joints are applied, is presented

in Fig. 4.7. Actually, the nominal programmed control signals calculated
by the centralized model of the robot, are added to signals from simple
local servo systems.

If we apply the control presented in Fig. 4.7, we may guarantee that the
approximative model of the robot dynamics (4.4.14), (4.4.15) is stabilized around the nominal trajectory $x^{oi}(t)$. However, it is quite obvious that this control does not guarantee the stabilization of the "actual" model of the robot (4.4.13). The actual coupling between the joints of the robot is not described by (4.4.15) (as assumed in synthesis of local servos), but actual coupling are total dynamic moment in
the joints:

$$\Delta P_i = H_i(q)\ddot{q} - H_i^o(q^o)\ddot{q}^o + h_i(q, \dot{q}) - h_i^o(q^o, \dot{q}^o) \qquad (4.4.17)$$

The total dynamic moment in the joint P_i is described by the dynamic
model of the mechanical part of the system (3.2.2).

The local servo systems stabilize the decoupled models of the joints
(and actuators) in which the actual coupling between the joints ΔP_i is
neglected. Therefore, the actual coupling between the joints ΔP_i (i.e.
the deviation of the actual coupling from the nominal coupling $P_i^o(t)$)
has not been compensated neither by local servo systems, nor by nominal
centralized control. Actually, the difference:

$$f^i(\Delta P_i - \Delta \tilde{P}_i) = f^i[H_i(q)\ddot{q} - H_i(q^o)\ddot{q}^o + h(q, \dot{q}) -$$
$$- h(q^o, \dot{q}^o) - \bar{H}_{ii}\Delta \ddot{q}^i] \qquad (4.4.18)$$

has not been taken into account in the synthesis of the control (4.4.16).
Therefore, we have to examine the effects of these factors upon the
stability of the robot system. This analysis is presented in Appendix
4.A.2. If the effects of these "factors" upon the stability of the entire system is relatively "weak", then the control (4.4.16) is sufficient to stabilize the robot and to ensure acceptable tracking of the
nominal trajectory.

By comparison of the equation (4.4.7) (representing the uncompensated
coupling when local controller (4.4.4) are applied) and the equation
(4.4.18), we can see that if the centralized nominal control (4.4.16)
is applied, then the uncompensated coupling is decreased by the value

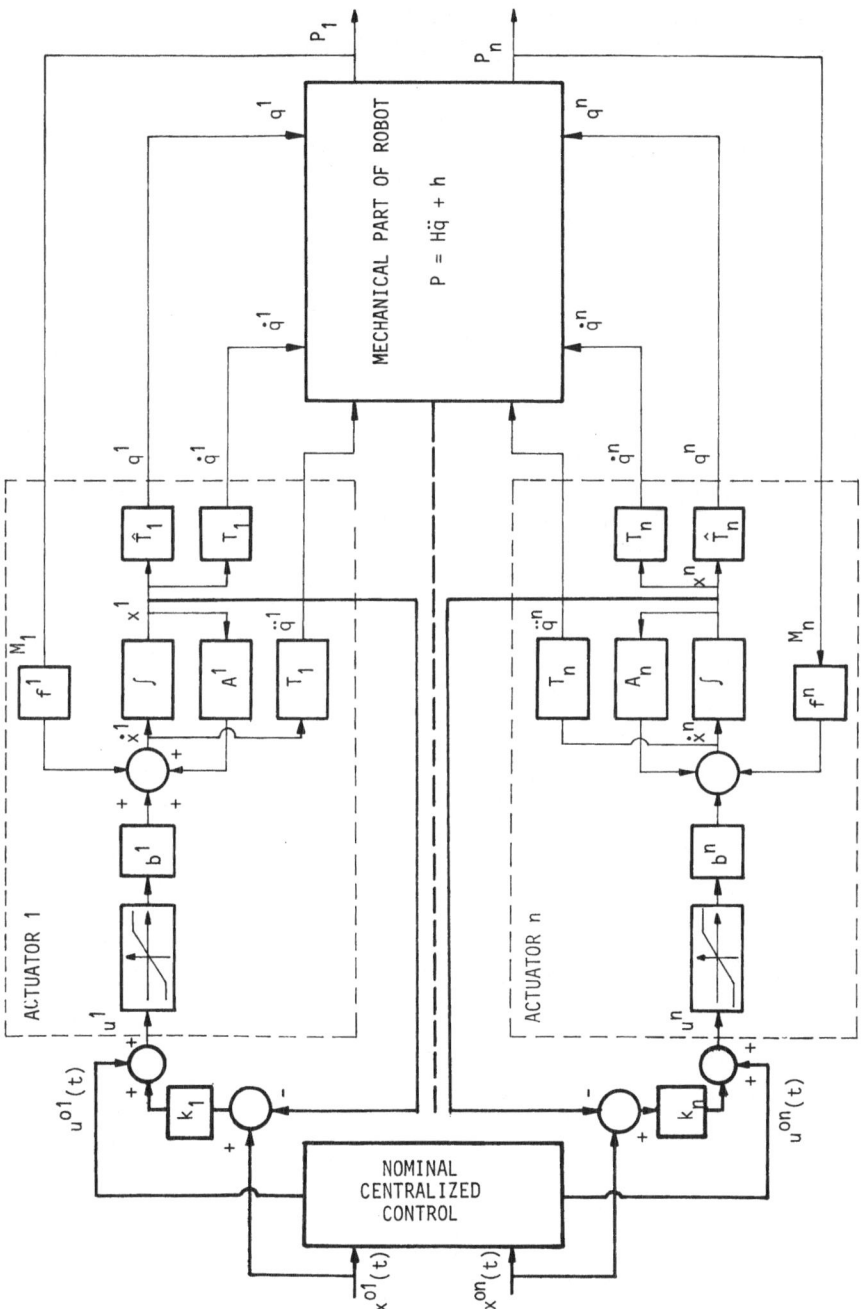

Fig. 4.7. Control scheme: local servo systems and centralized
nominal control

of nominal coupling $P_i^o(t)$. Namely, local controllers (4.4.4) do not compensate for nominal dynamic torques $P_i^o(t)$. Therefore, the effects of coupling (4.4.7) upon the system performance (when just local controllers (4.4.4) are applied), might be considerably stronger than the effects of the difference between the actual and nominal coupling (4.4.18) (which is uncompensated in the case when the centralized nominal control is added to local servos). In other words, the stabilization of the robot around the nominal trajectory is much efficient if centralized nominal control is applied than if just local controllers with local nominal control are applied.

However, the application of the centralized nominal control suffers from certain drawbacks. Let us consider how we can compute centralized nominal control. In Section 3.5. we have shown how we can compute local nominal programmed control. Let us assume that the nominal trajectory $x^o(t)$ of the system state is given, i.e. let us assume that the nominal trajectories $x^{oi}(t)$ of all coordinates of the states vectors of all subsystems (actuators) are given. Let us assume that all models of the actuators are of the second orders $n_i = 2$ and that the matrices of models (3.2.6) are in the form:

$$A^i = \begin{bmatrix} 0 & 1 \\ 0 & a_{22}^i \end{bmatrix}, \quad b^i = \begin{bmatrix} 0 \\ b_2^i \end{bmatrix}, \quad f^i = \begin{bmatrix} 0 \\ f_2^i \end{bmatrix} \tag{4.4.19}$$

The state vector of the system in this case is given by $x(t) = (x^{1T}(t), x^{2T}(t), \ldots, x^{nT}(t))^T = (q^1, \dot{q}^1, q^2, \dot{q}^2, \ldots, q^n, \dot{q}^n)^T$. If the trajectories of the joint angles (linear displacements) $q^{oi}(t)$ and of the velocities $\dot{q}^{oi}(t)$ are given, by differentation we can get the desired variation of the joint accelerations $\ddot{q}^{oi}(t)$ along the nominal trajectories. On the basis of the dynamic model of the mechanical part of the system (3.2.2), nominal driving torques $P_i^o(t)$ can be computed according to (4.4.12). These nominal driving torques have to be realized around the robot joints in order to ensure that the robot joints change their angles in time according to the desired functions $q^{oi}(t)$. If all three above listed conditions were fulfilled, the realization of nominal driving torques $P_i^o(t)$ would cause desired movement of all robot joints. To realize the nominal driving torques, we have to realize the nominal programmed signals $u^o(t) = (u^{o1}(t), u^{o2}(t), \ldots, u^{on}(t))^T$ at the inputs of actuators. These nominal signals have to satisfy (4.4.11) and, therefore, they are computed as (taking into account the models of actuators (3.2.6) and (4.4.19)):

$$u^{oi}(t) = (\dot{x}_2^{oi}(t) - a_{22}^i x_2^{oi}(t) - f_2^i p_i^o(t))/b_2^i \qquad (4.4.20)$$

where by $x_2^{oi}(t)$ is denoted the second coordinate of the state vector of the i-th actuator (i.e. $x_2^{oi}(t)$ denotes the nominal velocity $\dot{q}^{oi}(t)$ of the i-th joint), and $\dot{x}_2^{oi}(t)$ denotes the acceleration of the i-th joint $\ddot{q}^{oi}(t)$. Based on (4.4.20) we can compute the centralized nominal programmed control $u^{oi}(t)$. It can be seen that expression (4.4.20) for the centralized nominal control differs from expression (3.5.8) for the local nominal control, due to the term $f_2^i p_i^o(t)$ which represents the effects of nominal driving torques, i.e. of the nominal coupling caused by simultaneous motions of all joints. When $u^{oi}(t)$ is computed by (4.4.20), we have to test whether the nominal control is within the permitted amplitude limits:

$$|u^{oi}(t)| < u_m^i \qquad (4.4.21)$$

If (4.4.21) is not fulfilled for any joint, that means that the given nominal trajectory cannot be realized by the particular robot and its actuators. In that case we have either to slow down the desired motion of the robot, or, if the robot is under design, we may re-select actuators (see Sect. 3.5).

It is obvious that the centralized nominal control requires computation of the nominal driving torques $P_i^o(t)$. The nominal driving torques are computed on the basis of total dynamic model of the mechanical part of the robot. As we have already explained, the dynamic model of the robot might be extremely complex nonlinear equations. Therefore, computation of nominal driving torques for the given nominal angles, velocities and accelerations might require the control computer to perform a large number of adds and multiplies in a short sampling interval. If the centralized nominal control has to be computed *on-line* (during the execution of motions of the robot), then we have to ensure that the control computer computes the values of the nominal driving torques and the values of the nominal programmed control according to (4.4.20), at rate of 5-10 [ms]. This means that within 5-10 [ms] the microcomputer has to calculate once $P_i^o(t)$ for given $q^o(t)$, $\dot{q}^o(t)$, $\ddot{q}^o(t)$ and to calculate $u^{oi}(t)$ for all joints. Therefore, the microcomputer which is capable to achieve desired rate of computation must be relatively powerful and expensive (or, several microprocessor

might be applied in parallel, but this certainly complicates implementation of control system).

However, the nominal driving torques and centralized nominal programmed control are exclusively functions of nominal (imposed) trajectories
of the joints angles, velocities and accelerations, and they do not depend on actual (realized) coordinates of the robot. If the desired nominal trajectories of the joints are known in advance, then the nominal
driving torques and the nominal programmed control might be computed
off-line (in advance)and memorized in the control microcomputer. In some
industrial applications the process to be realized by the robot is performed in strictly pre-defined conditions, in defined environment and
in precisely pre-specified manner. In such cases the nominal trajectories of the robot hand (and corresponding nominal trajectories of the
joints) might be defined in advance. Then, we can compute nominal control off-line in the phase of robot teaching to perform the set task.
In this way we avoid computation of the nominal programmed control in
real time, during the task execution. The control computer may compute
nominal driving torques and control relatively slowly. During the execution of the movements of the robot, the control computer has just to
take, from its memory, prepared, computed values of the nominal control
signals and to send them at the actuators inputs (obviously, the control signals of local servo systems are added to these nominal control
signal according to the scheme in Fig. 4.7). In this way requirements,
regarding the necessary speed of computation, that are imposed upon the
control microcomputer are reduced.[*)]

However, such a solution has certain drawbacks. In modern industry
(specially if we think on flexible manufacturing systems) the robot
tasks are often executed in variable conditions, which cannot be strictly pre-defined, and therefore it is not possible to compute in advance
nominal trajectories of the robot. On the other hand, many tasks (movements), which robots have to realize, might be very complex, consisting of a large number of various elementary movements, so that the
ammount of data on nominal trajectories which have to be memorized
might be enormous (specially if we take into account that we must memorize nominal trajectories and off-line computed nominal programmed
control for n joints). To memorize so large ammount of data on nominal

---

[*)] The nominal control might be computed and memorized with lower sampling rate then 10-15 [ms], and then, in on-line control the computer can determine the nominal control signals by interpolation between the memorized values.

trajectories and on nominal control, large capacity of the computer memory is required, which also increases the price of the control microcomputer.

Due to these reasons, the application of off-line computed and memorized nominal programmed control is limited to only a narrow set of robots tasks in which all conditions are pre-defined and which require only a few short movements of the robot to be repeated for many times. However, the nominal programmed control suffers from some other drawbacks, too. The centralized nominal programmed control compensates just for so-called nominal coupling between the robot joints along the nominal trajectory. The actual coupling between the joints which appears when the joints coordinates deviates from the nominal trajectories is not compensated for by nominal control. This means that the effects of the centralized nominal control are limited. As we have explained above, application of the centralized nominal control together with the local servos (Fig. 4.7) does not always guarantee that the robot is stabilized around the nominal trajectory (see Appendix 4.A). It might be necessary to introduce, besides the centralized nominal control, an additional global control which has to compensate for the actual coupling between the joints (see Chapter 5).

The computation of the centralized nominal control is based on the total dynamic model of the robot. This means that the nominal control is efficient only if all parameters of the robot (geometric data, masses and moments of inertia, friction coefficients, data on actuators, etc.) are identified very accurately. This requirement cannot be often satisfied, and therefore the efficiency of the centralized nominal programmed control, computed on the basis of the total dynamic model of the robot, might be even unefficient regarding the compensation of the nominal coupling between the joints. Namely, nominal programmed control is not robust (see Chapter 6).

In modern industry it is usually required that the robot plans its movements in *real time* (using information, obtained from cameras or other sensors, on actual state in its workspace or using information from other subsystems, conveyers, other robots cooperating in the same process and so on). In such tasks many parameters are unknown in advanced and they might vary during tasks executions. In such tasks application of nominal centralized control is not efficient. The decentralized control (4.4.4) which includes local nominal control is much simpler for

application. The decentralized controller (4.4.4) can be easily implemented *on-line* and it is much more appropriate for the tasks which require on-line planning of the robot paths and on-line generation of the
joints trajectories. However, in comparison to the centralized nominal
control, local controllers are less efficient in reducing the effects
of coupling, since they do not even compensate for the nominal coupling.

Example 4.4.1. For the robot in Fig. 3.2. the nominal trajectories of
the joints have been given in Fig. 4.5. Based on these trajectories
$q^{oi}(t)$ we can easely obtain the velocities of the joints $q^{oi}(t)$ and
corresponding accelerations $q^{oi}(t)$. Based on dynamic model of the mechanical part of the robot (3.2.3) and corresponding data (given in
Table 3.2), we calculate the nominal driving torques which are presented in Fig. 4.8. Based on the models of actuators (D.C. electro-motors
– data on which are given in Table 3.1) we compute centralized nominal
programmed control for all three joints (eq. (4.4.20)). The centralized nominal control is presented in Fig. 4.9.

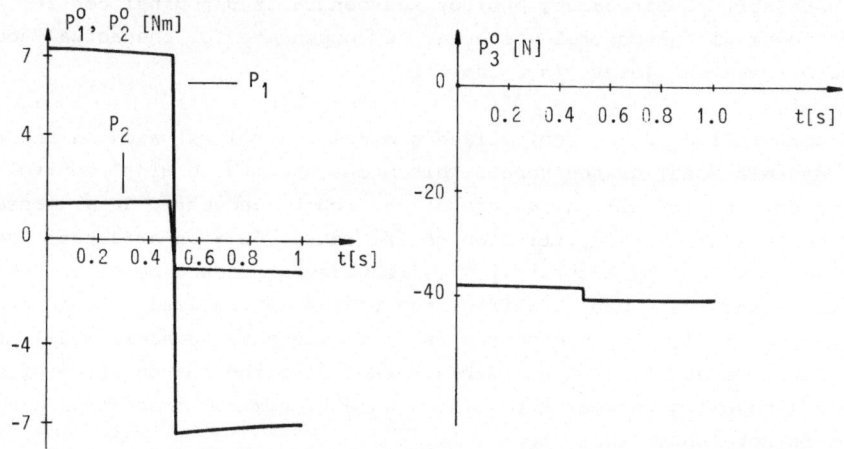

Fig. 4.8. Nominal driving torques for trajectories in Fig. 4.5.

The linearized model of this robot has been presented in Example 4.3.1.
The matrices of the linearized model are given by (4.3.13). The variations of the elements of the linearized model of the robot along the
nominal trajectories in Fig. 4.5. are presented in Fig. 4.10 (A_L =
$[a_{ij}]$, B_L = $[b_{ij}]$). The eigen-values σ of the matrices of the open-
-loop linearized model of the robot A_L also very along the nominal
trajectories as presented in Fig. 4.11. Let us assume that at the

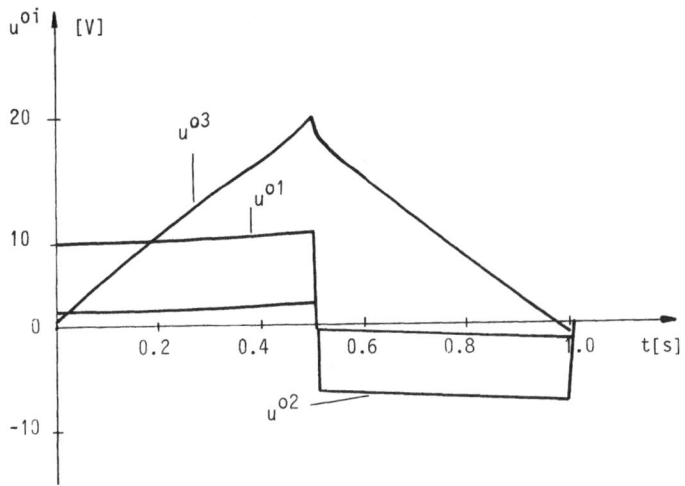

Fig. 4.9. Nominal programmed control (centralized)
for robot in Fig. 3.2.

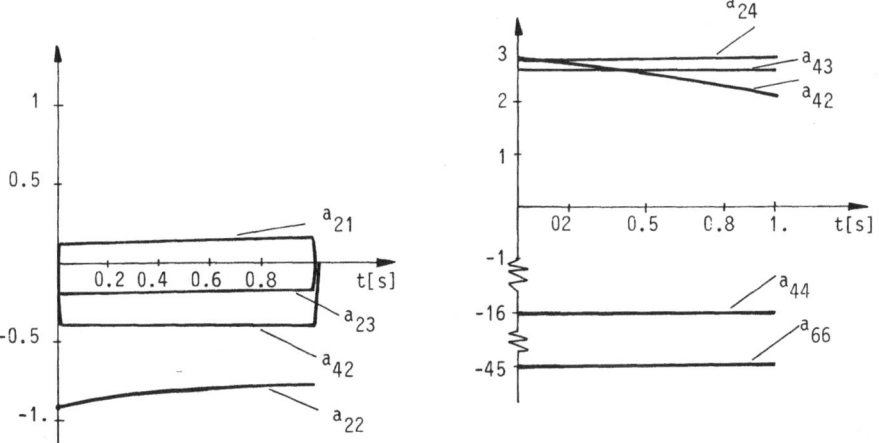

Fig. 4.10. Variations of elements of linearized model matrix of
robot along nominal trajectory (robot from Fig. 3.2)

inputs of the actuators are fed the nominal centralized control signals (from Fig. 4.9) and that the local servos with constant feedback gains are applied. If the local servo feedback loops are closed
with the feedback gains given in Table 4.1 (see Example 4.3.2), then
the eigen-values of the matrix of the linearized model of the robot
also varies along nominal trajectories as presented in Fig. 4.12.
However, based on Fig. 4.12 we can conclude that the eigen-values of
the matrix of the closed-loop system vary slightly and that they stay

at the left hand side from the line Re(s)=-3 along the entire nominal
trajectory. This means that the linearized model of the system around
the nominal trajectory is exponentially stable with stability degree
of α=3. Since all eigen-values of the closed-loop system matrix are
real along the nominal trajectory, this means that the local servos
together with the nominal programmed control u^{oi}(t) are sufficiently
robust to ensure satisfactory performances of the linearized model of
the robot along the nominal trajectory. However, the problem is whether such control can satisfy nonlinear model of the robot, or we
must introduce variable local gains, or additional global control
has to be applied.

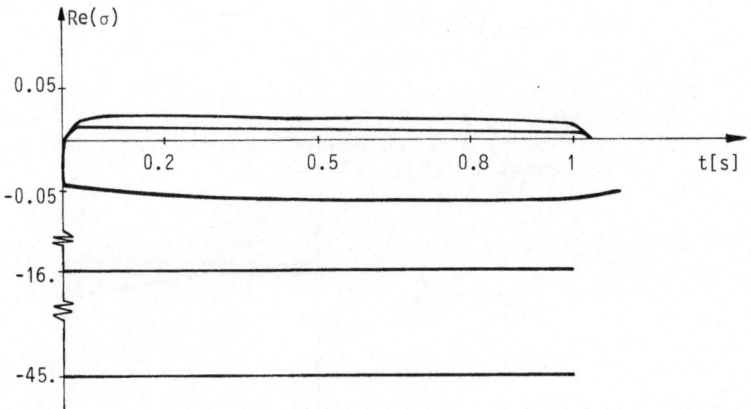

Fig. 4.11. Variation of eigen-values of matrix of open-loop
           linearized model of robot along nominal trajectory

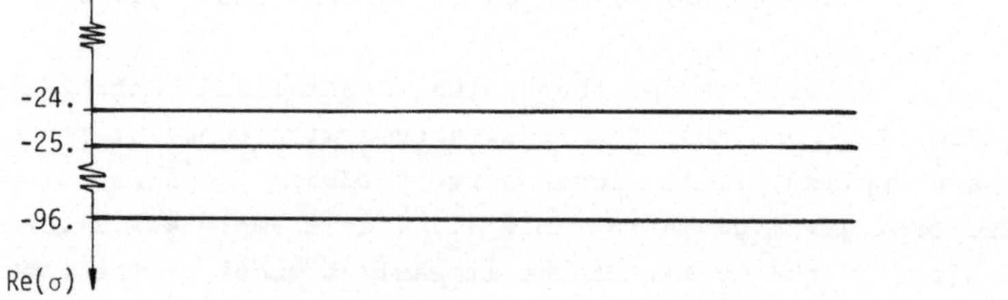

Fig. 4.12. Variation of eigen-values of matrix of closed-loop
           linearized model of robot along nominal trajectory

Exercises

4.17. For the robot in Fig. 2.5. the trajectories of the joints are
given by the following time functions:

$$q^{oi}(t) = \begin{cases} \dfrac{a^i t^2}{2} & 0 < t \leq 0.5[s] \\ \dfrac{a^i}{2}(2t\tau - \dfrac{\tau^2}{2} - t^2) & 0.5 < t \leq 1[s] \end{cases}, \text{ where } a^1 = 0.5 \ [\dfrac{rad}{s^2}],$$

$$a^2 = 0.2 \ [\dfrac{m}{s^2}], \qquad a^3 = 0.4 \ [\dfrac{m}{s^2}], \qquad \tau = 1[s]$$

Compute the nominal driving torques if all data on the mechanism are given in Table 3.4. and compute the centralized nominal control if the models of actuators are of the second order
and if data on D.C. electro-motors are given in Table 3.3.

4.18. For the robot considered in the previous exercise compute the
local nominal control and compare it with the centralized nominal control.

4.19. For the robot in Example 4.4.1. check whether the nominal driving torques are corectly computed (Fig. 4.8) in the following
time instants $t=0.$, $t=\tau/2$, and $t=\tau$. Also check the nominal control (Fig. 4.9).

4.20. For the third order models of actuators $n_i=3$ in the form
(3.2.6) where the matrices are in the form (3.2.7) show that
the centralized nominal control is computed according to the
following equations (assuming that q^{oi}, $\dot{q}^{oi}$, $\ddot{q}^{oi}$ are imposed
and that $x^i = (q^i, \dot{q}^i, i_R^i)^T = (x_1^i, x_2^i, x_3^i)^T$):

$$P_i^o(t) = H_i(q^o(t))\ddot{q}^o(t) + h_i(q^o(t), \dot{q}^o(t))$$

$$x_3^{oi}(t) = (\dot{x}_2^{oi}(t) - a_{22}^i x_2^{oi}(t) - f_2^i P_i^o(t))/a_{23}^i$$

$$\dot{x}_3^{oi}(t) \approx (x_3^{oi}(t) - x_3^{oi}(t-\Delta t))/\Delta t \quad (\Delta t - \text{small}), \quad x_3^{oi}(-\Delta t) \approx 0$$

$$u^{oi}(t) = (\dot{x}_3^{oi}(t) - a_{23}^i x_2^{oi}(t) - a_{33}^i x_3^{oi}(t))/b_3^i, \quad i=1,2,\ldots,n$$

4.21. Calculate the nominal programmed control for the robot in Example 4.4.1. if all models of actuators are of the third orders, using the expressions given in Exercise 4.20. Compare these results with the nominal control presented in Fig. 4.9.

4.22. Explain why in expressions (4.4.20) for the centralized nominal control (or in expressions in Exercise 4.20) we use matrices of actuators without taking into account the moments of inertia of the mechanism, while in the expressions for the local nominal control (3.5.8) (or, in Exercise 3.60) we use the matrices which include the moments of inertia of the mechanism around the corresponding joints axes.

4.23. Explain why in Example 4.4.1. the eigen-values of the open-loop linearized model of the robot (Fig. 4.11) vary much more along the nominal trajectory than the eigen-values of the matrix of the system with the closed local servo loops (Fig. 4.12)?

4.24.* Write in one of high-level programming language the programme for computation of the nominal driving torques (for the robot in Fig. 3.2) and for computation of the centralized nominal control (if the models of actuators are either of the second, or of the third order). Try to minimize the number of adds and multiplies required for these computations. The programme inputs are the nominal trajectories of the joints $q^{oi}(t)$, velocities $\dot{q}^{oi}(t)$, and accelerations $\ddot{q}^{oi}(t)$, and the outputs are the nominal driving torques $P_i^o(t)$ and the centralized nominal control $u^{oi}(t)$, i=1, 2, 3. Combining this programme with the programme written in Exercise 3.57 write the programme which will implement (by micro-computer) the control law presented in Fig. 4.7 (for particular robot in Fig. 3.2).

References

[1] Vukobratović M., _Applied Dynamics of Manipulation Robots: Modelling, Analysis and Examples_, Springer-Verlag, 1989.

[2] Vukobratović M., Stokić D., _Control of Manipulation Robots: Theory and Application_, Series: Scientific Fundamentals of Robotics Monograph, Springer-Verlag, Berlin, 1982.

[3] Medvedov S., Leskov A., Juschenko A., <u>Control Systems of Manipu-</u><u>lation Robots</u>, (in Russian), Monograph, "Nauka", Moscow, 1978.

[4] Vukobratović M., Kirćanski N., "Computer-Oriented Method for Li-nearization of Dynamic Models of Active Spatial Mechanisms", Me-chanism and Machine Theory, Vol. 17, No. 1, 1982.

[5] Chen C.-T., <u>Linear System Theory and Design</u>, Holt, Rinehart and Winston, New York, 1984.

[6] Takigaki M., Arimoto S., "A New Feedback Method for Dynamic Con-trol of Manipulators", Trans. of the ASME, Journal of Dynamic Systems, Measurement and Control, Vol. 103, No 2, 1981.

[7] Michel N.A., "Stability, Transient Behaviour and Trajectory Bounds of Interconnected Systems", Int. Journal of Control, Vol. 11, No. 4, pp. 703-715, 1970.

[8] Vukobratović M., Stokić D., Kirćanski N., <u>Non-Adaptive and Adap-</u><u>tive Control of Manipulation Robots</u>, Monograph, Series: Scienti-fic Fundamentals of Robotics 5., Springer-Verlag, Berlin, 1985.

[9] Asada H., Youcef-Toumi K., <u>Direct-Drive Robotics: Theory and</u><u>Practice</u>, The MIT Press, 1987.

Appendix 4.A
Stability Analysis of Nonlinear Model of Robot

In Chapter 4. we have shown how we can analyze the linearized model of robot in the case when the nominal centralized control is applied together with the local controllers (local servo systems) and the simultaneous movements of the robot joints are required. As we have already underlined, the analysis of linearized model gives necessary, but not sufficient conditions of stability. Nonlinear model of the robot should be analyzed. Therefore, here we shall consider the analysis of the nonlinear model of robot. Due to high nonlinearity of the dynamic model of robot, in general case, it is not possible to solve analytically the system of differential equations (which represents the model of the robotic system). Therefore, we have to apply methods for stability analysis which do not require explicit solution of the nonlinear model of the system. There is a number of methods for analysis of stability of large-scale nonlinear systems. These methods might be generaly classified into two groups: the input-output stability methods and the methods via Liapunov's stability [1-4]. Here, we shall consider the latter approach for stability analysis.

We shall analyze the stability of the nonlinear model of robot when the local servo systems are applied and when all joints might move simultaneously. First we shall consider the stability of the robot when it has to be positioned in various positions in work space. In doing this we assume that the control signals which compensate for gravity moments (forces) in the goal position, are implemented. Therefore, the imposed goal positions might be regarded as equilibrium points in the state space, and we may analyze the asymptotic (or exponential) stability of the robot around these imposed goal positions.

Next, we shall analyze the stability of the system around the nominal trajectory, i.e. we shall examine the robot capability to realize the desired trajectories. We shall assume that besides the local servo systems, the nominal centralized control is also applied. The application of the centralized nominal control (as we have explained in Section 4.4.1) ensures that the robotic system is in equilibrium along the nominal trajectory. Therefore in this case the asymptotic (or exponential) stability of the robot can be analyzed, too. If the nominal centralized control is not applied, but just the local nominal control and local servos are implemented, then the robotic is not in equilibrium along the nominal trajectory. Thus, in this case we cannot analyze asymptotic stability of the system around the equilibrium point, but we have to analyze the so-called practical stability of the system.

In presentation of the method for the stability analysis we shall not give rigorous proofs for each step of the procedure, but we shall try to present the method in a simple way. The more rigorous formal treatment of the subject might be found in the literature [1-9].

4.A.1 Analysis of asymptotic stability of robot position control

First, we shall consider the problem of stability of the nonlinear model of the robot in the position control. Let us assume that desired position of the robot is imposed q^{oi}, i=1,2,...,n, (i.e. the goal positions of all joints). The state vector x^o corresponds to the imposed position q^o (assuming that all joints velocities have to be zero in the goal positions $\dot{q}^o = 0$). We shall assume that control u^o is determined which satisfies (4.3.2), i.e. this control compensates the gravity moments in the imposed positions q^o (since when the robot stops, just the gravity moments are acting around the joints axes). If the control signals satisfying (4.3.2) are applied at the actuator inputs, then when the robot reaches the desired state x^o the first derivative of the state vector must be equal to zero $\dot{x}^o=0$, which means that the system is in the equilibrium state. Since the state x^o is the equilibrium state of the robotic system, we may analyze the *asymptotic stability* of the robot around this state. We have to examine whether the control ensures that for any initial state x(0) the system is driven to the imposed position q^o. As it is well known [2], the equilibrium state of the system x^o is asymptotically stable if for each number $\varepsilon > 0$, there exists a number $\delta > 0$, such that (here $||\cdot||$ denotes the norm of the vector)

$$||\Delta x(0)|| < \varepsilon \qquad (4.A.1)$$

where $\Delta x(t) = x(t)-x^o$ is the deviation of the system state from the equilibrium state x^o, implies

$$||\Delta x(t)|| < \delta \quad \text{for} \quad t \geq 0 \qquad (4.A.2)$$

and that there exists a number $\mu > 0$ such that $||\Delta x(0)|| < \mu$ implies

$$\lim_{t \to \infty} x(t) = x^o \qquad (4.A.3)$$

Since we are considering stationary system it is obvious that the initial moment might be any time instant (and therefore this is also *uniform asymptotic stability of the system*). It is obvious that if the robot is asymptotically stable around the imposed positions q^o, then the control system ensures positioning of the robot in the desired (goal) position. However, from (4.A.3) if follows that the robot might reach the desired position (state) in theoretically infinite time. The asymptotic stability does not say anything about the speed by which the robot approaches the desired position. Since it is necessary to ensure sufficiently fast positioning of the robot, we have to examine exponential stability of the system. It is well known [2] that the equilibrium state of the system x^o is exponentially stable, if there exist two numbers $\pi > 0$ and $\pi > 0$ which are independent from the initial state $x(0)$ and which satisfy

$$||x(t) - x^o|| \leq \Pi ||x(0) - x^o|| \exp(-\pi t) \qquad (4.A.4)$$

This practically means that the state x^o is exponentially stable if each motion of the system (starting from any initial state $x(0)$) converges towards x^o faster than the exponential function $\exp(-\pi t)$. Therefore, by examination of the exponential stability of the robotic system we estimate the rate of its positioning[*].

The analysis of the asymptotic (exponential) stability of the robotic system around the imposed position can be realized by application of the Liapunov's direct method [3]. The basic idea of the Liapunov's direct method is to analyze a nonlinear system of arbitrary high order by one scalar (positive) function of the system state. If it is possible to select such a continuously differentiable positive definite scalar function of the system state $v(\Delta x)$, that its first derivative by time $\dot{v}(\Delta x)$ is negative definite function (except for $v(0) = 0.$), then

---

[*] We have to note that we have already used these concepts in Chapter 3 concerning the local control synthesis. We also note that under the notion of the robot stability (asymptotic, exponential, practical) we actually assume the stability of the robot state, or trajectory.

the considered system is asymptotically stable around the equilibrium
state x^o. This means that for any initial state of the system x(0) the
selected function v(Δx) must continually decrease along the solution
of the model of the system x(t) until it reaches its minimum v(0), (since $\dot{v}$(0) = 0), which means that the system will reach its equilibrium
state x^o. Therefore, the method enables analysis of the stability of
the system without solving the differential equations which describe
the system behaviour. (Actually, the nonlinear differential equations
representing the model of the system cannot be solved analytically in
general case). Here, we shall not prove the validity of this method.
It should be underlined that the basic problem in application of this
method lies in the selection of the Liapunov's function for the considered system. There is no general procedure for unifold selection of
the Liapunov's function which will guarantee the fulfillment of the
stability conditions. In other words, the Liapunov's method gives just
sufficient but not necessary conditions of the system stability: if
the selected Liapunov's function satisfies that its derivative along
the system solution is negative definite, then we may guarantee that
the system is asymptotically stable. However, if this condition is not
fulfilled, we cannot say anything about the system stability. The system might be stable, although the selected Liapunov's function does
not fulfill the above mentioned conditions. Since the procedures for
selection of the Liapunov's function are missing, the stability analysis by Liapunov's method might be conservative. If we do not select
an adequate Liapunov's function we may get negative results of the
stability test although the system is actually stable. This means that
we would require the stronger conditions to ensure system stability
than they are actually necessary. (In other words, we would require
higher feedback gains than they are actually necessary to ensure the
system stability). Therefore, it is necessary to pay special attention
to selection of the Liapunov's function in order to minimize the conservatism of the stability test.

Here, we have presented just a few general notes on the direct Liapunov's method for the stability analysis. The basic problem in application of this method to analysis of stability of large scale nonlinear
systems lies in the selection of the adequate Liapunov's function: the
complex high order nonlinear system (which means the system with high
number of the state coordinates) is substituted by a scalar function.
To minimize the conservatism of this method, a number of aggregate-
-decomposition methods for stability analysis of large-scale systems

has been developed. The basic idea of all these methods is to ensure
the least conservative tests of the system stability by an insight in
the physical structure of the system. Actually, the idea is to "use"
the system structure in the stability analysis. In this it is assummed
that the majority of the large-scale systems might be decomposed to a
number of subsystems of the lower order. In essence all aggregate-de-
composition methods for stability analysis follows the same procedure
[1]:

a) The system is decomposed to a number of subsystems of lower order
   (in doing this it is the most important to observe the physical
   structure of the system and to decompose the system according to
   its physical features; in this way the subsystems are determined
   which physical interconections are weak - under assumption that such
   subsystems can be identified at all).

b) Each local subsystem is analyzed independently from the rest of the
   system (decoupled subsystems) and the "measure" of their local stability is estimated.

c) The quantitive estimates of the interconnections between the subsystems are determined.

c) The conditions for the stability of the entire (complex) system are
   determined on the basis of the quantitative estimates of the stability of the local subsystems and on the basis of the quantitative
   estimates of the interconnections between the subsystems.

It is clear that if there is no interconnections between the subsystems
(i.e. if the subsystems are completely decoupled), then the stability
of the entire large-scale system is guaranteed by the local stability
of the subsystems. If the coupling between the stable subsystems is
relatively weak, the entire system will be stable. Therefore, the methods for stability analysis are better if stronger coupling between
the subsystems is allowed but yet they can prove the stability of the
complete system. The majority of methods assume that the subsystems
are the stabilizing elements in the system while the coupling between
the subsystems is "the source" of instability. Concerning the application of the Liapunov's direct method, two approaches can be recognized:
by application of scalar Liapunov's function [4] and by concept of Liapunov's vector function [1, 5, 6]. The concept of the Liapunov's vector
function assumes that the stability of each subsystem is represented

by one scalar Liapunov's function, which can be used as a component of
the Liapunov's vector function. In this way an aggregate model of the
system is obtained as a vector differential inequality, the order of
which is equal to the number of subsystems. The stability of the subsystems and of the aggregate model ensures the stability of the entire
system. Here, we shall not consider various methods for analysis of
the large-scale systems. There is a number of papers which elaborate
these methods and also a number of very good survey papers on these
topics [1, 4].

We shall restrict ourselves to only one method which can be efficiently applied for the stability analysis of robotic systems. The robotic
systems in general meets the assumption that it can be considered as a
set of subsystems and these subsystems represent stabilizing elements
in the system. In principle, it is possible to decompose the robotic
system to subsystems in various ways. The most adequate decomposition
from the point of view of physical features of the system, is to consider each joint of the robot and its actuator as a local subsystem.
Namely, as a local subsystem we may adopt one joint with its actuator
as it has been considered in the previous chapter. The coupling between
such subsystems is represented by dynamic moments (forces) which are
produced due to simultaneous motions of several joints. Such decomposition is justified also from the point of view of the control synthesis. The local controllers have been synthesized based on these subsystems (see Chapter 3).

Let us consider the model of the robotic system in the form of the models of actuators (3.2.6) and the model of the mechanical part of the
system (3.2.2). Let us consider the model of deviation of the system
state around the desired (nominal) state x^o (i.e. around the desired
position q^o). Let us assume that the nominal control u^o, which satisfies (4.3.2), is applied. Therefore it is fulfilled that $\dot{x}^o = 0$. If the
control u^o satisfies (4.3.2), then it must also satisfies the models
of actuators:

$$0 = A^i x^{oi} + b^i u^{oi} + f^i P_i^o, \qquad i=1,2,\ldots,n \qquad (4.A.5)$$

where P_i^o is the moment around the i-th joint when the robot is in the
state $x^o = (x^{o1T}, x^{o2T}, \ldots, x^{onT})^T$:

$$P_i^o = h_i(q^o, 0) \qquad (4.A.6)$$

The model of deviation of the system state and control of the i-th actuator from the desired state x^o and the corresponding nominal control u^o can be written in the following form:

$$\Delta \dot{x}^i = A^i \Delta x^i + b^i N(u^{oi}, \Delta u^i) + f^i \Delta P_i, \quad i=1,2,\ldots,n \qquad (4.A.7)$$

where Δx^i is the deviation of the state vector of the i-th actuator from the nominal state (position) $\Delta x^i = x^i - x^{oi}$, Δu^i is the deviation of the i-th input from the nominal signal $\Delta u^i = u^i - u^{oi}$, ΔP_i is the deviation of the driving torque from its nominal value, $\Delta P_i = P_i - P_i^o$, and this deviation of the driving torque is described by the model of the mechanical part of the system:

$$\Delta P_i = H_i(q^o, \Delta q) \Delta \ddot{q} + h_i(q^o, \Delta q, \Delta \dot{q}), \quad i=1,2,\ldots,n \qquad (4.A.8)$$

where $\Delta q^i = q^i - q^{oi}$.

In the previous chapter we have considered the independent motion of each joint of the robot and we have synthesized local controller for each isolated actuator and joint. Each actuator and joint might be considered as a subsystem. The model of the i-th joint motion is given by (3.3.1). The model of the i-th actuator and the joint is given by (3.3.4). The model of deviation of the actuator and joint around the nominal point x^{oi}, u^{oi} can be written, analogously to (4.A.7), in the form:

$$\Delta \dot{x}^i = \hat{A}^i \Delta x^i + \hat{b}^i N(u^{oi}, \Delta u^i), \quad i=1,2,\ldots,n \qquad (4.A.9)$$

The model (4.A.9) describes the motion of the i-th actuator and the i-th joint when all the other joint are kept locked. When the joints are moving simultaneously the model of the state deviation around the nominal is given by (4.A.7) and (4.A.8). By combining (4.A.9), (4.A.7) and (4.A.8) the model of the deviation of the state of the robotic system from the nominal can be written in the following form:

$$\Delta \dot{x}^i = \hat{A}^i \Delta x^i + \hat{b}^i N(u^o, \Delta u^i) + \hat{f}^i \Delta \bar{P}_i, \qquad (4.A.10)$$

$$\Delta \bar{P}_i = H_i(q^o, \Delta q) \Delta \ddot{q} + h_i(q^o, \Delta q, \Delta \dot{q}) - \bar{H}_{ii} \Delta \ddot{q}^i \qquad (4.A.11)$$

$$i=1,2,\ldots,n$$

where $\bar{H}_{ii}$ has been already explained in relation to (3.3.4).

It is very simple to show that the models (4.A.7), (4.A.8) and (4.A.10), (4.A.11) are equivalent.

The robotic system might be "decomposed" to subsystems which corres- pond to "decoupled" joints and actuators. In other words, the system might be observed as a set of n subsystems (4.A.9) which are intercon- nected by $\hat{f}^i \Delta \bar{P}_i$ where $\Delta \bar{P}_i$ is given by (4.A.11). In this way we satisfy the assumption that the robotic system can be decomposed to subsystems, and therefore we may apply the above described procedure for stability analysis of the large-scale systems. As we shall show in the text to follow, it is also fulfilled the assumption that the local subsystems (4.A.9) are stabilizing elements in the system, while the coupling $\hat{f}^i \Delta \bar{P}_i$ is the destabilizing factor. However, on the contrary from some other large-scale systems (for example power systems, economic systems etc.) the subsystems of the robotic system might be "strongly" coupled.

Here, we shall present a method for analysis of asymptotic stability of the robotic systems [7, 8] at the finite regions in the state space, i.e. the method for estimation of regions of asymptotic stability of system, which has been developed for general large-scale systems by Weisenberger [9]. Namely, instead to examine the stability conditions in the entire state space, which in general case is very complex due to system nonlinearity, the system is considered at the finite regions in the state space. In other words, the robotic system is considered for the limited variations of the joints angles, velocities and the rotor currents (i.e. for limited variations of the state coordinates). Due to constraint upon the amplitude of the actuators inputs (4.3.4) the system can be stabilized just in some bounded regions in the state space (see Section 3.3.5).

We have assumed that the robot has to be positioned in the position q^o, i.e. we want to stabilize the robot around the state $x^o = (x^{o1T}, x^{o2T}, \ldots, x^{onT})^T$, where $x^{oi} = (q^{oi}, 0)^{T*)}$. In this we shall assume that we want to ensure positioning of the robot for some limited (allowed) vari- ation of the joints angles, i.e. we shall assume that the variation of the angle of the i-th joint must satisfy

---

*) If we consider the third order model of the actuator $n_i = 3$ the nomi- nal state of the actuator is defined by $x^{oi} = (q^{oi}, 0, i_R^{oi})^T$ where i_R^{oi} is the nominal value of the current in the rotor curcuit for the given nominal positions q^o of the robot joints.

$$|\Delta q^i| < \Delta q^i_{max} \tag{4.A.12}$$

where Δq^i_{max} is the maximal allowed deviation of the angle of the i-th joint from the given position q^{oi}. This also means that the initial position of the joint $q^i(0)$ must satisfy (4.A.12). Let us assume that we have synthesized local controller around the i-th joint in the form (static controller - see Section 3.3).

$$u^i = u^{oi} - K^i_p \Delta q^i - K^i_v \Delta \dot{q}^i \tag{4.A.13}$$

Due to amplitude constraint upon the input (4.3.4) and limited varia- tions of the joint angle (displacement) (4.A.12), on the basis of

$$|u^{oi} - K^i_p \Delta q^i - K^i_v \Delta \dot{q}^i| \leq u^i_m \tag{4.A.14}$$

we may determine the contraint upon the velocity of the i-th joint

$$\pm \Delta \dot{q}^i \leq [u^i_m \pm |u^{oi}| \pm K^i_p \Delta q^i_{max}]/K^i_v \tag{4.A.15}$$

Constraints (4.A.12) and (4.A.15) define the finite set, in the state space of the i-th subsystem, for which we have to ensure the robot po- sitioning. In other words, we have to ensure stability of the robot around the nominal point x^{oi} for all states in the finite region (in the i-th subsystem state space) which is defined by (4.A.12) and (4.A.15). This finite set of states (finite region) is presented in Fig. 4.A.1. (similar to Fig. 3.17).

Let us denote by X_i the set of states given by:

$$X_i = \{x^i: |\Delta q^i| \leq \Delta q^i_{max}, |u^{oi} - K^i_p \Delta q^i - K^i_v \Delta \dot{q}^i| \leq u^i_m\} \tag{4.A.16}$$

If we define such sets X_i for all n joints of the robot, then for the entire robot is defined the set X which represents the product of the sets X_i corresponding to subsystems, i.e. in the state space of the entire system, the region X is defined by:

$$X = X_1 \times X_2 \times \ldots \times X_n \tag{4.A.17}$$

The region (4.A.17) in the state space of the robot (the dimension of this space is N) can be written as:

$$X = \{x: \ |\Delta q^1| \ \leq \ \Delta q^1_{max}, \ \ |u^{o1}-K^1_p\Delta q^1-K^1_v\Delta\dot{q}^1| \ \leq \ u^1_m,$$

$$|\Delta q^2| \ \leq \ \Delta q^2_{max}, \ \ |u^{o2}-K^2_p\Delta q^2-K^2_v\Delta\dot{q}^2| \ \leq \ u^2_m, \dots$$

$$\dots, |\Delta q^n| \ \leq \ \Delta q^n_{max}, \ \ |u^{on}-K^n_p\Delta q^n-K^n_v\Delta\dot{q}^n| \ \leq \ u^n_m\} \qquad (4.A.18)$$

Our task is to examine whether the robot is asymptotically (exponentially) stable in the region X around the imposed state x^o if just the local controllers are applied.

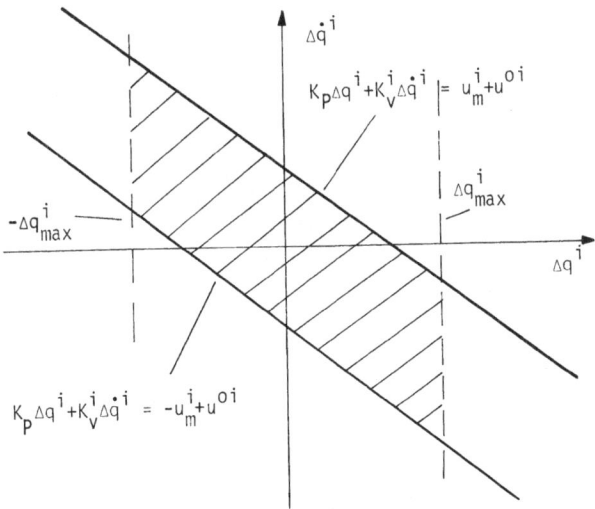

Fig. 4.A.1. Finite region in the state space of
the i-th local subsystem

We have adopted the robot decomposition into n-subsystems given by (4.A.9) which are interconnected by the factors $\hat{f}^i{}_\Delta\bar{P}_i$. By this we have performed the first step in the stability analysis of the entire non-linear model of the robot. The second step is to examine stability of the local subsystems if the interconnections between them are neglected. We have to examine whether or not the subsystem (4.A.9) is asymptotically stable around the point $\Delta x^i = 0$ in the region X_i if the local feedback loops are closed[*]:

$$\Delta\dot{x}^i = \hat{A}^i\Delta x^i-\hat{b}^ik^T_i\Delta x^i = (\hat{A}^i-\hat{b}^ik^T_i)\Delta x^i \qquad (4.A.19)$$

---

[*] Here, we shall restrict ourselves to consider the case when $n_i=2$ and when just the static controllers are applied. The cases when $n_i=3$ and when the dynamic controllers are introduced are considered in [7, 8].

where as before $k_i = (K_P^i, K_v^i)^T$ is the vector of the feedback gains. In (4.A.19) we have ommitted the nonlinear constraint upon the amplitude of the input due to assumption that the subsystem is observed at the region X_i in which (4.A.16), or (4.A.14), is fulfilled.

In the previous chapter we have synthesized the local feedback gains in such a way to ensure exponential stability of the subsystem (4.A.19). Namely, the local controller ensures that all the poles of the closed- -loop subsystem are on the left from the line $Re(s) = -\beta_i$ (in the s- -plane). Therefore, it is clear that the subsystem is exponentially stable with the exponential stability degree of β_i. However, to analyze the stability of the entire robotic system, we have to examine stability of the subsystem by introducing the *subsystems Liapunov's functions*.

Let us select as a candidate for the Liapunov's function of the i-th subsystem (4.A.19) the following function:

$$v_i = (\Delta x^{iT} \hat{H}_i \Delta x^i)^{1/2} \tag{4.A.20}$$

where $\hat{H}_i$ is the $n_i \times n_i$ positive definite matrix. Let us consider the derivative of this function along the solution of the closed-loop subsystem (4.A.19):

$$\dot{v}_i(\text{along solution of } (4.A.19)) = (\text{grad} v_i)^T \Delta \dot{x}^i =$$

$$= \frac{1}{2v_i} \Delta x^{iT} [\hat{H}_i^T (\hat{A}^i - \hat{b}^i k_i^T) + (\hat{A}^i - \hat{b}^i k_i^T)^T \hat{H}_i] \Delta x^i \tag{4.A.21}$$

It can be shown that for the stability analysis of the entire system it is the most convinient if the subsystems Liapunov's functions are selected in such a way that it is satisfied[*]:

$$\dot{v}_i(\text{along solution of } (4.A.19)) \leq -\beta_i v_i \tag{4.A.22}$$

Therefore, we have to select the matrix $\hat{H}_i$ in (4.A.20) in such a way that the derivative of the function v_i along the solution of (4.A.19), which is given by (4.A.21), satisfies the inequality (4.A.22). To fulfill this requirement we introduce a non-singular transformation $n_i \times n_i$ matrix T_i such that

---

[*] The proof of this statement can be found in [1].

$$T_i^{-1}(\hat{A}^i - \hat{b}^i k_i^T) T_i = \Lambda_i \qquad (4.A.23)$$

where Λ_i is the diagonal matrix given by

$$\Lambda_i = \begin{bmatrix} -\sigma_1^i & 0 \\ 0 & -\sigma_2^i \end{bmatrix} \qquad \text{or} \qquad \Lambda_i = \begin{bmatrix} -\sigma_1^i & \omega_1^i \\ -\omega_1^i & -\sigma_1^i \end{bmatrix} \qquad (4.A.24)$$

where $-\sigma_1^i$, $-\sigma_2^i$ or $-\sigma_1^i \pm \omega_1^i$ are the eigen-values of the matrix of the closed-loop subsystem, i.e. the eigen-values of the matrix $(\hat{A}^i - \hat{b}^i k_i^T)$. Since the closed-loop subsystem matrix $(\hat{A}^i - \hat{b}^i k_i^T)$ is stable (i.e. all its eigen-values are in the left part of the s-plane), it is possible to determine the matrix T_i which satisfies (4.A.23). For example, if the eigen-values of the matrix $(\hat{A}^i - \hat{b}^i k_i^T)$ are real and not equal, it can be shown that the transformation matrix T_i satisfying (4.A.23) is given by:

$$T_i = [s_1 \ s_2] \qquad (4.A.25)$$

where s_1, s_2 are the eigen-vectors (of dimensions $n_i \times 1$) of the matrix $(\hat{A}^i - \hat{b}^i k_i^T)$. The eigen-vectors satisfy the equations $(\hat{A}^i - \hat{b}^i k_i^T) s_j - \sigma_j^i s_j = 0$, $j=1,2$. Now, we can select the matrix $\hat{H}_i$ in the following way:

$$\hat{H}_i = (T_i^{-1})^T T_i^{-1} \qquad (4.A.26)$$

It is obvious that if we select the matrix $\hat{H}_i$ by (4.A.26), then it holds:

$$\hat{H}_i^T (\hat{A}^i - \hat{b}^i k_i^T) + (\hat{A}^i - \hat{b}^i k_i^T)^T \hat{H}_i = 2\Lambda_i \hat{H}_i \qquad (4.A.27)$$

If we substitute (4.A.27) into the expression for the first derivative of the Liapunov's function, along the solution of the subsystem, (4.A.21) we get

$$\dot{v}_{i \text{(along solution of (4.A.19))}} = \Delta x^{iT} \Lambda_i \hat{H}_i \Delta x^i / v_i \le$$

$$\le -\min |\sigma_j^i| v_i \le -\beta_i v_i \qquad (4.A.28)$$

since it holds $\beta_i \le \min_{j=1,2} |\sigma_j^i|$. Therefore, if we select the matrix $\hat{H}_i$ by (4.A.26), we can ensure that the first derivative of the Liapunov's function satisfies (4.A.22). This means that such selection of the

Liapunov's function ensures exact estimate of the subsystem stability. Namely, since the solution of the differential inequality (4.A.22) is given by $v_i < v_i(0) \exp(-\beta_i t)$, based at (4.A.22), we may state that, (taking into account obvious inequalities $\lambda_m^{1/2}(\hat{H}_i) ||\Delta x^i(t)|| < v_i(t)$ and $v_i(0) < \lambda_M^{1/2}(\hat{H}_i) ||\Delta x^i(0)||$):

$$||\Delta x^i(t)|| \leq \frac{\lambda_M^{1/2}(\hat{H}_i)}{\lambda_m^{1/2}(\hat{H}_i)} ||\Delta x^i(0)|| \exp(-\beta_i t) \tag{4.A.29}$$

where λ_M is the maximal, and λ_m is the minimal eigen-value of the matrix $\hat{H}_i$. The inequality (4.A.29) has the following meaning: the subsystem is exponentially stable with the exponential stability degree β_i. This means that we have determined the Liapunov's function by which the exponential stability degree of the subsystem is accurately estimated.

Since we have assumed that the subsystem state vector Δx^i belongs to region (finite set) X_i (and therefore in (4.A.19) and (4.A.21) we have not taken into account the constraints upon the input amplitude), we may state that the differential inequality (4.A.22) is satisfied for the region X_i. In other words, if the initial conditions are such that (4.A.16) is satisfied, then the local subsystem is exponentially stable with the stability degree β_i.

Next, we have to *estimate* the region X_i by the Liapunov's function. To do this, let us introduce the region (set) in the subsystem state space $\tilde{X}_i$ which is defined by:

$$\tilde{X}_i = \{\Delta x^i: v_i(\Delta x^i) \leq v_{io}\} \tag{4.A.30}$$

where $v_{io} > 0$ is the constant which has to be determined. In Fig. 4.A.2. the regions $\tilde{X}_i$ are presented for various values of v_{io}. The region $\tilde{X}_i$ represents an estimate of the region X_i if all points of the region $\tilde{X}_i$ belongs also to the region X_i. Therefore, all regions $\tilde{X}_i$ which are inscribed in the region X_i (Fig. 4.A.2) might be considered as estimates of the region X_i. The best estimate is the region which covers the largest "area" (but, still, is inscribed in the region X_i). Therefore, we have to determine the largest number v_{io} for which the region $\tilde{X}_i$ given by (4.A.30) completely belongs to the region X_i.

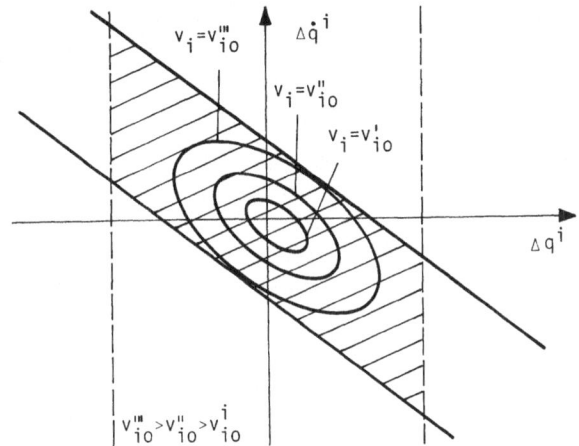

Fig. 4.A.2. Estimates of the stability regions
of the local decoupled subsystems

Now, let us proceed to the third step in the stability analysis of the
entire robotic system. The selected Liapunov's functions candidates v_i
guarantee stability of the local subsystems if the coupling between
them is completely ignored. (If the subsystems are actually decoupled
this will guarantee the stability of the entire system). We have to
estimate quantitatively the nonlinear coupling between the subsystems.
As we have already said, the coupling between the subsystems is given
by $\hat{f}^i \Delta \bar{P}_i$ where $\Delta \bar{P}_i$ is given by (4.A.11). According to the model of the
mechanical part of the system (4.A.11), the driving torques P_i are the
complex nonlinear functions of system state. Therefore to estimate the
coupling might be very complex job. It is obvious that:

$$\lim_{\Delta x \to 0} \Delta \bar{P}_i \to 0 \qquad (4.A.31)$$

since $\Delta x \to 0$ means that $\Delta q^i \to 0$, $\Delta \dot{q}^i \to 0$ and $\Delta \ddot{q}^i \to 0$ for all joints.
Therefore, we may determine the numbers ξ_{ij} which satisfy the inequalities:

$$(\mathrm{grad} v_i)^T \hat{f}^i \Delta \bar{P}_i \leq \sum_{j=1}^{n} \xi_{ij} v_j, \qquad i=1,2,\ldots,n \qquad (4.A.32)$$

for all values of the state vector Δx which belong to the region X
(4.A.18). We have to determine such number ξ_{ij} which ensures that the
inequalities (4.A.32) are fulfilled for all possible values of the
state vector of the robotic system x belonging to the region X. In
this, the numbers ξ_{ij} must satisfy:

$$\xi_{ij} \geq 0 \quad \text{for} \quad i \neq j \qquad (4.A.33)$$

Determination of the numbers ξ_{ij} in general case is not simple due to the fact that $\Delta \bar{P}_i$ is the complex nonlinear function of all state coordinates of the robotic system. It will be shown in the text to follow, that it is necessary to determine the least numbers ξ_{ij} which fulfill the inequalities (4.A.32). In the cases of simple robot structures the numbers ξ_{ij} might be estimated analytically. However, in the general case the determination of these numbers requires implementation of the digital computers. By application of computer programmes it is possible to examine the values of $\Delta \bar{P}_i$ for "all" points[*] in the region X in the state space of the system. In this way we can determine the numbers ξ_{ij} which fulfill inequalities (4.A.32). Various numerical procedures for searching along the region X and for determination of the minimal numbers ξ_{ij} which satisfy (4.A.32), might be applied.

The numbers ξ_{ij} represent the quantitative estimates of the coupling between the subsystems (i.e. between the robot joints motions).

At last, we proceed to the fourth step in the stability analysis - and that is establishment of the sufficient conditions for the stability of the entire robotic system using the estimates of the subsystems stability and quantitative estimates of coupling. To examine the stability of the entire system, let us select the Liapunov's function candidate for the entire (interconnected) system (4.A.7) in the following form [9]:

$$v = \max_{i=1,2,\ldots,n} (v_i/v_{io}) \qquad (4.A.34)$$

As we have explained before, the entire system (4.A.7) (or, (4.A.10), (4.A.11)) is asymptotically stable, if the first derivative of the selected Liapunov's function candidate is negative along the solution of the model of the system (i.e. for all states in the considered region). Let us consider the first derivative of our Liapunov's function candidate:

$$\dot{v} = \max_{i=1,2,\ldots,n} (\dot{v}_i/v_{io}) \qquad (4.A.35)$$

---

[*] It is obvious that the examining of $\Delta \bar{P}_i$ can be performed for a finite number of points in the region X, but since system is smooth such approach is quite acceptable.

where max denotes max v_i of all v_i, $i=1,2,\ldots,n$ (and not the maximum of the derivative $\dot{v}_i$).

The first derivative of the Liapunov's function for the i-th subsystem must satisfy:

$$\dot{v}_{i(\text{along solution of }(4.A.10))} = (\text{grad}v_i)^T \Delta \dot{x}^i =$$

$$= (\text{grad}v_i)^T (\hat{A}^i \Delta x^i + \hat{b}^i N(u^{oi}, \Delta u^i) + \hat{f}^i \Delta \bar{P}_i) =$$

$$= (\text{grad}v_i)^T [\hat{A}\Delta x^i + \hat{b}^i N(u^{oi}, \Delta u^i)] + (\text{grad}v_i)^T \hat{f}^i \Delta \bar{P}_i \qquad (4.A.36)$$

Taking into account that the local control has been introduced in the form (4.A.13) and that we consider only solutions of the system which completely belong to the finite region X given by (4.A.18), we may write:

$$\dot{v}_{i(\text{along solution of }(4.A.10))} = (\text{grad}v_i)^T [(\hat{A}^i - \hat{b}^i k_i^T) \Delta x^i] +$$

$$+ (\text{grad}v_i)^T \hat{f}^i \Delta \bar{P}_i \qquad (4.A.37)$$

Based on (4.A.28) and (4.A.32) we may write:

$$\dot{v}_{i(\text{along solution of }(4.A.10))} \leq -\beta_i v_i + \sum_{j=1}^{n} \xi_{ij} v_j \qquad (4.A.38)$$

Let us note that the expressions (4.A.38) are valid for all subsystems (for each $i=1,2,\ldots,n$) if the system state belongs to the region X. Let us restrict our consideration to the case when the system state belongs to the region $\tilde{X}$ given by

$$\tilde{X} = \tilde{X}_1 \times \tilde{X}_2 \times \ldots \times \tilde{X}_n \qquad (4.A.39)$$

As we have already shown, the regions $\tilde{X}_i$ are estimates of the finite regions X_i, and therefore the region $\tilde{X}$ is an estimate of the finite region X. All states belonging to the region $\tilde{X}$ must also belong to the region X (but, vice versa does not hold). The region $\tilde{X}_i$ is defined by (4.A.30). Therefore, if the state belongs to the region $\tilde{X}$ it must be fulfilled:

$$v_i < v_{io} \qquad \text{for} \qquad i=1,2,\ldots,n \qquad (4.A.40)$$

It can be proved that if the following inequalities are fulfilled:

$$-\beta_i v_{io} + \sum_{j=1}^{n} \xi_{ij} v_{jo} \leq 0 \quad \text{for each} \quad i=1,\ldots,n \quad (4.A.41)$$

then it must hold:

$$-\beta_i v_i + \sum_{j=1}^{n} \xi_{ij} v_j \leq 0 \quad \text{for} \quad v_i/v_{io} = \max_j (v_j/v_{jo}) \quad (4.A.42)$$

This means that if the conditions (4.A.41) are fulfilled, then the following relation is then the following relation is also fulfilled (taking into account (4.A.35) and (4.A.38)):

$$\dot{v}_{(\text{along solution of } (4.A.10))} = \max_{i=1,2,\ldots,n} (\dot{v}_i/v_{io}) \leq 0 \quad (4.A.43)$$

In other words, if the condition (4.A.41) is satisfied, then the first derivative of the selected Liapunov's candidate (4.A.34) along the solution of the model of the entire robotic system is negative for all states belonging to the finite region $\tilde{X}$. Therefore, if the conditions (4.A.41) are fulfilled the nonlinear model of the robotic system is asymptotically stable around the imposed position q^o, for all states whithin the frame of the finite region $\tilde{X}$. The finite region $\tilde{X}$ represents *an estimate of the asymptotic stability region of the robotic system*.

The conditions (4.A.41) might be written in a more convinient form. Let us introduce a matrix G of dimensions n×n whose elements are given by:

$$G_{ij} = -\beta_i \delta_{ij} + \xi_{ij} \quad (4.A.44)$$

where δ_{ij} is the Kronecker's symbol ($\delta_{ij}=0$ for i≠j, $\delta_{ij}=1$ for i=j). Let us introduce an n×1 vector v_o which is given by $v_o=(v_{1o}, v_{2o},\ldots,v_{no})^T$. Now, the conditions (4.A.41) might be written in the following form:

$$Gv_o \leq 0 \quad (4.A.45)$$

The matrix inequality (4.A.45) represents the *sufficient condition* for the robotic system to be asymptotically stable in the region $\tilde{X}$ defined by (4.A.30), (4.A.39), or, we may state that (4.A.45) is the sufficient condition for the region X to represent an estimate of the asymptotic stability region of the system. If the condition (4.A.45) is fulfilled we may *guarantee* that the model of the robotic system is asasymptotically stable around the desired set position q^o for all initial states belonging to the region $\tilde{X}$. This means that all solutions of the nonlinear model of the robot starting from any state which belongs to the finite region $\tilde{X}$ must terminate in the desired nominal state x^o, assuming that just the local controllers (4.A.13) are applied.

Even more, based on the above analysis we may estimate the speed by which the robot positioning will be realized (i.e. how fast will the robot be driven towards the desired position q^o). Based on (4.A.35), if follows:

$$v(x(t)) \leq v(x(0))\exp(-\eta t) \qquad (4.A.46)$$

where η is the degree of the exponential "shrinkage" of the region to which the system state must belong during the transient process (during positioning) and it is given by:

$$\eta = \min_{i=1,2,\ldots,n} \left| -\beta_i v_{io} + \sum_{j=1}^{n} \xi_{ij} v_{jo} \right| / v_{io} \qquad (4.A.47)$$

Based on (4.A.46) we may estimate that the robot state approaches the imposed state x^o by the speed which is higher than $\exp(-\eta t)$, where η is given by (4.A.47). Based on (4.A.46) it follows:

$$||\Delta x(t)|| \leq \frac{\max\limits_{i=1,2,\ldots,n} \lambda_M^{1/2}(\hat{H}_i)}{\min\limits_{i=1,2,\ldots,n} \lambda_m^{1/2}(\hat{H}_i)} ||\Delta x(0)||\exp(-\eta t) \qquad (4.A.48)$$

which means that the system is exponentially stable around the imposed desired position with the exponential stability degree which is higher or equal to η defined by (4.A.47). Now, we have just to check whether it is fulfilled that $\eta > \alpha$, where α is desired exponential stability degree.

Using the described procedure we are able to investigate whether the selected local control ensures positioning of the robot in the desired position when all joints are moving simultaneously. It is obvious that we may use this method to synthesize the decentralized control (i.e. for the synthesis of the local controllers) which will ensure the accurate positioning of the robot when all the joints are moving simultaneously. This procedure for control synthesis will be considered in Chapter 5.

If we consider the sufficient conditions (4.A.45) for the asymptotic stability of the robotic system we may conclude the following: the stability tests will be "relaxed" if the exponential stability degrees of the local subsystems β_i are greater (by absolute values), and if the numbers ξ_{ij}, which estimate the coupling between the subsystems (joints), are low. If we introduce the local controllers only, then we cannot affect the coupling between the subsystems, but we can increase the exponential stability degrees of the isolated subsystems. However, the increase of the exponential stability degrees require increase of the local feedback gains. The high feedback gains are unconvinient since they might excite resonant structural oscillations of the mechanism and since they amplify the noise in the system (see Chapter 3).

Regarding the presented method for the stability analysis, it is obvious that it is less conservative if we could obtain better (more precise) estimates β_i and ξ_{ij} (i.e. if we may determine β_i as high as possible and numbers ξ_{ij} as low as possible).

It should be noticed that the numbers ξ_{ij} which fulfill (4.A.32) have to be determined for the estimated regions $\tilde{X}$ (4.A.39) and not for the regions X (4.A.18), since the regions $\tilde{X}$ are "inscribed" in the regions X. By this we can obtain lower numbers ξ_{ij}, and the stability is anyway proved just in the estimated regions $\tilde{X}$.

The asymptotic stability of the position control of robots, if just local controllers are applied, might be proved in simpler way by application of the Liapunov's method. Namely, starting from the dynamic model of the robot it is possible to directly determine the Liapunov's function candidate for the entire system which satisfy that its first derivative is negative along the solution of the system, if just the local controllers are applied for the robot positioning [10, 11]. However, this approach doesnot work when tracking of the nominal trajectories is in question. Therefore, we presented the more complex agreggate-decomposition method for the stability analysis also in the case of position control of robots, since it can be easely extended to the trajectory servoing problem, as will be presented in the text to follow.

Let us briefly explain how we may directly prove asymptotic stability of the position control of the robot when only local servos are applied. Let us consider the dynamic model of the mechanical part of the robot (3.2.2) in the form:

$$P = H(q)\ddot{q} + \dot{q}^T C(q)\dot{q} + g(q)$$

Let us consider the second order models of the actuators which might be written in the following form:

$$N_V^i N_m^i J_M^i \ddot{q}^i = -(F_V^i + C_M^i C_E^i N_m^i N_V^i / r_R^i) \dot{q}^i + (C_M^i N_m^i / r_R^i) u^i - P_i$$

$$i = 1, 2, \ldots, n$$

All symbols have been described in Section 3.2.

Let us assume that local servo systems are applied which include perfect on-line compensation of gravity term, i.e. the actuator input signals are computed as:

$$u^i = [r_R^i / C_M^i N_m^i] g_i(q) - K_P^i \Delta q^i - K_V^i \dot{q}^i$$

If we combine the model of the mechanical part of the system, models of actuators and applied control we get model of the entire system in the following form:

$$[H(q) + J] \ddot{q} + [\dot{q}^T C(q) + \bar{D}] \dot{q} + \bar{K}_V \dot{q} + \bar{K}_P \Delta q = 0$$

where J is the n×n matrix defined by $J = \text{diag}(N_V^i N_m^i J_M^i)$, $\bar{D}$ is the n×n matrix defined by:

$$\bar{D} = \text{diag}(F_V^i + C_M^i C_E^i N_m^i N_V^i / r_R^i),$$

$\bar{K}_V$ is the n×n matrix:

$$\bar{K}_V = \text{diag}(C_M^i N_m^i K_V^i / r_R^i)$$

$\bar{K}_P$ is the n×n matrix:

$$\bar{K}_P = \text{diag}(C_M^i N_m^i K_P^i / r_R^i)$$

We have to analyze asymptotic stability of the system described by this model around the point q^o, i.e. around the point $\Delta q = 0$. Actually since $\dot{q}^o = 0.$, $\ddot{q}^o = 0.$ (position control is considered), we may write the model of the system deviation around the nominal (desired) position as:

$$[H(q) + J] \Delta \ddot{q} + [\Delta \dot{q}^T C(q) + \bar{D}] \Delta \dot{q} + \bar{K}_V \Delta \dot{q} + \bar{K}_P \Delta q = 0$$

In order to analyze stability of this system let us select the Liapunov function candidate for the entire system in the following form [10, 11]:

$$v(\Delta q, \, \Delta \dot{q}) = \frac{1}{2} [\Delta q^T \bar{K}_p \Delta q + \Delta \dot{q}^T (H(q) + J) \Delta \dot{q}]$$

Since both matrices $\bar{K}_p$ and $[H(q)+J]$ are positive definite, the Liapunov's function candidate is positive $v > 0$ for all Δq and $\Delta \dot{q}$, except for $\Delta q = 0.$, $\Delta \dot{q} = 0$. The first deviative of this function along the solution of the system model is given by:

$$\dot{v} = \Delta q^T \bar{K}_p \Delta \dot{q} + \Delta \dot{q}^T (H(q)+J) \Delta \ddot{q} + \frac{1}{2} \Delta \dot{q}^T \dot{H} \Delta \dot{q} =$$

$$= \Delta q^T \bar{K}_p \Delta \dot{q} - \Delta \dot{q}^T \{ [\Delta \dot{q}^T C(q) + \bar{D}] \Delta \dot{q} + \bar{K}_v \Delta \dot{q} + \bar{K}_p \Delta q \} + \frac{1}{2} \Delta \dot{q}^T \dot{H} \Delta \dot{q}$$

At this point we shall exploit the relation which is well known from theory of robot dynamics:

$$\Delta \dot{q}^T C(q) \Delta \dot{q} = \dot{H}(q) \Delta \dot{q} - \frac{1}{2} \Delta \dot{q}^T \frac{\partial H}{\partial q} \Delta \dot{q}$$

Using this relation, we get for the derivative of the Liapunov's function candidate:

$$\dot{v} = - \Delta \dot{q}^T (\bar{D} + \bar{K}_v) \Delta \dot{q} < 0$$

This means that the first derivative of the Liapunov's function of the entire system is negative for all system states (save for $\Delta \dot{q} = 0$, but $\Delta \ddot{q} \neq 0$ for $q \neq q^o$).

By this it is proved that the system is asymptotically stable if only local controllers are applied. This explaines why the robot work well whith the simple servo control. However, as already mentioned above, if trajectory control is in question this analysis cannot be applied.

4.A.2 Analysis of asymptotic stability of trajecotry control

We have considered the problem of stability of the entire nonlinear model of the robotic system around the prescribed position, i.e. in the position control of robot. Now, we shall consider the problem of trajectory control of robotic systems, i.e. the problem of the system stability around the specified trajectory. We shall assume that the robot is controlled by local controllers synthesized for each joint independently, as it has been explained in the previous chapter. However, there are two possible variants regarding the feedforward term which has to compensate for delay along the nominal trajectory. The first solution is to introduce the nominal programmed control synthesized on the basis of complete centralized dynamic model of the robot (Fig. 4.7), as it has been explained in Section 4.4.1. The second solution it to apply local feedforward terms in the form of the local nominal programmed control synthesized on the basis of the models of isolated actuator and joint (i.e. the model of the local subsystem), as it has been presented in Section 4.4. (Fig. 4.4).

Here, we shall consider the case when the centralized nominal programmed control is applied.

Let us define the stability conditions of the system around the imposed nominal trajectory. Here, we shall consider so-called *practical stability* of the robot around the specified nominal trajectory. We shall not present in detail various definitions of the practical stability of the system, since they can be found in the literature [12]. We shall adopt a definition which is adequate for requirements appearing in robotics. Let us assume that the nominal trajectory of the state vector of the robot $x^o(t)$, $t \in (0, \tau)$ is given. We assume that the trajectory $x^o(t)$ is continual with respect to all coordinates of the state vector x. Let us assume that the maximal deviation of the actual initial state $x(0)$ from the nominal initial state $x^o(0)$ is defined by the positive number $\bar{x}^I$ such that

$$||x(0) - x^o(0)|| \leq \bar{x}^I \qquad (4.A.49)$$

The robotic system is practically stable around the specified nominal trajectory $x^o(t)$ if for each $x(0)$ which fulfilles (4.A.49) the actual trajectory $x(t)$ fulfilles the following condition

$$||x(t) - x^o(t)|| \leq \bar{x}^t \exp(-\alpha t), \quad t \in (0, \tau) \qquad (4.A.50)$$

where $\bar{x}^t > \bar{x}^I$ and α are positive numbers independent of $\bar{x}^I$. This definition of the practical stability of the system in essence represents the exponential stability of the system around the nominal trajectory at the finite time interval and at the finite region in the state space. Our task is to examine if the synthesized control ensures that for all initial states of the system which fulfill (4.A.49) the actual trajectory of the system satisfies (4.A.50).

Let us consider the case when the nominal control $u^o(t)$, $t \in (0, \tau)$ is applied, which is synthesized based at the centralized model of the robot, i.e. let us apply the control $u^o(t)$ which satisfies (4.4.11). The task is to examine the stability of the model of deviation (4.4.13). The practical stability of the robot around the specified trajectory, when the nominal centralized control is applied, might be considered as the problem of the exponential stability of the deviation model. Namely, for the model of the state deviation (4.4.13), the following statement holds: $\Delta x=0$ and $\Delta u=0$ implies $\Delta \dot{x}=0$. This means that $\Delta x=0$ is the equilibrium point of the model of the state deviation. The conditions of the practical stability (4.A.49), (4.A.50) of the robotic system around the nominal trajectory $x^o(t)$ correspond to the following conditions of the exponential stability of the model of deviations. If the model of deviations (4.4.13) is exponentially stable around the point $\Delta x=0$ with the exponential stability degree greater or equal to α for all initial conditions which satisfy

$$||\Delta x(0)|| \leq \bar{x}^I \qquad (4.A.51)$$

then the actual trajectory of the robot meet the condition (4.A.50) for all initial states fulfilling (4.A.49), and therefore the robot is practically stable around the specified nominal trajectory. This means that if we prove that the model of the state deviations around the nominal (4.4.13) is exponentially stable around the point $\Delta x=0$, then we may state that the robot is practically stable around the imposed nominal trajectory in the sense of conditions (4.A.49), (4.A.50). We have to note that, in the general case, the exponential stability of the model of deviations is stronger condition than the practical stability of the robot around the specified trajectory. However, it is simpler to examine the exponential stability of the model of the state deviations, and, therefore, we shall apply this procedure: we shall examine the exponential stability of the model of the state deviations (4.3.22) and by this we shall check if the robot is practically stable around the set nominal trajectory.

Therefore, we have to examine the exponential stability (with the prescribed stability degree α) of the model of the state deviation around the point $\Delta x=0$ for all initial conditions which belong to the finite region defined by (4.A.51). We shall examine the system stability if the nominal centralized programmed control and the local servo systems, synthesized for each actuator and joint independently (see Chapter 3), are applied:

$$u^i = u^{oi} + \Delta u^i = u^{oi}(t) - k_i^T \Delta x^i \qquad (4.A.52)$$

Let us consider the case when the static local servos are applied. The stability analysis of the deviation model can be carried on analogously to the procedure presented in Appendix 4.A.1. for the case of robot position control. The application of the agreggation-decomposition method for stability analysis of the complex nonlinear system is performed according to the steps described in the previous section.

First, we shall consider the model of deviation (4.4.13) in an approximative form as a set of decoupled subsystems each corresponding to one joint and its actuator, in the form (4.A.9), where, now, Δx^i represents the deviation of the subsystem state vector from the nominal trajectory $\Delta x^i(t) = x^i(t) - x^{oi}(t)$, and $u^{oi}(t)$ denotes the nominal programmed control synthesized at the basis of the complete model of the system (4.A.21):

$$\Delta \dot{x}^i = \hat{A}^i \Delta x^i + \hat{b}^i N(u^{oi}(t), \Delta u^i) \qquad (4.A.53)$$

Let us examine the stability of the decoupled subsystems (4.A.53) (in which couplings are ignored) if the static controllers are applied

$$\Delta u^i = -k_i^T \Delta x^i(t) \qquad (4.A.54)$$

As in the previous case, the nonlinearity of the amplitude saturation type upon the actuator input $N(u^{oi}, \Delta u^i)$ bounds the region in the state space in which the system can be stabilized. The amplitude constraint upon the input Δu^i defines the set of points in the state space (the region):

$$|k_i^T \Delta x^i| \le u_m^i \mp u^{oi}(t) \qquad (4.A.55)$$

The bounderies of this region are obviously time dependent. Instead "the time-varaying region" defined by (4.A.55) we shall adopt the region which is inscribed in (4.A.55) and which is defined as

$$X_i = \{\Delta x^i : |k_i^T \Delta x^i| \leq u_m^i \dotplus \max_{t \in (0, \tau)} u^{oi}(t)\} \qquad (4.A.56)$$

However, it should be remembered that we want to examine exponential stability of the system for the finite region of the initial states which is defined by (4.A.51). The condition (4.A.51) bounds the region by all state coordinates of the robotic system. We may assume that by the condition (4.A.51) the finite region of the subsystem (4.A.53) is also defined, so that the subsystem is considered at the region given by:

$$x_i^I = \{\Delta x^i : ||\Delta x^i|| \leq \bar{x}_i^I\}, \qquad i=1,2,\ldots,n \qquad (4.A.57)$$

where the constraint $\bar{x}_i^I$ might be adopted as $\bar{x}_i^I = \bar{x}^I/\sqrt{n}$.

Similarly we may assume that the region (4.A.50) might be presented in the "decoupled" form, i.e. by the regions in the subsystems state space:

$$X_i(t) = \{\Delta x^i : ||x^i(t)-x^{oi}(t)|| \leq \bar{x}_i^t \exp(-\alpha t)\}$$

where $\bar{x}_i^t$ are the numbers which might be calculated as $\bar{x}_i^t = \bar{x}^t/\sqrt{n}$.

The set of states x_i^I represents the region of the initial conditions for which the subsystem stability should be guaranteed. Let us assume that x_i^I is the subset of X_i given by (4.A.56). In other words, let us assume that for all initial states of the subsystems for which we want to examine the system stability, the control u^i (4.A.54) does no violate the constraint upon the actuator input amplitude[*]. This means that for all initial states belonging to the region (4.A.57) we may assume that the subsystem behaves as linear and that the nonlinearity $N(u^{oi}(t), \Delta u^i)$ might be ignored:

$$\Delta \dot{x}^i = (\hat{A}^i - \hat{b}^i k_i^T) \Delta x^i \qquad (4.A.58)$$

The stability of the subsystem (4.A.58) will be examined as presented in the previous section. The Liapunov's function candidate for the subsystem (4.A.58) is adopted in the form:

$$v_i = (\Delta x^{iT} \hat{H}_i \Delta x^i)^{1/2} \qquad (4.A.59)$$

---

[*] The case when the set of states (4.A.57) is not the subset of (4.A.56) we shall not consider. The solution to this case can be found in [8].

where the matrix $\hat{H}_i$ can be computed according to (4.A.27). In this case the first derivative of the Liapunov's function for the decoupled subsystem (4.A.58) satisfies the condition:

$$\dot{V}_{i\text{(along the solution of (4.A.58))}} \leq -\beta_i V_i \qquad (4.A.60)$$

where β_i is the exponential stability degree of the decoupled subsystem (i.e. the eigen-value of the closed-loop subsystem matrix ($\hat{A}^i - \hat{b}^i k_i^T$) which real part has the least absolute value. The espression (4.A.60) is valid for all initial states of the subsystem which belong to the finite set (4.A.57).

The set of initial states of the subsystems X_i^I might be estimated by the subsystems Liapunov's functions, as described in the previous section. The region $\tilde{X}_i$ which estimates the region X_i^I is given by (4.A.20), where the highest numbers v_{io} have to be determined, such that all points in the region $\tilde{X}_i$ belong also to the region X_i^I (i.e. the region $\tilde{X}_i$ must be "inscribed" into the region X_i^I).

The next step in the stability analysis is to examine the coupling between the subsystems (4.A.53). The coupling between the subsystems is given by $\hat{f}^i \Delta \bar{P}_i$ where $\Delta \bar{P}_i$ is given by (4.A.11), i.e.

$$\Delta \bar{P}_i = P_i(q, \dot{q}, \ddot{q}) - P_i^o(q^o, \dot{q}^o, \ddot{q}^o) - \bar{H}_{ii} \Delta \ddot{q}^i \qquad (4.A.61)$$

It is obvious that it holds:

$$\lim_{\Delta x \to 0} \Delta \bar{P}_i \to 0 \qquad (4.A.62)$$

The condition (4.A.62) is fulfilled since we assume that the applied nominal centralized control $u^{oi}(t)$ "perfectly" compensates for the nominal moments P_i^o, and therefore the coupling between the subsystem (4.A.53) represents the deviation of the actual driving torques from the nominal moments. When the actual state of the robot approaches the nominal trajectory $x^o(t)$ (i.e. when $\Delta x \to 0$), then the actual moments P_i must convergate to the values of the nominal moments $P_i^o(t)$, and, therefore, the coupling $\Delta \bar{P}_i$ must convergates towards zero. Due to (4.A.62), we can determine the numbers ξ_{ij} which satisfy the following inequalities:

$$(\text{grad} v_i)^T f^i \Delta \bar{P}_i \leq \sum_{j=1}^{n} \xi_{ij} v_j, \qquad \text{for} \quad t \in (0, \tau), \; i=1,2,\ldots,n \quad (4.A.63)$$

This means that the numbers ξ_{ij} have to be determined such that the inequalities (4.A.63) are fulfilled for all deviations from the nominal trajectory $x^o(t)$ which satisfy the conditions (4.A.30). In determining the numbers ξ_{ij} we have to examine the deviation of the moment P_i from $P_i^o(t)$ for "all" points along the nominal trajectory $x^o(t)$, i.e. not just for the initial state $x^o(0)$, but along the "whole" trajectory, for "each" $t \in (0, \tau)$.

We must keep in mind that the practical stability of the robot requires that the system state belongs to the region given by (4.A.50), (i.e. the region to which the actual state of the robot must belong is exponentially "shrinking" around the nominal trajectory). Therefore, the inequalities (4.A.63) must be fulfilled for all states around the nominal trajectory which belong to the region (4.A.50). This region might be estimated by the following region (expressed by the subsystems Liapunov's functions):

$$\tilde{X}_i(t) = \{\Delta x^i : v_i(\Delta x^i) \leq v_{io} \exp(-\alpha t)\}$$

$$\tilde{X}(t) = \tilde{X}_1(t) \times \tilde{X}_2(t) \times \ldots \times \tilde{X}_n(t) \tag{4.A.64}$$

It is obvious that $\tilde{X}(t)$ (4.A.64) is an estimate of the region (4.A.50), since the numbers v_{io} are determined in (4.A.30) so that all points of $\tilde{X}_i$ must belong to X_i^t. On the other hand, it must be satisfied that $\bar{X}^t > \bar{X}^I$, and therefore $\bar{X}_i^t > \bar{X}_i^I$, and $\tilde{X}_i(t)$ is the subset of $X_i(t)$.

It is clear that if ξ_{ij} are determined numerically by digital computer, it is possible to examine the deviation of P_i from $P_i^o(x^o(t))$ for a finite number of points at the nominal trajectory. In other words, instead to determine the numbers ξ_{ij} by examination of the moments in the robot joints around the specified position, as we did in the previous case, here we have to examine the deviations of the moment from the nominal moments along the nominal trajectory. This might be performed by determination of the coupling $\Delta \bar{P}_i(\Delta x)$ for "the sufficient number of states" Δx (fulfilling the condition (4.A.64)), and around "the sufficient number of points" at the nominal trajectory $x^o(0)$, $x^o(t_1)$, $\ldots, x^o(t_\ell)$. It is possible to establish an algorithm which will search for the least numbers ξ_{ij} which satisfy the inequalities (4.A.63), by determination of the values of the coupling $\Delta \bar{P}_i(\Delta x)$.

Once the numbers β_i, estimating the exponential stability degrees of the decoupled subsystems, and the numbers ξ_{ij}, estimating the actual coupling between the subsystems, are determined, we may apply the same procedure as before for the stability analysis. It is easy to show that if the following condition is fulfilled

$$Gv_o < 0 \qquad\qquad (4.A.65)$$

where the n×n matrix G is given by:

$$G_{ij} = -\beta_i \delta_{ij} + \xi_{ij} \qquad\qquad (4.A.66)$$

and the n×1 vector v_o is given by $v_o = (v_{1o}, v_{2o}, \ldots, v_{no})^T$, then the model of the deviations (4.4.13) is asymptotically stable around the point $\Delta x = 0$ for all states belonging to the finite region $\tilde{X}$. Even more, if the test (4.A.65) is fulfilled then we may say that the model (4.4.13) is exponentially stable around the point $\Delta x = 0$ for all states in the region $\tilde{X}$, and the exponential stability degree can be estimated by:

$$\eta = \min_{i=1,2,\ldots,n} \left| -\beta_i v_{io} + \sum_{j=1}^{n} \xi_{ij} v_{jo} \right| / v_{io} \qquad\qquad (4.A.67)$$

If $\eta \geq \alpha$ is fulfilled we may state that the model of deviation is exponentially stable around the point $\Delta x = 0$ with the prescribed exponential stability degree in the finite region $\tilde{X}$. As we have presented above, the exponential stability of the model of deviation is sufficient condition for the robotic system to be practically stable around the specified nominal trajectory. This means that the analysis of the practical stability of the robot around the nominal trajectory has been reduced to testing of the condition (4.A.65) and determination of the number η which fulfilles (4.A.67). If the test (4.A.65) is satisfied and if $\eta \geq \alpha$, then it can be guaranteed that the model of the robot is practically stable around the nominal trajectory, but just in the estimated regions (4.A.64). The procedure for the stability analysis might be used to establish an algorithm for iterative estimation of the region $\tilde{X}$ in which the system is practically stable. The numbers v_{io} in (4.A.64) and ξ_{ij} in (4.A.63) have to be iteratively determined until "the largest" region in which the robot is stable is determined. If the estimated region of the practical stability $\tilde{X}(0)$ completely "covers" the region (4.A.51) we may guarantee the practical stability of the nonlinear model of the robot.

We have to note once again that the basic drawback of the presented procedure for the stability analysis of the robot lies in its conservativity. The condition (4.A.65) might be too conservative. If the test (4.A.65) is not fulfilled, the robot still might be practically stable around the specified trajectory. Therefore, in analyzing the robot stability by the presented procedure we must carefully interpret the results of the test. In estimating of the stability degrees of the local subsystems (the numbers β_i) and in estimating of the coupling between the subsystems (the numbers ξ_{ij}) we must try to get the best possible estimates (i.e. to determine the largest numbers β_i satisfying (4.A.60), and the least numbers ξ_{ij} satisfying (4.A.63)).

References

[1] Šiljak D.D., <u>Large Scale Dynamic Systems</u>, North Holland, 1978.

[2] Chen C.T., <u>Linear System Theory and Design</u>, Holt, Rinehart and Winston, New York, 1984.

[3] Liapunov A.I., <u>Selected Papers</u>, (in Russian), edition of Academy of Science SU, editor V.I. Smirnov, 1948.

[4] Voronov A.A., "The State of the Art and Problems of the Theory of Stability", (in Russian), Automatica and Remote Control, No. 1, 1983.

[5] Bellman R., "Vector Liapunov Functions", SIAM Journal of Control, 1, 32-34, 1962.

[6] Bailey F.N., "The Application of Liapunov's Second Method to Interconnected Systems", SIAM Journal of Control, 3, 443-461 1966.

[7] Vukobratović M., Stokić D., "Contribution to the Decoupled Control of Large-Scale Mechanical Systems", Automatica, No. 1, 1980.

[8] Vukobratović M., Stokić D., Kirćanski N., <u>Non-adaptive and Adaptive Control of Manipulation Robots</u>, Monograph, Series: Scientific Fundamentals of Robotics 5, Springer-Verlag, Berlin, 1985.

[9] Weissenberger S., "Stability Regions of Large-Scale Systems", Automatica, Vol. 9, 653-663, 1973.

[10] Takegaki M., Arimoto S., "A New Feedback Method for Dynamic Control of Manipulators", Trans. of the ASME, Journal of Dynamic Systems, Measurement and Control, Vol. 102, No. 2, 1981.

[11] Arimoto S., Miyazaki F., "Stability and Robustness of PID Feedback Control of Robot Manipulators of Sensory Capability", The First Intern. Symposium on Robotics Research, 1983.

[12] Michel N.A., "Stability, Transient Behaviour and Trajectory Bounds of Interconnected Systems", Int. Journal of Control, Vol. 11, No. 4, pp. 703-715, 1970.

Chapter 5
Synthesis of Robot Dynamic Control

5.1 Introduction

As we have explained in previous chapter, during simultaneous motions
of robot joints there appear dynamic forces (torques) which act around
the joints axes and affect the performances of local servo systems. To
ensure accurate tracking of nominal trajectories (imposed by the higher tactical control level) we have to compensate for the effects of
these dynamic forces. If the executive control level has to ensure only
positioning of the robot hand in various positions in work space, or to
ensure tracking of "slow" trajectories of the robot hand, then the effects of these dynamic forces might be relatively weak and they might
be ignored. Therefore, the local servos might accomplish such control
tasks. In the previous chapter we have shown how we can examine whether
or not the local servos could ensure accurate tracking of desired trajectories. If local servo systems can not "overcome" the effects of dynamic forces, we must introduce additional control loops which take
into account these dynamic forces. Such control law, which accounts
for dynamic characteristics of robotic systems is called *dynamic control of robots*.

In the previous chapter we have already considered one possibility to
introduce dynamic control law. We have introduced centralized nominal
control which takes into account dynamics of robot, but only at the nominal level. We have considered adventages and drawbacks of this type
of dynamic control.

One of the problems in introducing of dynamic control lies in complexity of dynamic model of robot, and therefore, implementation of control which has to compensate for dynamic effects, might be complex,
too. Due to this, our aim is to synthesize dynamic control which is
the simplest possible, but which meets imposed requirements.

In this chapter we shall consider several forms of dynamic control
laws. First, we shall consider synthesis of global dynamic control
which has to compensate directly for the effects of dynamic forces
upon performance of local servos. We shall consider various forms of

such dynamic control. However, the synthesis of dynamic control is complex due to complexity of dynamic model of robot, and, therefore, it is difficult to realize this synthesis without the aid of a digital computer. The computer-aided synthesis of control for robot is often applied today. These algorithms, set at a computer, have to help a robot designer to select the simplest control law which still satisfies the requirements regarding the accuracy of trajectory tracking. In this chapter we shall consider such a computer-aided procedure for synthesis of dynamic control of robots.

In this chapter we shall consider yet another approach to dynamic control of manipulation robots. We shall consider well-known procedure for control synthesis using so-called *"computed torque method"*. We shall also consider *Cartesian based control* of robots. In Appendix we shall briefly present a well-known procedure for synthesis of control for complex systems by centralized optimal regulator which is interesting more from the theoretical standpoint than for its real applicability in robotics.

5.2 Synthesis of Global Control

In the previous chapter we have presented how, by stability analysis, we can test whether local servo systems can meet the requirements regarding the accuracy of tracking of imposed nominal trajectory. If local servos can meet the imposed requirements, there is no need to introduce any additional control loops. However, if local servos do not guarantee sufficiently accurate positioning, or tracking of trajectories we must either re-select local servo systems, or we must introduce additional global control. The re-selection of local servos ussualy means increasing of the stability degrees of the local subsystems. To meet requirements regarding the stability of the over-all system, it is required to increase the stability degrees of subsystems β_i. However, increase of the stability degrees β_i means that the servo feedback gains are to be increased. The high feedback gains are not acceptable since they might excite resonant structural oscillations and due to effects of noice in servo systems (see Section 3.2.2). Therefore, this solution (by increasing of β_i) might be applied up to certain limit. If we cannot increase any more the local feedback gains, and the required accuracy in tracking of nominal trajectories is not achieved yet, we must introduce additional global feedback loops.

Local servo systems are supplied with information (feedback loops) on
the state (position, velocity) of the corresponding local subsystem
(joint) only. Dynamic forces (moments) which act upon the servo are
functions, in general case, of all state coordinates (angles, velocities and accelerations of all joints). To compensate for these forces we
have to introduce not only the local feedback loops, but also *global
cross-feedback loops*. The purpose of these cross-feedback loops (from
one subsystem (joint) to another) is to compensate for dynamic forces
and ensure satisfactory tracking of the nominal trajectories of all
robot joints.

First, let us consider the case when centralized nominal control, synthesized using the complete model of robot (Fig. 4.7), it introduced.
The control already compensates for the nominal dynamic moments $P_i^o(t)$
$(4.4.12)^{*)}$.

However, during the realization of the imposed trajectories $q^{oi}(t)$ there appear deviations of the actual trajectories of joints from nominal,
and dynamic moments also deviates from nominal driving torques $P_i^o(t)$.
As we have already explained, the model of deviation of the dynamics of
the mechanical part of the system from "the nominal dynamics" might be
written in the form:

$$\Delta P_i = H_i(q)\Delta\ddot{q} + [H_i(q) - H_i(q^o)]\ddot{q}^o + h_i(q, \dot{q}) - h_i(q^o, \dot{q}^o) \qquad (5.2.1)$$

The model of deviation of the i-th actuator from the nominal dynamics
(i.e. around the nominal trajectory $x^o(t)$, and nominal centralized control u^o (4.4.11)) might be written in the form:

$$\Delta\dot{x}^i = A^i\Delta x^i + b^i N(u^{oi}, \Delta u^i) + f^i\Delta P_i, \qquad i=1,2,\ldots,n \qquad (5.2.2)$$

where $\Delta x^i = x^i - x^{oi}$ is the deviation of the actual state vector of the
i-th actuator around the nominal trajectory $x^{oi}(t)$, $\Delta u^i = u^i - u^{oi}$ is
deviation of the i-th input signal from the nominal control signal
$u^{oi}(t)$, and $\Delta P_i = P_i - P_i^o$ is deviation of the i-th driving torque from
the nominal driving torque.

In chapters 3 and 4 we have synthesized local controllers in the form
of local servo systems around the individual joints which stabilize

---

$^{*)}$ We shall consider tracking of the nominal trajectories, and it can
be easely reduced to the case of position control of robots.

subsystems (actuator and joint) given by (3.3. 4). The models of deviations around the nominal trajectory and control $x^{oi}(t)$, $u^{oi}(t)$ for such decoupled subsystems (in which actual coupling has been neglected) are given by:

$$\Delta \dot{x}^i = \hat{A}^i \Delta x^i + \hat{b}^i N(u^{oi}, \Delta u^i) \tag{5.2.3}$$

Local controller has been selected in the form of linear feedback control (in the case of static controllers):

$$\Delta u_L^i = -k_i^T \Delta x^i \tag{5.2.4}$$

where k_i is the vector of feedback gains in the i-th servo. Local servos (5.2.4) ensure stabilization of local decoupled subsystems (5.2.3), i.e. they guarantee that the closed-loop local subsystems:

$$\Delta \dot{x}^i = (\hat{A}^i - \hat{b}^i k_i^T) \Delta x^i \tag{5.2.5}$$

are exponentially stable around the nominal trajectory $x^{oi}(t)$, $u^{oi}(t)$. (In Eq. (5.2.5) the amplitude constraint upon the actuator input has been ignored).

Thus, the local servos guarantee stability of subsystems (5.2.5) if they were decoupled from each other. However, it is obvious that the actual subsystems are not decoupled. The models (5.2.3) do not represent accurate model of the complete actual system, but just very approximative model. The model of the actual system (5.2.2) might be written in the form

$$\Delta \dot{x}^i = (\hat{A}^i - \hat{b}^i k_i^T) \Delta x^i + \hat{f}^i \Delta \bar{P}_i \tag{5.2.6}$$

where $\Delta \bar{P}_i$ is given by

$$\Delta \bar{P}_i = H_i(q) \Delta \ddot{q} + (H_i(q) - H_i(q^o)) \ddot{q}^o + h_i(q, \dot{q}) - h_i(q^o, \dot{q}^o) - \bar{H}_{ii} \Delta \ddot{q}^i \tag{5.2.7}$$

In (5.2.7) the dynamics of the joint (represented by $\bar{H}_{ii} \Delta \ddot{q}^i$) has been "submitted" from ΔP_i since it has been included in the model of subsystem (5.2.3) and compensated by local control. The difference between the accurate model (5.2.6) and approximative model formed of decoupled subsystems (5.2.3), is in the term $\hat{f}^i \Delta \bar{P}_i$. Obviously, the nominal programmed control (4.4.11) and the local control (5.2.4) do not compensate for this term. This term represents interconnection (coupling) between the subsystems (5.2.3) which has not been taken into account neighter in nominal programmed control synthesis, (4.4.11), nor in synthesis of local control (5.2.4). If the effects of this term are strong, we have to introduce additional global control in order to compensate for these effects.

Therefore, we introduce dynamic control in the following form:

$$u^i = u^{oi}(t) + \Delta u_L^i + \Delta u_i^G \qquad (5.2.8)$$

where Δu_i^G is the scalar global control which has to compensate for coupling between subsystems (5.2.3), i.e. to compensate for the effects of the term $\hat{f}^i \Delta \bar{P}_i$. This term is given by (5.2.7) which obviously represents dynamic forces which act upon the i-th joint due to the movements of the other joints of the robots. Actually, this term represents deviation of the dynamic forces from their nominal values. Since these dynamic forces $\Delta \bar{P}_i$ enter linearly into the model (5.2.6), the global control can be introduced in the following way. Let us assume that ΔP_i^* represents some scalar function, measured or calculated, which corresponds to deviation of the dynamic forces from nominal forces ΔP_i, i.e. which satisfies the following inequality, (for all points in the state space around the nominal trajectory $x^o(t)$):

$$|\Delta P_i^*(x)| \leq |\Delta \bar{P}_i(x)|,$$

$$\text{sign}(\Delta P_i^*(x)) = \text{sign}(\Delta \bar{P}_i(x)) \qquad (5.2.9)$$

where sign() denotes the sign of the value in brackets. Thus, the global control in the i-th joint (servo system) might be introduced in the form [1]:

$$\Delta u_i^G = -\bar{K}_i^G(\Delta x^i) \Delta P_i^* \qquad (5.2.10)$$

where by $\bar{K}_i^G(\Delta x^i)$ is denoted the scalar function of the state vector of the i-th subsystem Δx^i. The function $K_i^G(\Delta x^i)$ might be selected in various ways.

The aim is to ensure compensation for the coupling, i.e. for the dynamic forces which act upon the i-th joint. If this global control is applied in the i-th controller, then the model of deviation of the system state from the nominal becomes:

$$\Delta \dot{x}^i = \hat{A}^i \Delta x^i - \hat{b}^i k_i^T \Delta x^i + \hat{f}^i \Delta \bar{P}_i - \hat{b}^i \bar{K}_i^G (\Delta x^i) \Delta P_i^* \qquad (5.2.11)$$

where, for the sake of simplicity, the amplitude constraint upon the input signal has been ignored. It is obvious that $\bar{K}_i^G(\Delta x^i)$ has to be selected in such a way as to minimize the term

$$|\hat{f}^i \Delta \bar{P}_i - \hat{b}^i \bar{K}_i^G (\Delta x^i) \Delta P_i^*|$$

Let us assume that the model of the i-th actuator is of the second order ($n_i = 2$). Then, the vectors $\hat{f}^i$ and $\hat{b}^i$ are in the following forms:

$$\hat{f}^i = (0, \bar{f}^i)^T, \qquad \hat{b}^i = (0, \bar{b}^i)^T \qquad (5.2.12)$$

If we consider the model of the i-th subsystem (5.2.11) with the applied global control, it is obvious that the effects of coupling $\hat{f}^i \Delta \bar{P}_i$ is compensated for, if the global control is introduced in the following form:

$$\Delta u_i^G = -K_i^G (\bar{f}^i / \bar{b}^i) \Delta P_i^* \qquad (5.2.13)$$

where $K_i^G > 0$ is scalar global feedback gain. The selection of this global gain will be adressed in the text to follow. Now, the system model becomes

$$\Delta \dot{x}^i = (\hat{A}^i - \hat{b}^i k_i^{iT}) \Delta x^i + \hat{f}^i (\Delta \bar{P}_i - K_i^G \Delta P_i^*) \qquad (5.2.14)$$

where amplitude constraint upon the input signal has been again ignored. The effect of coupling $\hat{f}^i \Delta \bar{P}_i$ has been reduced to $\hat{f}^i (\Delta \bar{P}_i - K_i^G \Delta P_i^*)$, taking into account that the function ΔP_i^* fulfilles (5.2.9). The scheme of such control (5.2.8), (5.2.13) is presented in Fig. 5.1.

However, such global control directly compensates for the coupling between the subsystems (i.e. the dynamic forces acting upon the i-th joint), only if we adopt the second order models of actuators as sufficiently accurate. The second order model of actuator is approximative since it ignores the delay between the actuator input and the driving torque (force produced by the actuator). If this delay cannot be ignored, the introduced global control cannot efficiently compensate for

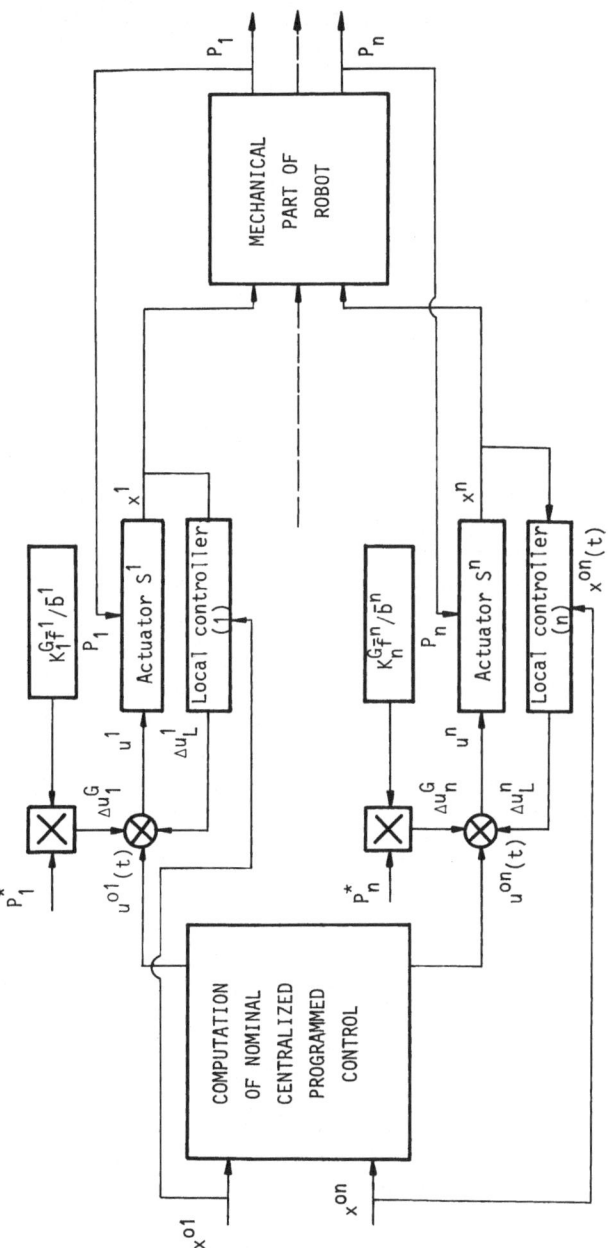

Fig. 5.1. Control scheme including nominal centralized
control, local controllers and global control

dynamic forces. Therefore, in the synthesis of global control we have to take into account this delay between the input signal u^i and the driving torque $P_i^{*)}$. One of methods to solve this problem is considered in Appendix 5.A.

In the synthesis of the global control we have to select the scalar gain K_i^G in (5.2.13). This is the global feedback gain. It is obvious that the compensation for the coupling between the subsystems (i.e. the compensation for the dynamic forces acting upon the i-th subsystem), depends on the selection of K_i^G. It is clear that if we select higher K_i^G then the global control will compensate for "larger portion" of coupling $\hat{f}^1 \Delta \bar{P}_i$. However, if we assume that we have realized the function ΔP_i^* in such a way that it satisfies $\Delta P_i^* = \Delta \bar{P}_i$, then it is obvious that the global gain K_i^G must be less than 1 (i.e. $K_i^G \leq 1$), otherwise an "overcompensation" will appear. If we select K_i^G to be equal 1 ($K_i^G = 1$), the global control would compensate for total coupling (in ideal case, if the delay in actuator can be ignored and if it is fulfilled $\Delta P_i^* = \Delta \bar{P}_i$). However, since we can never ensure such an ideal compensation that $\Delta P_i^* = \Delta \bar{P}_i$, the global control might cause oscillatory behaviour of the system (specially in close suroundings of the nominal trajectory when ΔP_i is very small). On the other hand, too high feedback gains are not acceptable, due to reasons already explained in Chapter 3. Therefore, we must select the global gain to be as low as possible, but to ensure the stabilization of the robot. In the global control synthesis we have to analyze the stability of the robot with the selected global control (see Appendix 5.A), and determine the minimal gains K_i^G which ensure desired stability of the robot.

Up to this point, we have considered the case when the global control is applied together with local servo systems and the centralized nominal control. We have explained that the centralized nominal control suffers from certain drawbacks. Thus, instead the centralized nominal control we may apply local nominal control together with local servos, which is much simpler for implementation (Fig. 4.4). However, the local nominal control does not compensate even for the nominal dynamic forces. Therefore, the effects of these forces upon the servo systems performances are considerably stronger than in the previous case. This

---

$^{*)}$ In D.C. electro motors this delay is defined by the electrical constant T_R^i of the rotor curcuit. This constant is often very small but it cannot be always ignored, specially in the case of large D.C. motors having massive windings with high inductances.

means that it is much "harder" to ensure accurate tracking of nominal trajectories (or positioning), if just local nominal control is applied. Further, it is necessary to introduce global control which has to compensate not only for the difference between the actual dynamic coupling and the nominal coupling, but for the complete dynamic forces (moments) which load the local servo systems.

The global control might be introduced in an analogous way to the previous case. The only difference is that instead of function ΔP_i^* which corresponds to the deviation of the dynamic forces from the nominal forces $\Delta \bar{P}_i$ (5.2.7), here we must introduce a function P_i^* which corresponds to total actual dynamic forces P_i (3.2.2). We select the function P_i^* to satisfy the following inequality for all states around the nominal trajectory:

$$|P_i^*(x)| \leq |P_i(x) - \bar{H}_{ii}\ddot{q}^i| = |\bar{P}_i(x)|$$

$$\text{sign}(P_i^*(x)) = \text{sign}(\bar{P}_i(x))$$

(5.2.15)

If such function $P_i^*(x)$ is introduced, the global control might be represented in the form (5.2.10):

$$\Delta u_i^G = -\bar{K}_i^G(\Delta x^i)P_i^*(x)$$

(5.2.16)

The global control might get particular form which is analogous to (5.2.13):

$$\Delta u_i^G = -K_i^G(\bar{f}^i/\bar{b}^i)P_i^*(x)$$

(5.2.17)

Here $K_i^G > 0$ again denotes scalar global gain. Next, we have to examine directly the practical stability of the system if the following control is applied:

$$u^i = u_L^{oi} - k_i^T\Delta x^i - \bar{K}_i^G(\Delta x^i)P_i^*(x)$$

(5.2.18)

i.e. if together with local nominal control and local servo feedback loops we apply the global control (5.2.17). As we have already said, the analysis of the practical stability of the robotic system is out of the scope of this book, and it can be found elsewhere [2].

So far we have not explained yet how we can realize the functions ΔP_i^* or P_i^*, which play the central roles in the global control law. The realization of these functions represents the basic problem in global control synthesis. We shall consider two possible realizations of these functions, i.e. of the global control: via force feedback and by on-line computation of dynamic couplings. We shall consider the case when the local servo systems, the feedforward terms in the form of the local nominal control and the global control in the form (5.2.16) are applied, since such control is much more applicable for industrial robots. However, all considerations can be easely extended to the control law which includes the feedforward terms in the form of the centralized nominal control and the global control in the form (5.2.10).

5.2.1. Force feedback as global control

The function P_i^* has to correspond to dynamic forces P_i, which act upon the servo system as coupling from other joints. Dynamic force (moment) P_i which acts upon the i-th joint can be directly measured using force transducers. If we implement force transducer in the joint we may obtain direct information on the effects of coupling upon this joint. In that case the function P_i^* can be made equal to the actual dynamic moment (force) P_i, taking into account "joint dynamics" $\bar{H}_{ii} \ddot{q}^i$

$$P_i^* = P_i - \bar{H}_{ii} \ddot{q}^i$$

In other words, the global control directly compensates for dynamic forces, which are measured by sensors. If we introduce global control in the form (5.2.17) and if we assume that the actuator is of the second order $n_i = 2$, and if we adopt $K_i^G = 1$, then by force feedback we would obtain the total compensation of coupling, i.e. the system would be ideally *decoupled to subsystems* (local servos). By introducing force feedback we achieve that the system behaves as a set of local subsystems among which there are no interconnections (i.e. the subsystems would behave as we have considered in Chapter 3. when we have assumed that, except the i-th joint, all the other joints are locked). Actually, even force feedback cannot completely decouple robotic system: due to an amplitude constraint upon the actuator inputs, the global control by force feedback is also constrained, and therefore for large deviations from the nominal trajectories it cannot completely compensate for the actions of dynamic moments. On the other hand, information from sensors are not perfectly equal to actual dynamic moments since they always include noice.

The control scheme which includes, becides local nominal control and local servo systems, the global control in the form of force feedback is presented in Fig. 5.2. The force feedback, as global control, has certain adventages:

a) *Simple control structure.* By introducing of force feedback we do not complicate significantly the control structure in respect to *decentralized structure* of control (Fig. 4.4). Practically in the joint for which the global control is required, we introduce only one new feedback loop-a feedback by force measured by force transducers.

b) *Minimal computation.* Force feedback does not require additional computation since information from the force transducers gives total information on coupling. It is sufficient to multiply information from force transducer by global gain. (It is also required to submitte from measured moment P_i the value $\bar{H}_{ii}\ddot{q}^i$). This means that such global control does not require additional computation time in a control microcomputer.

c) *Robustness to parameter variations.* Since force transducers measure the total dynamic forces, this information does not depend on the parameters of the robot mechanism and actuators. If some parameters of the mechanism vary, this global control ensures compensation for dynamic forces regardles to parameters variations. In other words, the force feedback guarantees robustness of the control system to variations of parameters of the mechanism (for example, to variation of payload parameters, see Chapter 6), and to uncertainites of robot modelling.

However, the application of force feedback also suffers from certain drawbacks:

a) *Technical problems of force transducers implementation.* To measure directly the forces acting in the joint, we have to build the force transducers directly in a shaft of actuator (reducer), or at some other place on the mechanism. Low stiffness of the force transducers might produce serious problems since it reduces the structural stiffness of the mechanism but it has been shown that these problems can be successfully solved [3, 4].

244

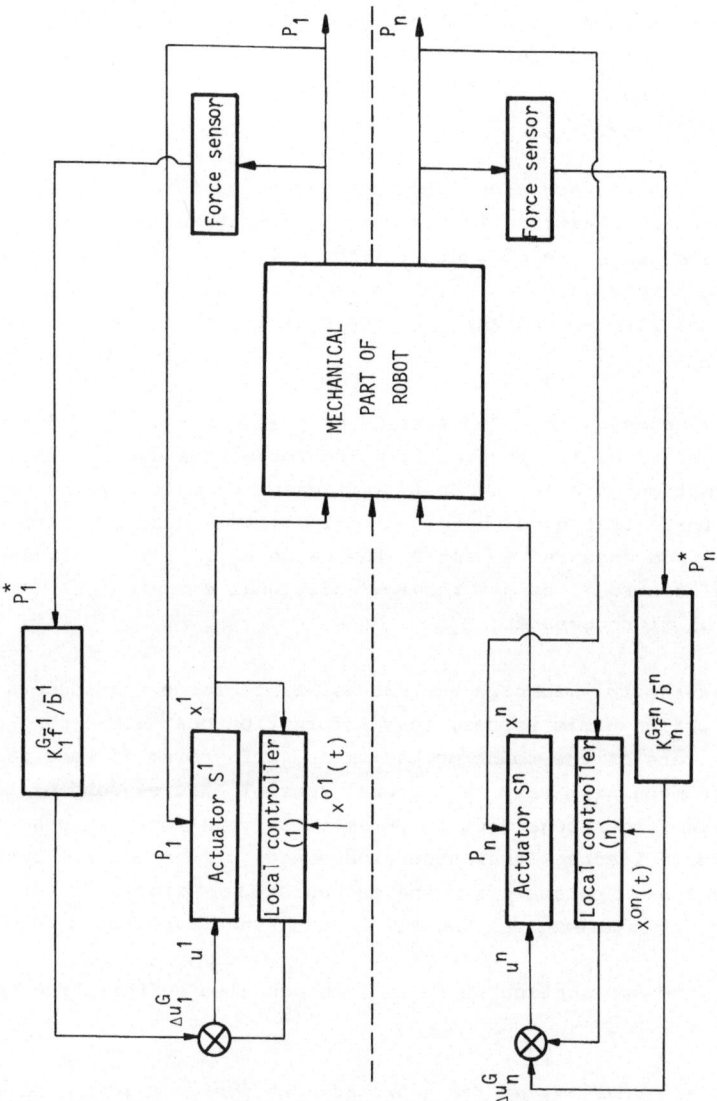

Fig. 5.2. Control scheme including global control in the form of force feedback (trajectory generation level and "joint dynamics" $\bar{H}_{ii}\ddot{q}^i$ are omitted for simplicity)

b) *Price of force transducers.* The high-quality force sensors, which can precisely measure the forces might be relatively expencive. Instead to apply faster (and expencive) microprocessors, in this case we use force transducers the price of which might be also high and increase the total cost of the control system. Therefore, the robot designer has to consider which of these two options is from the techno-economical aspects more acceptable: force feedback, or on--line computation of coupling (see Section 5.2.2).

c) *Noice at force transducers.* The force sensors, as well as all other sensors, in their output information always include certain noice. Most often this noice is negligable and can be ignored, specially if high-quality force transducers are used. However, the problem lies in the fact that these transducers are built directly in the mechanisms shaft which might vibrate due to their elastic modes. These vibrations are transfered to the sensors outputs (force information), and then they are amplified in the control system. This might lead to resonant vibrations of the mechanism. Therefore the global gain K_i^G must be limited, and filters have to be applied to filter the signals from the force sensors [3, 4].

Example 5.2.1. For the robot in Fig. 3.2. we shall show the effects of application of global control in a form of force feedback. Let us assume that we have to ensure tracking of the nominal trajectories presented in Fig. 4.5. If we apply just local programmed control and local feedback loops (local feedback gains are given in Table 4.1. - Example 4.3.2), then tracking of the trajectories is quite unsatisfactory (Example 4.4). Therefore, it is required to apply global control which compensates for effects of coupling. Obviously, it is sufficient to apply global control just in joint 1 and 2. Let us apply force feedback in these joints. If we put force transducers in these joints, then it can be shown that, to stabilize the robot, we have to apply global gains which are at least:

$$K_1^G = 0.5; \qquad K_2^G = 0.5$$

Thus, the global control is applied in the form (5.2.17):

$$\Delta u_1^G = 0.5 \cdot (0.28/0.26)(P_1 - 3.49 \cdot \ddot{q}^1)$$

$$\Delta u_2^G = 0.5 \cdot (5.28/4.96)(P_2 - 0.16 \cdot \ddot{q}^2)$$

where P_i are the dynamic moments in the joints which are measured by force sensors built in the actuator shafts. In Fig. 5.3 the simulation of tracking of nominal trajectories from Fig. 4.5. is presented, if the local nominal control, local servos and the global control in the form of force feedback are applied.

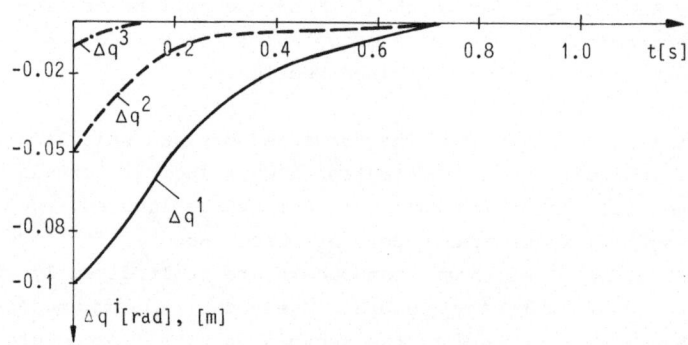

Fig. 5.3. Trajectory tracking for robot in Fig. 3.2. with force feedback as global control

5.2.2. On-line computation of dynamic forces for global control

The second possible implementation of function P_i^* for global control is by on-line computation of dynamic moments in joints, i.e. of the coupling between joints. By potentiometers in the joints (or by some other sensors) we get information on the actual positions of all joints q^i. By tachogenerators we obtain information on the joints velocities $\dot{q}^i$. In this we get information required to compute actual values of matrix of inertia $H(q)$ and of vector of centrifugal, Coriolis and gravity moments $h(q, \dot{q})$. If we forward the information on q^i, $\dot{q}^i$ from sensors to a control microcomputer, it can compute these elements of the dynamic model of mechanism of robot (3.2.2). On the basis of models of actuators (3.2.6) and of the mechanism (3.2.2), we can determine accelerations of the joints $\ddot{q}^i$, or they might be computed by numerical differentation of the joints velocities $\dot{q}^i$, or they can be measured by accelerometers.

In this way a microcomputer computes values of the driving torques $P_i(x)$, or the coupling between the servo systems $\bar{P}_i(x)$. Therefore, a microcomputer generates function $P_i^*(x)$ which is, theoretically, equal to actual coupling.

However, this solution has certain drawbacks. The basic problem related to on-line computation of the coupling lies in complexity of dynamic model of robot. In general case the model of the mechanism dynamics (3.2.2) might be very complex and it requires a large number of adds and multiplies to be performed by a microcomputer in order to compute matrix $H(q)$ and vector $h(q, \dot{q})$. Various computer-oriented methods for computation of dynamic models of robots have been developed. These methods differ in their efficiencies from the standpoint of the amount of computation necessary to compute matrix H and vector h. Here, we shall not consider various methods for computation of dynamic model of robot [5, 6]. Let us mention that the methods for computer generation of dynamic models of robots in symbolic forms are the most efficient for computation of dynamic moments P_i [7].

If the robot is moving fastly, the dynamic moments P_i, i.e. matrix H and vector h, also might vary fastly. Therefore, the microcomputer must compute their values at each 10-20 [ms]. This means that a control microcomputer has to be fast enough to compute new values of matrix H and vector h, (i.e. of driving torque $P_i(x)$), at rate of 10 - 20 [ms]. Since for execution of each computational operation a microprocessor requires some period of time, this means that to compute driving torque P_i a microprocessor might require much longer time than 10-20 [ms]. To achieve *sampling period* which is compatible with dynamic characteristics of the robot, a control microprocessor should be very fast, and that means that it might be relatively expencive. For certain robot structures a number of computational (real) operations, to be performed at each sampling period in order to compute coupling $\bar{P}_i(x)$, might be so high that even the current microprocessors hadrly can achieve it [6]. In such case we may apply several microprocessors *in parallel*, which means that the required computation load has to be distributed over two or more microprocessors. Such solution, obviously causes some new technical problems (for example, how to distribute computations over microprocessors, how to exchange data between microprocessors, how to synchronized microprocessors execution, etc.).

However, it is not always necessary to compute complete (total) dynamic models of robots, i.e. it is not always necessary to compute all components of dynamic moments (forces). As we have explained in Sect. 4.2, the effects of all components of dynamic moments upon performances of servos is not equal. Therefore, it is not necessary to compute all components of dynamic coupling, but in the function P_i^* we may include only those components whose effects upon the system performance is significiant. In other words, the global control might use just *approximative models of dynamics of robot*. Let us consider some approximative models of robots which might be used for the global control [8]:

a) If relatively slow motions of robot are considered, whith which velocity and inertia terms are not significiant, the global control might include only gravity moments, so that:

$$P_i^* = g_i(q) \qquad (5.2.19)$$

Computation of gravity moments usually is simple and if we compensate for these moments, we may eliminate static error in the robot positioning as well the error during trajectory tracking (see Sect. 3.3.5).

b) In Sections 3.3.5 and 4.2. we have considered the effects of variable inertia of the mechanism around each joint $H_{ii}(q)$. These eigen-inertia terms may be compensated for by the global control, if on-line computation of eigen-moments of inertia of mechanism and of corresponding accelerations is implemented at control microcomputer:

$$P_i^* = H_{ii}(q)\ddot{q}^i - \bar{H}_{ii}\ddot{q}^i + g_i(q) \qquad (5.2.20)$$

In this way we may achieve more uniform performance of local servo systems since the effects of variations of moments of inertia of mechanism are compensated for.

c) If we consider fast motions of the robot, with high accelerations and deaccelerations, cross-inertia terms produce significant dynamic load upon the servo systems and, by this, errors in trajectory tracking appear. If we introduce real-time computation of these cross-inertia terms, then we may achieve compensation of their effects. In that case approximative model of the robot dynamics is given by:

$$P_i^* = H_{ii}(q)\ddot{q}^i + \sum_{\substack{j=1 \\ j \neq i}}^{n} H_{ij}(q)\ddot{q}^j + g_i(q) - \bar{H}_{ii}\ddot{q}^i \qquad (5.2.21)$$

Computation of cross-inertia terms often requires a large number of real operations and therefore it should be carrfully studied whether all cross-inertia terms are necessary to be computed.

d) At last, if very fast motions of robot are considered, with which
it is necessary to achieve high accuracy in tracking of trajectories (along entire trajectory, not only when the robot reaches its
final position at the end of the motion), then we have to compensate even for Coriolis and centrifugal forces, which depend on joint velocities. Often the computation of these dynamic forces is
the most complex, and, therefore, we have to pay special attention
to analysis whether it is justified to compute them on-line. If we
include Coriolis and centrifugal forces in the model (5.2.21) we
obtain complete "exact" model of the robot mechanism dynamics:

$$P_i^* = \sum_{j=1}^{n} H_{ij}(q)\ddot{q}^j + h_i(q, \dot{q}) - \bar{H}_{ii}\ddot{q}^i \qquad (5.2.22)$$

As it follows from the above considerations, we have to carrfully study which components of dynamic moments we have to compensate for depending on specific robot and specific task. The adventage of such approach to synthesis of global control lies in the fact that it enables
compensation for only those components of moments which are essential
in particular case. In this way the global control can be simplified
and unenecessary computations in a control microcomputer are avoided. A
decision which approximative model is acceptable for particular robot
and task, might be made by analysis of the robot stability, as explained in Appendix 5.A. In doing this, for the function P_i^* we should use
various approximative models starting from the simplest one (5.2.19) up
to the complete model (5.2.22).

It should be also noted, that all components of dynamic moments do not
require the same computational rate. Some of them vary faster then
others, so it is not necessary to compute them at the same rate. Generally speaking, it is often sufficient to compute dynamic components at
the rate which are from 10 to 20 [ms], and this is for several time less
than servo rate which for industrial robots must be between 1 and 5[ms].
(i.e. the rate at which local servo control must be computed). It is
also necessary to analyze which is the lowest rate at which each component of dynamic moment might be computed.

However, on-line computation of dynamic forces suffers from some other
drawbacks, too. This form of global control assumes that all parameters of the robot are precisely identified and that they do not vary,
which is not true in general case. In industrial applications of robots, this assumption is often valid, since the processes are often

known in advance. However, as we shall consider in Chapter 6, with
current robots parameters of a payload are often not known in advance.
Therefore, the compensation for the effects of payloads upon the robot
dynamics is not simple if we apply on-line computation of dynamic moments. Actually, it is necessary to implement on-line identification
of unknown parameters of a payload. In other words, the global control
by on-line computation of dynamic coupling is not robust to parameters
variations, and to uncertainities of the robot modelling.

The control scheme which includes on-line computation of dynamic moments is presented in Fig. 5.4. It can be seen that the structure of
this control scheme is much more complex than in the case of force feedback. This considerably complicates the implementation of control
and its maintaince, and reduces the reliability of the system.

Example 5.2.2. In Example 5.2.1. we have synthesized global control in
the form of force feedback for the robot presented in Fig. 3.2. Now,
we shall present global control for this robot which applies on-line
computation of the robot dynamics. We shall use approximative model of
the robot dynamics. As in the previous case, we have to ensure accurate tracking of trajectories in Fig. 4.5. We shall apply local servo
systems and local nominal control. By analysis of practical stability
of the robot, the simplest approximative model of the robot dynamics
is determined which can be used for computation of couplings in the
robot joints, and to ensure accurate tracking of the nominal trajectories. It can be shown that the global control might be introduced by on-
-line computation of inertia moments in the first and in the second
joint (for the third joint there is no need to apply global control).
The minimal global feedback gains which ensure robot stabilization in
this case are $K_1^G = 0.8$ and $K_2^G = 0.8$, so that the global control gets
the following form:

$$\Delta u_1^G = 0.8 \cdot (0.28/0.26) \{ [J_{1z} + m_1 \ell_1^{*2} + J_{3z} + m_3 (\ell_1^2 + \ell_3^{*2} + 2\ell_1 \ell_2^* \cos q^2] \ddot{q}^1 +$$

$$+ (J_{3z} + m_3 \ell_3^{*2} + m_3 \ell_1 \ell_3^* \cos q^2) \ddot{q}^2 - 3.49 \ddot{q}^1 \}$$

$$\Delta u_2^G = 0.8 \ (5.28/4.96) \cdot (J_{3z} + m_3 \ell_3^{*2} + m_3 \ell_1 \ell_3^* \cos q^2) \ddot{q}^1$$

The tracking of trajectories in Fig. 4.5. with such global control
(together with the local servos and the local nominal control) is presented in Fig. 5.5.

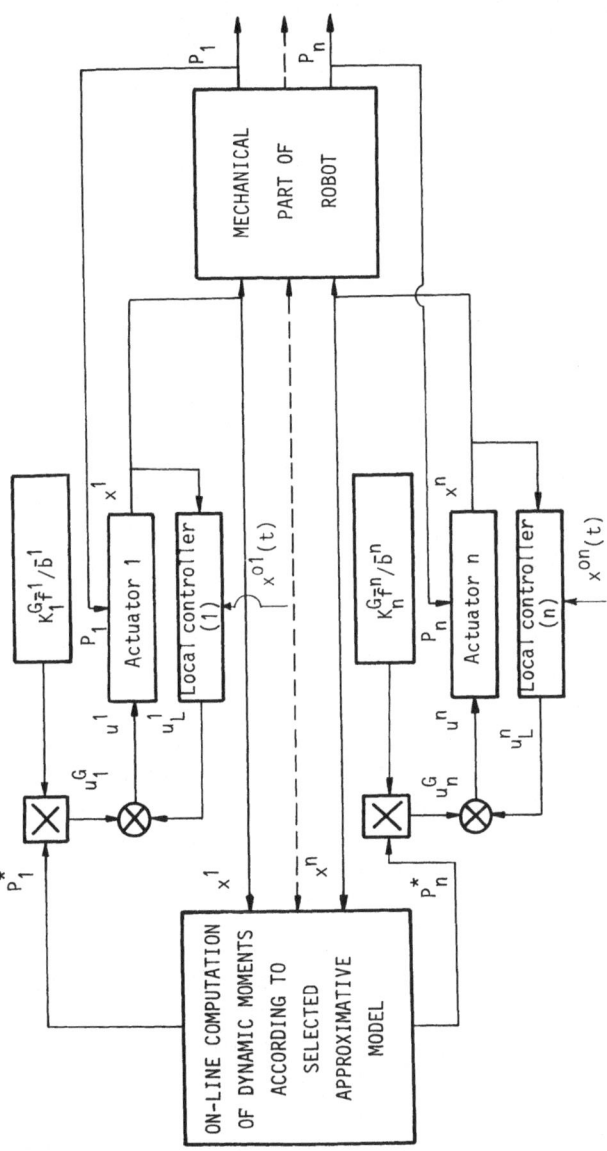

Fig. 5.4. Control scheme including global control via on-line computation of dynamic moments of robot (trajectory generation level is omitted for simplicity)

252

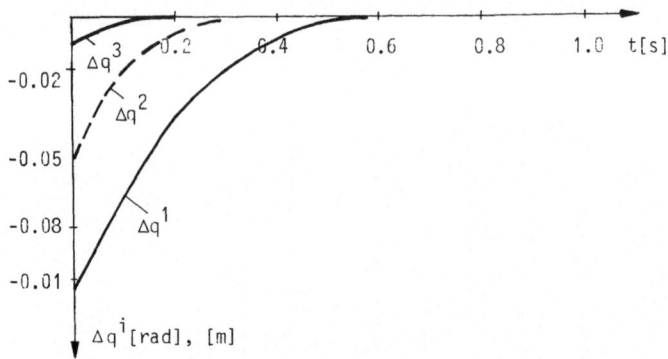

Fig. 5.5. Tracking of nominal trajectories for robot in Fig. 3.2.
with global control in form of on-line computation of
dynamic moments

Exercises

5.1. Why the global control (5.2.13) is approximative? Explain the statement that such control ignores delay in actuator.

5.2. Draw the control scheme for the robot in Fig. 3.2. if the control
considered in Example 5.2.1. is applied. (Local programmed control, local servo feedback loops, global control in the form of
force feedback). Determine the minimal number of real operations
(adds and multiples) that have to be performed at each sampling
interval in order to compute this control (the local nominal control is computed as in Example 3.5).

5.3. Estimate number of microprocessors that have to be applied in parallel in order to implement the control considered in Exercise
5.2, and to achieve a servo rate of 5 [ms] (for computation of local servos, local nominal control and global control) if we use:

a) microprocessors of type INTEL-80-80 (which takes for one floating-point multiply 1.5 [ms], and for one floating-point add
0.8 [ms]).

b) microprocessors of type INTEL-80-87 (which takes for one floating-point multiply 65 [µs], and for one floating-point add

35 [µs]). (Instruction: According to the number of multiplications and additions determined in previous exercise, determine the processor time required for computation of control at one sampling interval, assuming that the processor time is taken only for multiplies and adds (overhead is neglected), and then determine the number of processors required to achieve sampling rate of 5 [ms]).

5.4. For the robot in Fig. 3.2. determine approximative models in forms (5.2.19)-(5.2.22). For each approximative robot determine the number of real operations necessary to compute dynamic forces. Try to minimize these numbers. Assume that accelerations of the joints are given (that they are obtained by differentation of velocities). Assume that for computation of sinus and cosinus it is required 1 multiple and 2 adds.

5.5. Using results obtained in Exercise 5.4. determine the total number of real operations which have to be performed at each sampling interval in order to compute local nominal control, local servos (see Examples 5.2.1. and 5.2.2) and global control in the form (5.2.17) if various approximative models (5.2.19)-(5.2.22) are used. Estimate the number of microprocessors that have to be applied in parallel, for each of four control versions, in order to achieve sampling rate for local servos and local nominal control of 5 [ms] and for global control of 10 [ms]. Assume that we use the same microprocessors as in Exercise 5.3. Compare the results to those in Exercise 5.3.

5.6.* In one of high-level programming language write the programme for implementation of global control for the robot in Fig. 3.2. The global control is in a form: a) of force feedback as in Example 5.2.1 (assuming that the measured values of forces P_i are inputs for programme), or b) of on-line computation of dynamic moments as in Example 5.2.2. - using approximative dynamic models. Try to minimize the number of multiples and adds. (Inputs for programme are actual values of joint angles q^i, velocities $\dot{q}^i$ and accelerations $\ddot{q}^i$, and outputs are global control signals).

5.7.* For the robot in Fig. 3.2. determine expressions for calculation of joints accelerations $\ddot{q}^i$ as function of q^j, $\dot{q}^j$, j=1,2,3, using the second order models of actuators (n_i=2) and approximative

dynamic models (5.2.19)-(5.2.22) - see Exercise 5.4. If we use
this procedure for determination of joints accelerations, then
determine for how much increases the number of real operations
necessary to on-line compute dynamic moments by approximative models (5.2.19)-(5.2.22), in respect to the previous case (considered in
Exercise 5.4) in which we have assumed that accelerations are given. (Given accelerations means that they are either directly measured by accelerometers, or they are computed by numerical diffe-
rentation of velocities $\dot{q}^i$ according to equation $\ddot{q}^i \approx (\dot{q}^i(t+T_D)$ -
$\dot{q}^i(t))/T_D$ - where T_D is the sampling interval). In which way we
can reduce the number of real operations necessary to compute approximative models, if the accelerations are obtained by models of
actuators and mechanism. Solve the above problem (determination of
joints accelerations on the basis of actual joints angles and velocities using the second order models of actuators and approximative models of dynamics of robot mechanism) in general case, for
a robot of an arbitrary structure with n joints.

5.3 Computer-Aided Synthesis of Robot Control

From our previous discussion we may conclude that the selection of control law at the executive control level is strongly dependent on type
and structure of a robot and on a task which is assigned to it. If we
consider simple control tasks which can be reduced to position control
or to tracking of slow trajectories (requiring no high accuracy), then
we may adopt only local servo systems (eventually, we may also introduce local feedforward terms in a form of local nominal control). In
such cases effects of dynamics of the entire robotic system are weak
and therefore, we may ignore them. This approach has been applied in
control of many robots at the market. However, if very accurate tracking of fast trajectories is required, then the effects of dynamic forces upon the performances of servo systems might be strong, and then we
have to apply *dynamic* control which compensates for these forces. We
have seen that the dynamic control might be introduced in various ways:
by nominal centralized control, by global control in the form of force
feedback, or via on-line computation of dynamic forces using various
approximative models of robot dynamics. In dependance on complexity of
the robot structure and on requirements which are imposed before the
robot, we may introduce dynamic control of various degree of complexity.
For example, the global control need not to be applied in all joints.

In some joints just gravity moments have to be compensated for, in the other inertia moment require to be compensated for, etc. The aim is to determine the simplest dynamic control which still satisfy requirements imposed before the robot in particular tasks. However, this problem is not simple since it demands analysis of the robot dynamic performances. In Chapter 4 we have shown how we can analyze the stability of the robot and in Appendix 5.A. we shall present how we can synthesize global control by robot stability analysis. Obviously, such analysis is related to large amount of computations, which means that we have to use digital computer. Therefore, the computer-aided synthesis of control for manipulation robot is needed.

Application of computers for control synthesis of large-scale dynamic system is broadly encountered, today. Here, we shall briefly describe a *software package for computer-aided synthesis of control* for manipulation robots, which is based upon the above considered methods for control synthesis[*]. Since in the previous text we have explained in detail theoretical background for analysis of the robot performance and for synthesis of dynamic control, here we shall just briefly describe basic structure (flow-chart) of this software package [9].

The software package enables synthesis of the executive control level for robots of arbitrary structure with various number of joints and actuators. The user of the package has to impose data on robot and the package helps him to synthesize the control. Namely, the control synthesis is *interactive*. Instead of completely automatic synthesis of control, here interactive synthesis of control is applied in which the computer only helps the designer of the robotic system to select the most appropriate control law for the specific robot and task. The software package might be regarded as an assistive device which enables the designer to use his experience in the control system design [10].

The package is modular, i.e. it consists of several modules which the user can easely combine. We shall briefly describe the basic modules. The flow-chart of the package is presented in Fig. 5.6.

---

[*] The package has been developed in Institute Mihailo Pupin, Beograd.

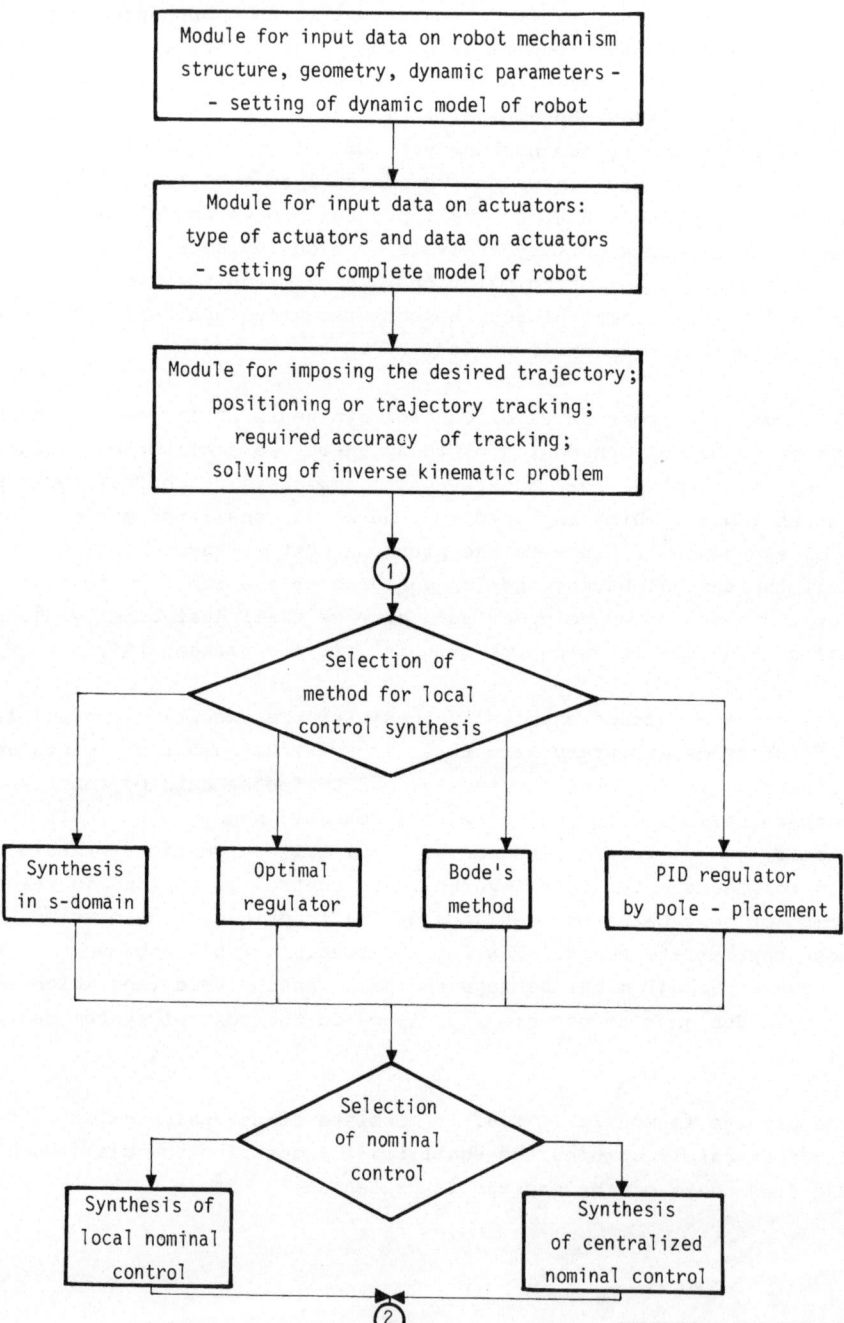

Fig. 5.6. Flow-chart of software package for control synthesis of robots

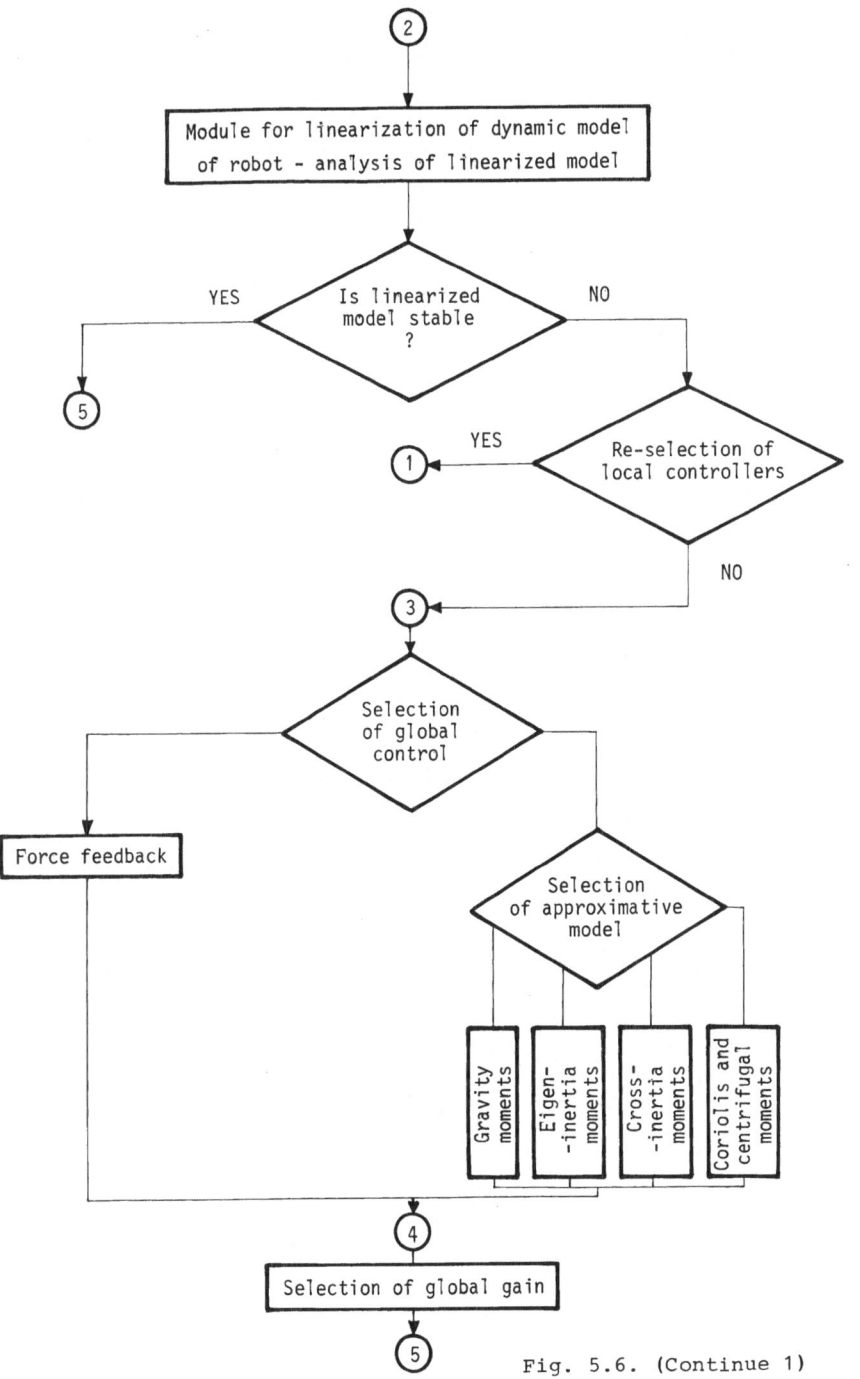

Fig. 5.6. (Continue 1)

258

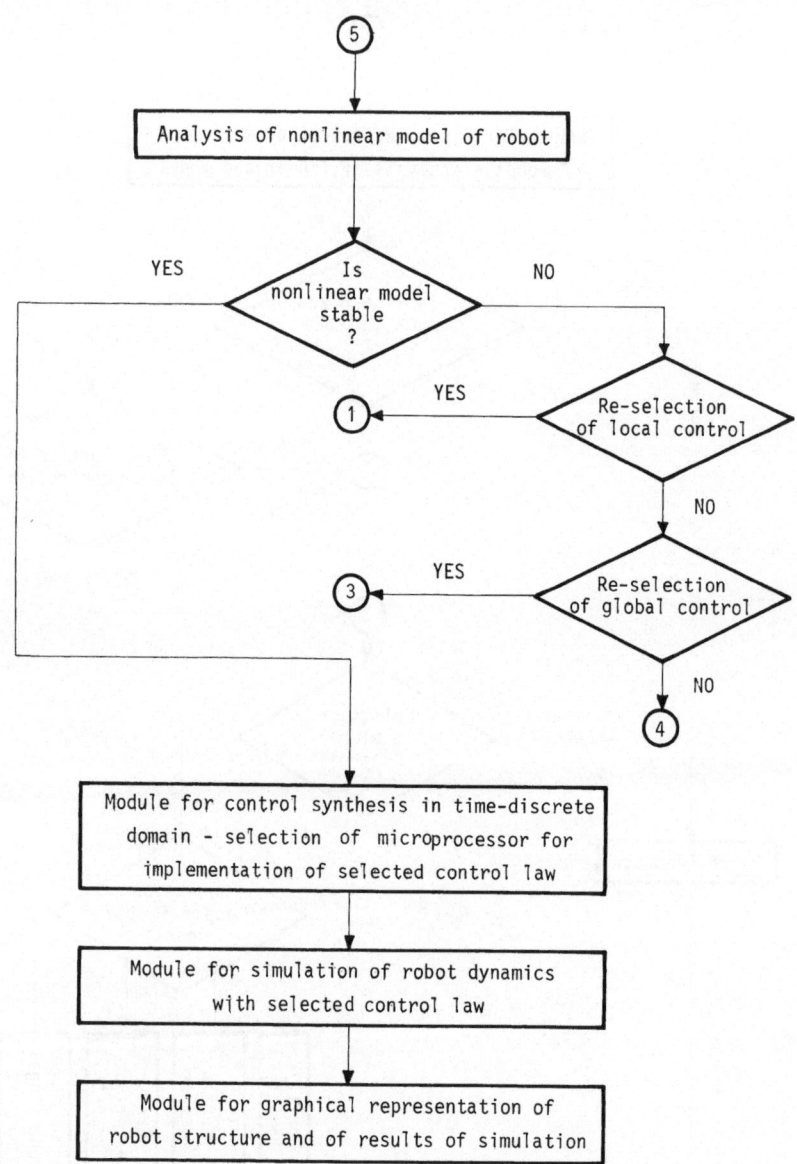

Fig. 5.6. (Continue 2)

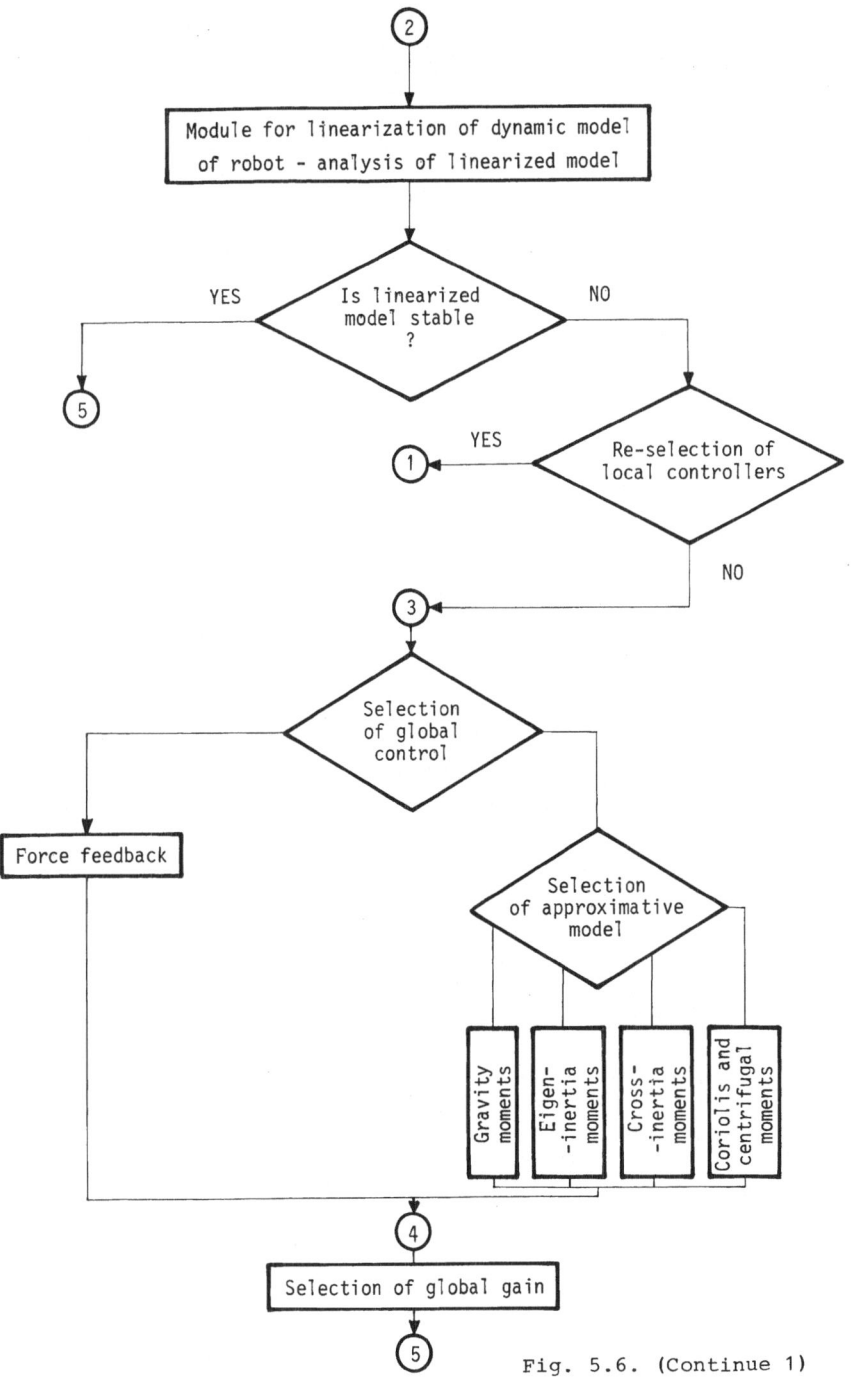

Fig. 5.6. (Continue 1)

258

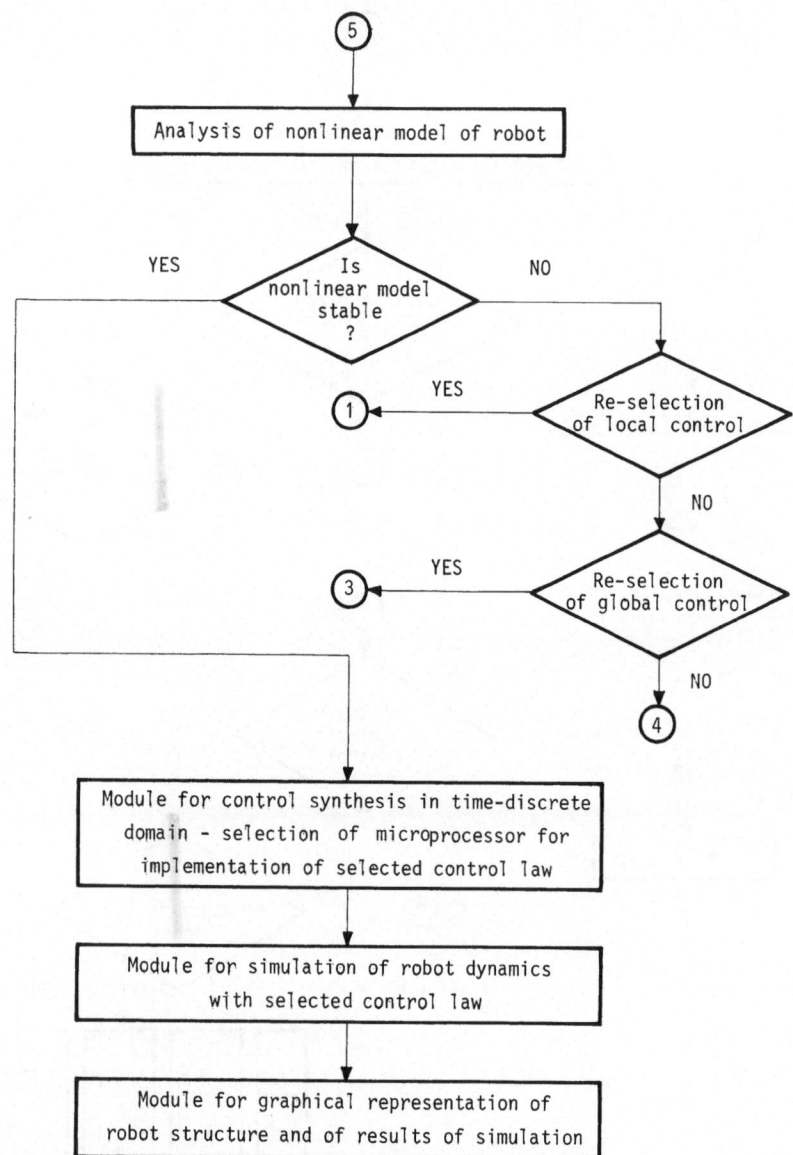

Fig. 5.6. (Continue 2)

1. Module for input data on robot mechanism. This module enables the user to impose data about geometry and structure of the robot mechanism for which he wants to synthesize the control. The user has to impose data on lengths of the links, orientation of the joint axes, types of the joints (rotational or linear). The user also has to specify data on masses and moments of inertia of the links. Based on these data, the software package automatically sets kinematic and dynamic models of robot, which are used in the synthesis of control. The user need not to take care directly on these models, but the package uses them automatically in control synthesis whenever they are needed. In this particular package, the procedure for automatic setting of dynamic models of robots based on D'Alambert's principle has been used [11].

2. Module for input data on actuators. In this module the user has to select type of actuator which he wants to apply for his robot (or, which has been applied to drive already designed robot), and the basic data on actuators (moment of inertia of the rotor, moment and electromotor constants of the motor, electrical resistance of the rotor circuit, if D.C. motors are applied as actuators). Based on these data the package automatically sets the models of actuators. Package includes D.C. electro-motors and hydraulic actuators.

3. Module for imposing the desired trajectory. The user has to specify the task which the robot has to implement, i.e. the user has to choose whether the robot has just to be positioned in various positions in its work space, or whether it should move "point-to-point", or if it should track the desired trajectory. The user has to impose data on the desired positions, or trajectories of the gripper of the robot. The trajectories might be imposed in various ways. Based on these data, the algorithm, using the kinematic model of the robot, determines the corresponding positions or trajectories of the robot joints, i.e. it solves the so-called inverse kinematic problem (see Chapter 2). In this way the desired (goal) positions, or trajectories of the robot joints are defined, and the executive control level has to implement them.

4. Module for synthesis of local servo systems. The package can synthesize the local servo systems using various methods in accordance with the user option. The package selects servo feedback gains for all joints of the robot.

5. <u>Module for nominal control synthesis</u>. The package enables automatic computation of the nominal driving torques which correspond to the imposed paths of the robot gripper, and synthesis of the nominal programmed control based on the complete dynamic model of the robot. The package also includes computation of the local nominal control on the basis of local subsystems models.

6. <u>Module for synthesis of global control</u>. The user is able (if he wishes so) to synthesize the global control using this software package. The global control has to compensate for dynamic coupling between the joints and, by this, to increase accuracy of tracking of desired trajectories. The global control can be introduced either as force feedback, or as on-line computation of dynamic moments. In the latter case the package helps the user to select the approximative model of the robot dynamics which can be used for computation of the global control. Namely, the package enables iterative selection of the simplest approximative model by which imposed requirements can be satisfied [10].

7. <u>Module for analysis of robot performance</u>. This module analyzes the the stability of the robot with the selected control law. First the stability of the linearized model of the system is analyzed, and then, the stability of the entire nonlinear model of the robot is verified. The software package automatically linearizes the dynamic model of robot and by testing eigen-values of the linearized model with the selected control, it examines whether the linearized model of the system is stable or not. If the linearized model is stable, then the nonlinear model is examined. The analysis of the stability of the nonlinear model of the robot is performed by aggregation-decomposition methods for analysis of stability of large-scale systems (Appendix 4.A). However, the analysis is performed automatically, and, thus, the user need not to learn much about these rather complex algorithms. For the user it is quite sufficient to obtain an answer whether the robotic system is stable with the selected control law, or not. If the answer is positive, the selected control can be accepted. Otherwise, he has to re-select either local servos, or global control (see flow-chart in Fig. 5.6).

8. <u>Module for simulation of robot dynamics</u>. The module for *simulation* of robot dynamics enables verification of robot behaviour with various control laws. Using this module, the designer of the robot obtains precise intight into the performance, quality and speed of the robot which helps him to make a final decision in selecting the

most appropriate control law. The package also permits simulation of only local isolated subsystems (joints) in order to get better insight into effects of dynamic interactions between the joints of the robot.

9. <u>Module for synthesis of control in time-discrete domain</u>. This module enables synthesis of control in time-discrete domain. Such control synthesis is necessary since the control of current robotic systems is implemented exclusively by microprocessors. This means that the feedback gains have to be synthesized taking into account a *sampling period*, i.e. the period which is taken by microprocessor to compute new values of control signals according to the selected control law. This module also helps the user to select microprocessor achitecture which is required for implementation of the synthesized control (taking into account computational capability of microprocessors). The control synthesis in time-discrete domain and microprocessor implementation of various control laws are out of the scope of this book.

Besides above listed modules, the package includes some auxiliary modules, for example module for graphic presentation of a robot structure and graphic representation of results of trajectory synthesis, nominal control synthesis, simulation etc.

At the end of this brief description of the package we have to make two notes. As we have explained before, in stability analysis of the robot with the selected control, first the linearized model of the robot is analyzed, and then the nonlinear model is analyzed. One of the reasons for such approach is that analysis of the linearized model including automatic linearization of dynamic model of the robot consumes much less computation time than the analysis of the nonlinear model. The analysis of the stability of the nonlinear model (due to estimation of dynamic coupling between joints in defined surroundings of nominal trajectory - see Appendix 4.A) might, in general case, require relatively long computation time. In order to increase speed of computer-aided analysis of the robot performances, first the fast analysis of the linearized model is performed. The analysis of the linearized models gives sufficient, but not necessary conditions of the system stability. This means, that if the linearized model is stable (according to adopted definition of stability), then we should proceed to stability analysis of the nonlinear model. If the analysis of the stability of the nonlinear model of the robot gives positive answer, than we can

guarantee that the robot is stabilized by the selected control law.
However, if the linearized model is not stable, there is no need to
analyze the nonlinear model, since, in that case, the robot is obviously
unstable. In that case, we have to re-select control (or, feedback
gains) and repeat analysis of the linearized model. This procedure
should be repeated until (desired) stability of the linearized model
is achieved. After that we have to proceed to the analysis of nonlinear model. We have to keep in mind that the analysis of the stability
of the nonlinear model gives sufficient but not necessary conditions
for the system stability (see Appendix 4.A), and if the answer of the
computer is that the stability tests are not fulfilled, we must not
state that the robot is not stable. In other words, the user has to combine
the results of stability analysis of the linearized and of the nonlinear
models, using his own experience, and in that way to get an answer
whether or not the robot is stabilized by the selected control, or re-
-selection of control law and parameters has to be done.

The second note concerns the simulation of the robot dynamics. The
simulation of the robot reduces to numerical integration of dynamic
model of robot (i.e. of the set of nonlinear differential equations
(3.2.27)). For the numerical integration we may use various algorithms,
which can be found in literature. The most simple procedure, (but also
the least accurate method) is *Euler's method*, which reduces to computation of the next system state (the state of the system at the next
sampling interval $t+\Delta T$) on the basis of the state at the instant t,
according to simple equation:

$$x(t+\Delta T) = x(t) + \Delta T \cdot \dot{x}(t) = x(t) + \Delta T \cdot [\hat{a}(x(t)) + \hat{B}(x(t))N(u(t))] \qquad (5.3.1)$$

where ΔT is the *integration interval* (i.e. the time interval at which
we compute a new state). For the given instant state $x(t)$ we have, in
accordance to a selected control law, to compute values of the control
signals, i.e. the values of the elements of the input vector $u(t)$.
Then, we have to compute expression at the right-hand side of the equation (3.2.27), and, then, based on expression (5.3.1) to compute the
state in the next sampling interval $x(t+\Delta T)$. Then, the procedure is
repeated at the next interval. Obviously, the user must impose the
initial state $x(0)$, and the integration interval ΔT (which must be
sufficiently short to ensure that the numerical integration (5.3.1)
gives state trajectory which is close to "actual" trajectory of the
system). This most simple procedure for numerical integration of the
differential equations (the model of the robot dynamics) suffers from

drawbacks regarding numerical accuracy, and therefore it is often necessary to apply more complex (and less approximative) procedures for numerical integration.

Exercise

5.8.[*] Write in a high-level programming language a programme for simulation of the dynamics of the robot in Fig. 3.2. if local nominal control, local servo systems and global control in the form of on--line computation of inertia moments in the first and in the second joints are applied (as in Example 5.2.2). Repeat simulation results presented in Example 5.2.2. Use the programmes written in Exercises 4.16 and 5.6.

5.4 Computed Torque Method for Robot Control Synthesis

To this point we have considered the dynamic robot control based upon the *decentralized approach* to control [12]. In this approach, the local servo systems at particular robot joints are synthesized first. Such control has a *decentralized structure:* each of the local servo systems possesses information only about the state of its own subsystem (actuator and joint) which it controls. As we have already seen, to ensure the accurate positioning and tracking of fast trajectories, such decentralized control is not always sufficient, so that it is necessary to introduce the control which takes into account the dynamics of the system as a whole. For this purpose, certain control signals have been "added" to the decentralized control structure to compensate the effect of the coupling between the subsystems. Such control, called the *nominal programmed control* is calculated on the basis of the complete, centralized robot model. The global control using either force feedbacks, or the on-line calculated moments has also the role to compensate the dynamic forces acting as the coupling between the subsystems. Both kinds of dynamic control are the deviation from the decentralized structure, because the control signals for the particular actuators are calculated as a function of states of all subsystems (not only of the corresponding subsystem). Therefore, such control possesses a *centralized structure.* However, the main aim in this approach is to retain, to the most possible extent, the decentralized structure, and

introduce the global, *cross* feedback loops only there where it is necessary. In doing so, as we have already shown (Section 5.3) the starting
point is always a decentralized structure whose main features are simplicity and reliability, and the centralized control components are
introduced later on.

Here, we shall consider an opposite approach: starting from the global
system model, the control is synthesized with the *centralized structure* and then, the unnecessary feedbacks are eliminated. Such a centralized approach is theoretically consistent but is much more complicated
and might lead to complex control laws.

It should be pointed out that the majority of the present day commercial robots possess a decentralized structure: each joint is controlled by a separate local servosystem. Centralized control schemes are
rarely used with robots because the majority of the commercial robots
have not been intended for solving *dynamic tasks* (accurate tracking of
fast trajectories), for which the centralized structure is justified.

The *"computed torque method"* or *"inverse dynamics"* method is one of the
centralized approaches to the robot control synthesis. The idea is to
include, in a direct way, the mathematical model of the robot dynamics
into the control law [13-15].

For the simplicity sake, let us consider first the model of the dynamics of the mechanical robot part, while the models of the actuators
will be introduced afterwards. The dynamics of the mechanical part of
the robot is described by the model in the form of (3.2.2)[*)]:

$$P = H(q)\ddot{q} + h(q, \dot{q}) \tag{5.4.1}$$

To ensure tracking of the given trajectories of the robot joints given
by $q^{oi}(t)$, let suppose the control system generates the driving torques by calculating them from the following control law:

$$P(t) = H_s(q)[\ddot{q}^o(t)+K_1(q(t)-q^o(t))+K_2(\dot{q}(t)-\dot{q}^o(t))]+h_s(q, \dot{q})$$

$$\tag{5.4.2}$$

[*)] For an explanation of the notations see Section 3.2.1.

where $H_s(q)$ is the inertia matrix of dimensions $n \times n$, calculated on the basis of the actual (measured) values of the joints angles q, $h_s(q, \dot{q})$ the vector of Coriolis', centrifugal and gravitational moments, calculated on the basis of the actual values of angles q, and velocities $\dot{q}$, K_1 is the matrix of the position feedback gains, of order $n \times n$, and K_2 is the matrix of velocity feedback gains, $(n \times n)$. The control system (computer), on the basis of sensory information about positions q and velocities $\dot{q}$ of the joints, should calculate H_s, h_s and P from (5.4.2). When the torques thus calculated are realized at the robot joints, it is obtained (by introducing (5.4.2) into (5.4.1)):

$$H_s(q) [\ddot{q}^o(t) + K_1(q(t) - q^o(t)) + K_2(\dot{q}(t) - \dot{q}^o(t))] +$$

$$+ h_s(q, \dot{q}) = H(q)\ddot{q} + h(q, \dot{q}) \tag{5.4.3}$$

If the calculation of the matrix H_s and vector h_s is perfectly accurate, then

$$H_s(q) = H(q), \ h_s(q, \dot{q}) = h(q, \dot{q}) \tag{5.4.4}$$

On the basis of (5.4.3) and (5.4.4) we obtain:

$$\ddot{q}^o(t) + K_1(q(t) - q^o(t)) + K_2(\dot{q}(t) - \dot{q}^o(t)) = \ddot{q}(t) \tag{5.4.5}$$

If we introduce the vector of deviation of the actual robot coordinates $q(t)$ from the nominal trajectory $q^o(t)$, $\Delta q(t) = q(t) - q^o(t)$, then (5.4.5) can be written as

$$\Delta\ddot{q}(t) = K_1\Delta q(t) + K_2\Delta\dot{q}(t) \tag{5.4.6}$$

If the gains K_1 and K_2 have been chosen such to ensure the solution of the system of differential equations (5.4.6) is asymptotically stable, the actual robot coordinates $q(t)$ will asymptotically approach the nominal trajectories $q^o(t)$. In other words, if K_1 and K_2 are chosen to be such that the linear model (5.4.6) is asymptotically stable, it will also guarantee an asymptotic stability of the nonlinear robot model around the nominal trajectory $q^o(t)$.

In order to ensure the linear model (5.4.6) is asymptotically stable, the gains K_1 and K_2 can be selected in various ways. One of the possible solutions is to select these gain matrices in the diagonal forms,

$K_1 = \text{diag}(K_1^{ii})$, $K_2 = \text{diag}(K_2^{ii})$. Then, the model (5.4.6) splits into a set of n independent second-order differential equations:

$$\Delta \ddot{q}^i(t) = K_1^{ii}\Delta q^i(t) + K_2^{ii}\Delta \dot{q}^i(t), \qquad i=1,2,\ldots,n \qquad (5.4.7)$$

The choice of gains K_1^{ii} and K_2^{ii} is now simple. It should be ensured the roots of the equation (in the s-domain)

$$s^2 - K_2^{ii}s - K_1^{ii} = 0, \qquad i=1,2,\ldots,n \qquad (5.4.8)$$

are in the left half of the complex plane[*]. Instead of the asymptotic stability of the model (5.4.6), (i.e. of the decoupled equations (5.4.7)), an exponential stability may be required: it suffices the solutions of equations (5.4.8) be in the complex plane on the left to the straight line $\text{Re}(s) = -\alpha$, where α is the required degree of exponential stability. In that case, the nonlinear robot model is also exponentially stable around the nominal trajectory. It is easy to show that, in this way, the practical robot stability can also be guaranteed.

The control scheme corresponding to the control law (5.4.2) is presented in Fig. 5.7. It is quite obvious that this scheme represents a combination of the control system with closed feedback loops (with respect to q and $\dot{q}$) and nonlinear control signals calculated on the basis of the nonlinear robot model. In this scheme, the compensation of the variable gravitational, centrifugal and Coriolis forces is ensured. The feedback gains are directly adjusted in accord with the changes in matrix H(q), i.e. in accord with the change of the moments of inertia of the mechanism; a precompensating signal is introduced with respect to the nominal acceleration $\ddot{q}^o(t)$, to compensate for the delay along the nominal trajectory.

The main problem in the implementation of the dynamic control (5.3.2) is in the requirement that the complete model of the robot dynamics is calculated on-line, which, as we explained in Section 5.2.2, is generally very difficult to attain, even by emploing very fast microprocessors. Here, similar to global control, instead of an exact calculation of the matrix H_s and vector h_s, different approximative models can be used. All the approximate models of the robot dynamics (5.2.19)

---

[*] Obviously, it is desirable to require the roots are real, so that the damping ratio would be $\xi \geq 1$.

- (5.2.22) that have been mentioned in relation to the on-line calculation of robot dynamics in the global control can also be applied here. For example, if the approximate dynamic model (5.2.20) is adopted,
which includes only the diagonal elements of the matrix $H_s(q)$ (the own
inertia moments) and gravitational moments, then, the calculation of
the driving moments is reduced to:

$$P_i(t) = H_{ii}(q)[\ddot{q}^{oi}(t) + K_1^{ii}(q^i(t)-q^{oi}(t)) +$$

$$+ K_2^{ii}(\dot{q}^i(t) - \dot{q}^{oi}(t)] + g_i(q) \qquad (5.4.9)$$

In this way, the calculation can be substantially simplified, though,
for certain robot types, it may be still complex. On the other hand,
in each specific case (for the specific robot type and its structure,
as well as the specific class of task) it should be examined if such
simplified control can satisfy the required stability conditions (as
relations (5.4.4) do not, obviously, hold any more).

In order to avoid too complicate calculation of the matrix H_s and vector h_s in real time, the elements of matrices H and C, and gravitational moments g_i can be calculated in advance (off-line) for the different values of angles q, and stored in a computer [16]. Here, C represents the matrix of the centrifugal and Coriolis effects $C = C_{jk}^i$ - see
the model (3.2.1). During the robot's motion, for the actual instantaneous values of joints angles q, the corresponding values for H_s, C
and g_i are taken from the memory and, on the basis of the measured
velocities $\dot{q}$, vector h_s is calculated:

$$h_s(q, q) = \dot{q}^T C\dot{q} + g(q) \qquad (5.4.10)$$

where $g(q) = [g_i(q)]$ is the vector of gravitational moments (n×1). In
this way, the amount of calculation in real time is substantially reduced (it is only necessary to calculate (5.4.10) and multiply the
stored matrices H_s by the term in the square bracket in (5.4.2)). This
approach has all the shortcomings of the off-line calculation of nominal control. First of all, it requires a large memory capacity. The
scheme for realization of such control is shown in Fig. 5.8.

Apart from the problems related to large amount of calculation, the
computed torque method suffers also from other shortcomings. This is
related to the problem of the robustness of such control to variations

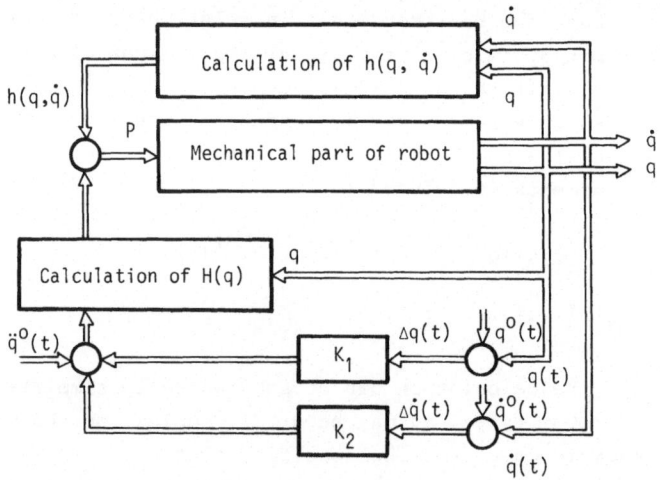

Fig. 5.7. Scheme of control by computed torque method (acc. [13])

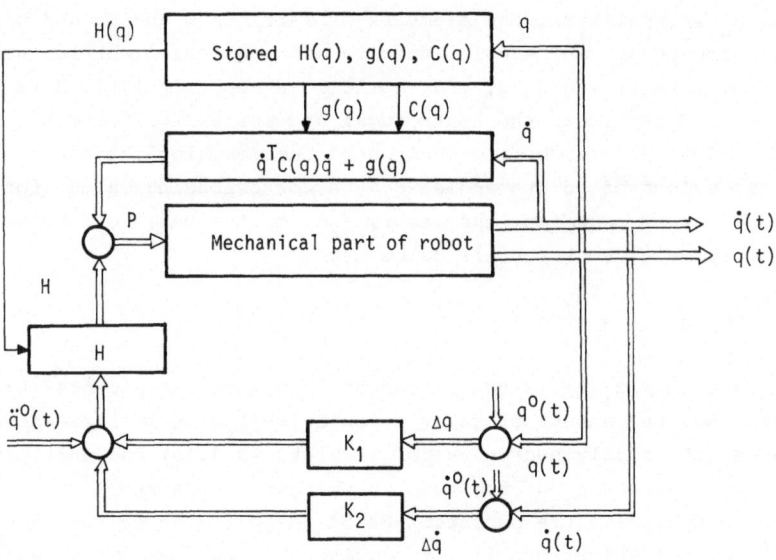

Fig. 5.8. Control scheme (acc. [16])

of the robot (and payload) parameters. The assumption in this control synthesis has been that all the robot parameters were ideally known in advance - see (5.4.4). If these parameters are changed it is necessary to examine robustness of such a solution. The robustness investigation is complicated by the complex centralized structure of the control. It has been shown that this control scheme is rather sensitive to both pa-rameter variations and to model uncertainities. Namely, it is evident that the control scheme is developed under assumption of perfect model-ling of the robot and actuators dynamics. Due to presence of unmodel-led high frequency modes of the systems (i.e. flexibility of the links, joints actuators and reducers which are not included in the rigid body model of the mechanism and of the actuators) the presented control scheme is valid only if the feedback gains are properly adjusted to prevent excitation of these high frequency modes. Therefore, the feed-back gains must not be too high, but this might cause that the system becomes sensitive to parameter variations (since robustness to parame-ter uncertainties is generally achieved by high feedback gains, i.e. by high control bandwidth). This method can be applied for the synthe-sis of adaptive control, but its realization is complex [17].

The main problem in this approach is in the complex, a priori chosen, centralized control structure which is making difficult the synthesis and choice of a simplest acceptable solution. Because of that, in the centralized control, it is much more difficult to develop a synthesis procedure which would utilize the robot designer's experience (see Section 5.3).

Finally, let us note that in considering the "computed torque" method we have neglected the dynamics of the actuators. If the models of the actuators have been chosen in the form of (3.2.11), (second-order ac-tuators models) it is easy to show that the input signals to the actu-ators to realize the driving torques (5.4.2) are $(J_R^i = J_M^i N_v^i N_m^i)$:

$$u^i = \frac{r_R^i}{C_M^i N_m^i} \{ (J_R^i I_n^i + H_s^i(q)) [\ddot{q}^o(t) + K_1(q - q^o(t)) + K_2(\dot{q} - \dot{q}^o(t))] +$$

$$+ h_s^i(q, \dot{q}) + (F_v^i + \frac{C_M^i C_E^i N_v^i N_m^i}{r_R^i}) \dot{q}^i \} \qquad (5.4.11)$$

where H_s^i and I_n^i are the i-th rows of matrix H_s, i.e. of the unit mat-rix I_n, h_s^i the i-th element of vector h_s, while the other symbols,

related to the D.C. electro-motors, have been explained for equations
(3.2.4) and (3.2.5).

It should be noticed that the problem of variable and unknown parameters in the "computed torque" method can be overcome by using the *robust control* [18] which represents a combination of the computed torque approach and the so-called *"sliding mode"* [19]. However, this attempt has not been developed sufficiently to be realized in practice,
and it is still at the stage of theoretical investigations.

Example 5.4. For the manipulator in Fig. 3.2, we shall write the expression for the driving torques in the "inverse dynamics" method
(5.4.2). The gains K_1^{ii} and K_2^{ii} have been chosen such that the roots of
(5.4.8) are both -5 (for all three joints, i=1,2,3). The obtained feedback gains are $K_1^{ii} = -25$, $K_2^{ii} = -10$, i=1,2,3. The driving torques
should be calculated (using the model of the robot dynamics (3.2.3))
from the following expressions:

$$P_1(t) = (J_{1z} + m_1 \ell_1^{*2} + m_3 \ell_1^2 + J_{3z} + m_3 \ell_3^{*2} + 2m_3 \ell_3^* \ell_1 \cos q^2) \cdot$$

$$\cdot [\ddot{q}^{o1}(t) - 25 \cdot \Delta q^1(t) - 10\Delta \dot{q}^1(t)] + (J_{3z} + m_3 \ell_3^{*2} + m_3 \ell_3^* \ell_1 \cos q^2) \cdot$$

$$\cdot [\ddot{q}^{o2}(t) - 25\Delta q^2(t) - 10\Delta \dot{q}^2(t)] - m_3 \ell_1 \ell_3^* \dot{q}^2 (\dot{q}^2 + 2\dot{q}^1) \sin q^2$$

$$P_2(t) = (J_{3z} + m_3 \ell_3^{*2} + m_3 \ell_3^* \ell_1 \cos q^2) [\ddot{q}^{o1}(t) - 25\Delta q^1(t) - 10\Delta \dot{q}^1(t)] +$$

$$+ (J_{3z} + m_3 \ell_3^{*2}) [\ddot{q}^{o2}(t) - 25\Delta q^2(t) - 10\Delta \dot{q}^2(t)] +$$

$$+ m_3 \ell_3^* \ell_1 (\dot{q}^1)^2 \sin q^2$$

$$P_3(t) = m_3 [\ddot{q}^{o3}(t) - 25\Delta q^3(t) - 10\Delta \dot{q}^3(t)] - m_3 g$$

Exercises

5.9. Prove that expression (5.4.11) represents the voltages at the inputs of D.C. motors which would generate the driving torques
(5.4.2) if the models of the motors had been given in the form
(3.2.6), (3.2.11).

5.10. For the manipulator in Fig. 3.2, the driving torques using the "computed torque" method are given as in Example 5.4. Write the expressions for the input signals to the actuators (5.4.11) corresponding to these driving torques for particular robot. If the approximative models (5.2.19) - (5.2.22) are used instead of the exact model of the robot dynamics, write the expressions for the driving torques (i.e. inputs to the actuators) using the "computed torque" method. Minimize the number of numerical operations which should be carried out.

5.11. For each of the set of voltage inputs to the actuators (5.4.11) written in the preceeding exercise for the robot in Fig. 3.2. determine, using the approximate models (5.2.19) - (5.2.22), the number of operations (multiplies and adds) needed for their calculation (for the given q and $\dot{q}$). Determine the minimal number of microprocessors which should work in parallel to calculate these input signals (for all three actuators) every 10 [ms], if the microprocessor used is:

a) INTEL-80-80 (one addition operation lasts 0.8 [ms], and one multiplication 1.5 [ms]),

b) INTEL-80-87 (addition 35 [μs], multiplication 65 [μs]). Suppose the processor time is used only for the operations of addition and multiplication (the calculation of a sine or cosine function is equivalent to two adds and one multiply operation).

5.12. For the control laws in Exercise 5.10, determined on the basis of the approximate models (5.2.15) - (5.2.22) for the robot in Fig. 3.2, draw the control scheme taking into account the model of the actuators.

5.13.* Write in a high programming language the programme for the on--line calculation of the input signals (5.4.11) to the actuators of the robot in Fig. 3.2. which would correspond to the driving torques in Example 5.4. (using the "computed torque" method). The programme input data are the robot actual and desired cordinates q^i, q^{oi} and velocities $\dot{q}^i$, $\dot{q}^{oi}$ and desired accelerations $\ddot{q}^{oi}$, and the output data are the input signals to the actuators u^i, i=1, 2,3 (the data about the robot are also inputs).

5.14.[*] Explain what is the similarity and the difference between the control law (5.4.11) according to the computed torque method and the control law presented in Section 5.2.2. which is composed of the local nominal control, local servo systems feedbacks, and global control in the form of on-line calculation of the complete model of the robot dynamics.

5.15. If in Example 5.7. the poles of equation (5.4.8) are set (both) at the position -20, instead of -5, what would be the gains K_1^{ii} and K_2^{ii}, i=1,2,3. What are the real gains in the feedback loops with respect to the position and velocity (the gains from the error Δq^i and $\Delta \dot{q}^i$ to the actuators inputs u^i). Are these gains acceptable, and how the poles of equation (5.4.8) should be chosen to constraint these gains?

5.16. Explain what would be the role of variation of the gains in the feedback loops with respect to the position and velocity as a function of the inertia of the mechanism. What would be achieved in respect of the uniformity of functioning of the system?

5.5 Cartesian Based Control of Robot

As we have already pointed out (Chapter 1), the task assigned to the robot is usually described in the so-called external (Cartesian) coordinates. In other words, the task is defined by assigning the path (or position) to be realized by the robot hand together with the workpiece. If the task is defined by the operator, by the direct lead-through method, or using the corresponding programming (robotic) language, it is much simpler to define the path, i.e. the action to be performed by the robot hand. Similarly, if the task is assigned by higher control levels, it is usually defined "what the gripper should do with the workpiece". On the other hand, the robot's joints are powered by the actuators, and they should be controlled to realize the desired positions or trajectories of the robot hand. To this point we have assumed the hierarchical robot control: at the higher, tactical level, the trajectories of the external (Cartesian) coordinates are transformed into the trajectories (positions) of the robot joints, while at the lower, executive level, the calculated joints coordinates are realized. In the previous chapters we considered the problems of realization of the given joints trajectories (positions), assuming they had been calculated on the basis of the gripper trajectories (Chapter 2). A global scheme of such control is shown in Fig. 5.9.

Such hierarchical solution to the robot control is practically used with all present day commercial robots. As we have seen, to realize the obtained joints coordinates, different control laws can be applied. Such control is called *joint-based control* [20]. It is obvious that this control requires the solving of the inverse kinematic problem, which is usually reduced to the inversion of the Jacobi matrix, and this may require a great number of numerical operations (i.e. a powerful microprocessor is needed to attain the sufficiently short sampling period). It is intuitively clear that another solution to the control problem is possible. As the task is usually defined in terms of external (hand) coordinates it would be possible to control the robot using directly these coordinates. The joints positions (and velocities), measured by the corresponding sensors (potentiometers, shaft-encoders, technogenerators, and the like) are transformed into the external coordinates (the hand coordinates) $s(t)$ and compared with the desired (nominal) hand trajectories $s^o(t)$, so that the errors are also obtained in terms of external coordinates. On the basis of the error in the hand coordinates, the control signals to the robot actuators are generated. A block-scheme of such control is presented in Fig. 5.10. This control is called the *Cartesian coordinates based control* (or the *control in the task-space coordinates* [20, 21], because the control is based on the errors in the coordinate frame in which the task is assigned, e.g. the frame related to tool, or workpiece, etc.).

After the microprocessor has calculated the error in the hand coordinates, it is possible to apply different control laws to generate the appropriate control signals. It should be borne in mind that the job to be done is to generate the inputs to the actuators, by which particular robot joints are powered. One of the possible solutions is to transform the error (using the inverse Jacobian), calculated in terms of Cartesian coordinates $\Delta s = s(t) - s^o(t)$ into the error of the internal coordinates Δq, and then, to generate the inputs to the actuators according to one of the control laws considered (e.g. for the local servo systems, the control is generated acc. to $u^i = -K_p^i \Delta q^i - K_v^i \dot{q}$, and the like). In this way, it is possible to calculate not only the error with respect to the position but also the errors with respect to the joints velocities. However, this solution requires the application of the inverse Jacobian whose calculation usually requires a large amount of calculation so that it has no substantial advantage over the solution presented in Fig. 5.9. On the other hand, the inverse transformation from the external into the internal coordinates is in this case carried out within the control loop, so that it may be required that the inverse Jacobian is calculated during a much shorter sampling period than in the scheme in Fig. 5.10.

Another solution to the Cartesian based control is, starting from the error in the hand coordinates Δs, to calculate the forces (moments) which should be produced on the hand so that the hand would follow the desired trajectory $s^o(t)$ (i.e. the given position s^o). Let denote by F the m×1 vector which is composed on the force components and corresponding moments acting at the gripper mass centre (i.e. around the hand mass centre)[*]. The components of the force and moment at the hand represent their projections to the axes of the Cartesian (or task specific) coordinate frame with respect to which the external coordinates have been defined. Therefore, the control law in terms of hand coordinates defines the link between the vector F and the error in Cartesian coordinates (as well as, the error in the Cartesian coordinates velocities):

$$F = f(\Delta s, \Delta \dot{s}, s^o(t)) \qquad (5.5.1)$$

The Cartesian forces and moments thus obtained should be applied to the robot hand. In order to realize these forces and moments, the corresponding torques (forces) to be produced by the actuators about the joints axes, should be determined. The link between these torques about the joints axes P and the vector of hand forces F is (in the static case) defined as:

$$P = J^T(q)F \qquad (5.5.2)$$

where $J(q)$ is the Jacobi matrix (m×n), defined in Chapter 2. The torques to be realized about the joints are given by (5.5.2). It is easy to determine the voltage signals to be applied to the actuators inputs to realize these torques. The global Cartesian control thus obtained is schematically represented in Fig. 5.11. The main advantage of this

---

[*] If the task is assigned with respect to the mass centre or the tip of the workpiece, and the external coordinates are refferred to this point, then the vector F is also defined with respect to this point. Actually, F should be reffered to the same coordinate frame to which s is reffered.

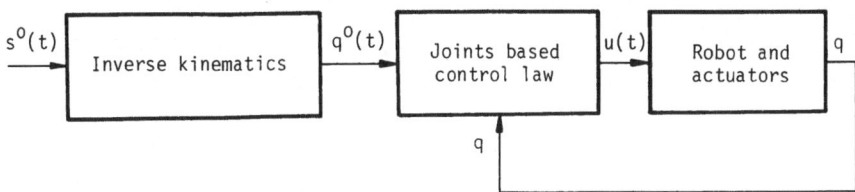

Fig. 5.9. Conceptual scheme of joint based control

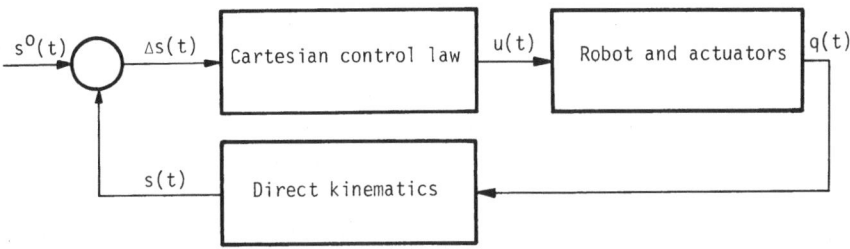

Fig. 5.10. Conceptual scheme of Cartesian based control

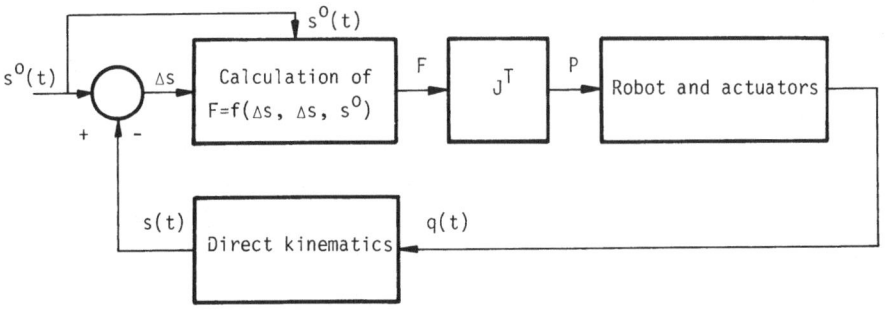

Fig. 5.11. Conceptual scheme of Cartesian based control
(for simplicity, velocity feedbacks and calculation of signals to actuators omitted)

control structure over the scheme in Fig. 5.10. is in that the calculation of the inverse kinematics is avoided. Instead of the inverse
Jacobian, a transposed Jacobian is used, and the direct robot kinematics is calculated (i.e. the hand coordinates are calculated on the
basis of the joint coordinates). These calculations require an incomparably smaller number of numerical operations.

The control law expressed in external (hand) coordinates (5.5.1) might be of different nature. Practically, all the control laws expressed in the joint coordinates, considered in this chapter and the preceeding ones, might be, (after cartain modifications) also applied for the Cartesian control. Thus, for example, the *decentralized control* might also be applied in the external coordinates [21]. In that case, each component of the vector F (the Cartesian force and moment) is obtained only through the errors in corresponding external coordinates:

$$F_i = -K^i_{pe} \Delta s^i - K^i_{ve} \dot{s}^i \qquad (5.5.3)$$

where F_i denotes the i-th component of vector F, to which corresponds the i-th component (s^i) of the vector s, K^i_{pe}, K^i_{ve} are the position and velocity gains in feedback loops by hand coordinates. It is obvious that (5.5.3) represents the control law which should ensure the robot positioning at the desired position s^o, while for the case of tracking the gripper trajectory $s^o(t)$ a precompensating term, and the like (analogous to the solution explained in Section 3.5), should be introduced. Similarly, it is possible to adopt the control law in the Cartesian coordinates which would correspond to the optimal regulator (Appendix 5.B), or which would include the complete dynamic model of the robot, i.e. which would represent the computed torque method applied in Cartesian coordinates [22].

Apart from the obvious advantages of such control schemes in respect of the reduction of the number of numerical operations to be carried out by the microprocessor to generate the control signals (due to the avoidance of the Jacobian inversion), the Cartesian control has also certain disadvantages. The feedback gains synthesis in these schemes is extremely complicated, because the analysis of stability in external coordinates is made difficult. It appears that, in order to achieve satisfactory positioning (or tracking of trajectories) using the decentralized control in external coordinates (5.5.3), it is not possible to adopt the fixed gains K^i_{pe}, K^i_{ve}, but these have to vary in dependence of the robot position. It should be borne in mind that the calculation of hand coordinates on the basis of the measured joint coordinates q and the transposed Jacobian in the scheme in Fig. 5.12. are within the control loop, so that it is possible to require their calculation much more frequently that in the case of the scheme shown in Fig. 5.9. (where the calculation of the inverse Jacobian is required, but outside the control loop). It would be possible to avoid the calculation of the hand coordinates on the basis of the joint coordi-

nates if the hand coordinates were measured directly, which would require the use of the corresponding sensors.

It should be noted that the Cartesian robot control is still at the
stage of theoretical and laboratory investigations so that the present
day commercial robots are controlled solely in the joint coordinates.

Exercises

5.17. Explain how the position and velocity gains in (5.5.3) should be
chosen in the decentralized control on the basis of the Cartesian coordinates, to ensure the robot exhibits no overshoots, nor
the oscillatory motion, and the characteristic frequency is not
close to the resonant frequency of the structure.

5.18.* Draw the control scheme for the robot in Fig. 3.2. which corresponds to the computed torque method in hand coordinates [22]. For
this purpose write the dynamic model of the robot in the Cartesian coordinates, assuming the external robot coordinates are given
by $s = (x_c, y_c, z_c)^T$, where x_c, y_c, z_c are the robot tip coordinates with respect to the coordinate frame fixed to the robot base
(at the first joint). Determine the minimal number of numerical
operations (adds and multiplies) to be carried out on each sampling interval to calculate input signals to the actuators using
the computed torque method in external coordinates for the given
robot (use expressions (5.5.1), (5.4.2) (5.5.2), (5.4.11)). Assume the adopted models of actuators are of second order.

5.19.* Repeat the same as in the preceeding problem for the robot in
Fig. 3.3, assuming the robot is in the horizontal plane, so that
the external robot coordinates are given by $s = (x_c, y_c)^T$, x_c,
y_c are the coordinates of the robot tip with respect to the coordinate frame fixed at the first robot joint. Determine the
number of operations needed to control this robot (to calculate
inputs to the actuators on each sampling interval) using the computed torque method in the joint coordinates, assuming the nominal trajectory is given in terms of Cartesian coordinates; these
Cartesian coordinates are first transformed (using the inverse
kinematic model) into the joint coordinates, and then, the control signals are calculated according to (5.4.2) and (5.4.11).

Compare the obtained number of operations to that needed to calculate the control (input) signals to the actuators using the
computed torque method on the basis of the Cartesian coordinates. (Assume that actuators are D.C. motors whose models are of
the second order, so that (5.4.11) holds).

5.20.* It has been shown that the dynamics of the robot hand might be
expressed by the model written in Cartesian (hand) coordinates
[22]. The dynamic model of the robot hand in Cartesian space
might be written as:

$$F = \lambda(s)\ddot{s} + \mu(s, \dot{s}) + p(s)$$

where F is 6×1 Cartesian force-moment vector, s is 6×1 vector of
Cartesian coordinates of the robot hand (positions and orienta-
tions), $\lambda(s)$ is the 6×6 inertia matrix in Cartesian space, $\mu(s, \dot{s})$ is the 6×1 vector of the hand centrifugal and Coriolis moments,
and p(s) is the 6×1 vector of gravity forces. Compare this model
to the model (3.2.2) in joint coordinates and show that the following relations hold:

$$\lambda(s) = J^{-T}(q)H(q)J^{-1}(q)$$

$$\mu(s, \dot{s}) = J^{-T}(q)\dot{q}^T C(q)\dot{q} - \lambda(q)\dot{J}(q)\dot{q}$$

$$p(s) = J^{-T}(q)g(q)$$

$$P = J^T(q)F$$

(Instruction: Use the fact that

$$\dot{q}^T C(q)\dot{q} = \dot{H}(q)\dot{q} - \frac{1}{2}\dot{q}^T \frac{\partial H}{\partial q}\dot{q}).$$

Starting from the above written model in the Cartesian space,
show that the Cartesian control on the basis of computed torque
method might be written as:

$$P = J^T(q)F = J^T(q)\{\lambda(q)[\ddot{s}^o + K_{pe}(s(t) - s^o(t)) +$$

$$K_{ve}(\dot{s}(t) - \dot{s}^o(t))] + \mu(s, \dot{s}) + p(s)\}$$

where $s^O(t)$ is desired hand path, K_{pe} is matrix of position feedback gains, K_{ve} matrix of velocity feedback gains. Try to rearange last expression to obtain more convinient form for computation, using above relations between models in joint space and in Cartesian space. Draw the corresponding control scheme.

References

[1] Vukobratović M., Stokić D., Control of Manipulation Robots: Theory and Application, Series: Scientific Fundamentals of Robotics 2., Monograph, Springer-Verlag, 1982.

[2] Stokić D., Vukobratović M., "Practical Stabilization of Robotics Systems by Decentralized Control", Automatica, Vol. 20, No. 3, 1984.

[3] Luh Y.S.J., Fisher D.W., Paul C.P.R., "Joint Torque Control by a Direct Feedback for Industrial Robots", IEEE Trans. on Automatic Control, Vol. AC-28, No. 2, 1983.

[4] Tanie K., Yokoi K., Kaneko M., Fukuda T.: "A Position Sensor Based Torque Control Method for a DC Motor with Reduction Gears", IEEE Int. Conference on Robotics and Automation, pp. 1867-1880, 1988

[5] Hollerbach M.J., "A Recursive Langrangian Formulation of Manipulator Dynamics and a Cooperative Study of Dynamics Formulation Complexity", IEEE Trans. on Systems, Man and Cybernetics, Vol. SMC-10, 730-736, November, 1980.

[6] Bejczy K.A., Paul P.R., "Simplified Robot Arm Dynamics for Control", Proc. of IEEE Conf. on Automatic Control, 261-262, 1981.

[7] Kirćanski M., Vukobratović M., Kirćanski N., Timčenko A., "A New Program Package for the Generation of Efficient Manipulator Kinematic and Dynamic Equations in Symbolic Form", Robotica, July, 1988.

[8] Vukobratović M., Stokić D., "Contribution to Suboptimal Control of Manipulation Robots", IEEE Trans. on Automatic Control, June, 1983.

[9] Vukobratović M., Stokić D., "A Procedure for Interactive Dynamic Control Synthesis of Manipulators", Trans. on Systems, Man, and Cybernetics, Sept./Oct. issue, 1982.

[10] Vukobratović M., Stokić D., "Is Dynamic Control Needed in Robotic Systems, and if so, to What Extent?", International Journal of Robotic Research, Vol. 102, Juni, 1982.

[11] Vukobratović M., Applied Dynamics of Manipulation Robots: Modelling, Analysis and Examples, Springer-Verlag, 1989.

[12] Vukobratović K.M., Stokić M.D., Kirćanski M.N., Non-adaptive and Adaptive Control of Manipulation Robots, Monograph, Series: Scientific Fundamentals of Robotics 5., Springer-Verlag, 1985.

[13] Paul R.C., <u>Modelling, Trajectory Calculation and Servoing of a Computer Controlled Arm</u>, A/i. Memo 177, Stanford Artificial Intelligence Laboratory, Stanford University, Septembar, 1972.

[14] Bejczy K.A., Robot Arm Dynamics and Control, Technical Memorandum 33-669, Jet Propulsion Laboratory, February, 1974.

[15] Saridis N.G., Lee G.S.C., "An Approximation Theory of Optimal Control for Trianable Manipulators", IEEE Trans. on Systems, Man, and Cybernetics, Vol. SMC-9, No. 3, March, 1979.

[16] Raibert H.M., Horn P.K.B., "Manipulator Control Using the Configuration Space Method", The Industrial Robot, Vol. 5, No. 2, 69--73, Juny, 1978.

[17] Timofeev, A.V., Ekalo, Yu.V., "Stability and Stabilization of the Programmed Motions of Robot", (in Russian), Automatic and Remote Control, No. 20, 1976.

[18] Asada H., Slotine E.J., <u>Robot Analysis and Control</u>, John Wiley and Sons, New York, 1986.

[19] Young K.K.D., "Controller Design for a Manipulator Using Theory of Variable Structure Systems", IEEE Trans. on Systems, Man, and Cybernetics, Vol. SMC-8, No. 2, 1978.

[20] Craig J.J., <u>Introduction to Robotics: Mechanics and Control</u>, Addison-Wesley Publishing Company, 1986.

[21] Takegaki M., Arimoto S., "A New Feedback Method for Dynamic Control of Manipulators", Trans. of the ASME, Journal of Dynamic Systems, Measurement and Control, Vol. 103, No. 2, 119-125, 1981.

[22] Khatib O., "A Unified Approach for Motion and Force Control of Robot Manipulators: The Operational Space Formulation", IEEE Journal of Robotics and Automation, Vol. RA-3, No. 1, February, 1987.

Appendix 5.A
Stability Analysis of Robot with Global Control

The analysis of asymptotic (and exponential) stability of the robot around the nominal trajectory $x^o(t)$ if just the nominal centralized control and decentralized controller (local servos around individual joints) are applied, has been presented in Appendix 4.A. Here, we shall briefly consider stability analysis when global control is also applied.

Let us consider the case when the centralized nominal control and local servos are applied. In Section 5.2. we have introduced global control in the form (5.2.13). However, as we have explained, this global control directly compensates for coupling between the joints (subsystems) only if we may ignore delay between actuator input and driving torque produced by the actuator. If this delay can not be neglected, this dynamic control cannot completely compensate for dynamic moments (which act as an external load upon the actuator). Therefore, in global control synthesis we have to take into account this delay between the input signal u^i and the driving torque P_i.

In stability analysis of the nonlinear model of the system we have introduced Liapunov function of subsystem $v_i(\Delta x^i)$ as a scalar function of the subsystem state Δx^i (see Appendix 4.A.1). It can be shown that, in order to directly compensate for the effects of coupling upon the system stability, the global control can be introduced in the following form:

$$\Delta u_i^G = -K_i^G[(\text{gradv}_i)^T \hat{b}^i]^{-1}(\text{gradv}_i)^T \hat{f}^i \Delta P_i^*$$ (5.A.1)

where K_i^G is scalar *global gain* (constant), while gradv_i denotes the derivative of function $v_i(\Delta x^i)$ by the coordinates of the vector Δx^i.

However, the Eq. (5.A.1) is defined only if $(\text{gradv}_i)^T \hat{b}^i \neq 0$. Since v_i is the function of the subsystem state Δx^i, then in some points in the subsystem state space $(\text{gradv}_i)^T \hat{b}^i$ might be equal to 0 (or, it might get values close to 0). In these points, the global control signals (5.A.1) would become extremely high (theoretically infinite). However, the amplitude of the total actuator input signal is constrained $N(u^i)$, and, therefore, in the points in which $(grad v_i)^T \hat{b}^i \to 0$, the global control is not realized according to (5.A.1), but some limited input signal is realized. We may introduce a number $\varepsilon_i > 0$ which satisfies the following inequality:

$$|K_i^G \varepsilon_i^{-1} (grad v_i)^T \hat{f}^i \Delta P_i^*| \leq |u_m^i \pm u^{oi} \pm k_i^T \Delta x^i| \qquad (5.A.2)$$

for all points Δx in the surrounding of the nominal trajectory $x^o(t)$, assuming that it is already fulfilled that $|u^{oi}(t) - k_i^T \Delta x^i| \leq u_m^i$. Now, the global control might be introduced in the form:

$$\Delta u_i^G = \begin{cases} -K_i^G [(grad v_i)^T \hat{b}^i]^{-1} (grad v_i)^T \hat{f}^i \Delta P_i^* & \text{for } |(grad v_i)^T \hat{b}^i| > \varepsilon_i \\ -K_i^G \varepsilon_i^{-1} sign[(grad v_i)^T \hat{b}^i] (grad v_i)^T \hat{f}^i \Delta P_i^* & \text{for } |(grad v_i)^T \hat{b}^i| \leq \varepsilon_i \end{cases}$$

$$(5.A.3)$$

The global control (5.A.3) is defined for all points in the state space of the i-th subsystem. The scheme of this global control is presented in Fig. 5.A.1. In (5.A.3) ΔP_i^* denotes function which satisfies (5.A.9) and which can be implemented in one of two ways described in Section 5.2: either by the force feedback, or by on-line computation of dynamic moments (using approximative models of robot dynamics).

Thus, the robot control is given by (5.2.8) and (5.A.3). It is obvious that the global control (5.A.3) directly reduces the effects of dynamic coupling upon the asymptotic (exponential) stability of robot. Namely, if we apply control (5.2.8), (5.A.3) the first derivative (by time) of Liapunov function $v_i(\Delta x^i)$ of the i-th subsystem gets the following form:

$$\dot{v}_{i (\text{along solution of } (5.2.11), (5.A.3))} = (grad v_i)^T \Delta \dot{x}^i =$$

$$= (grad v_i)^T [(\hat{A}^i - \hat{b}^i k_i^T) \Delta x^i] + (grad v_i)^T \hat{f}^i \Delta \bar{P}_i + (grad v_i)^T \hat{b}^i \Delta u_i^G$$

$$(5.A.4)$$

Now, the analysis of the stability of the nonlinear model of the robot is performed in completely analogous way to that one presented in Appendix 4.A.

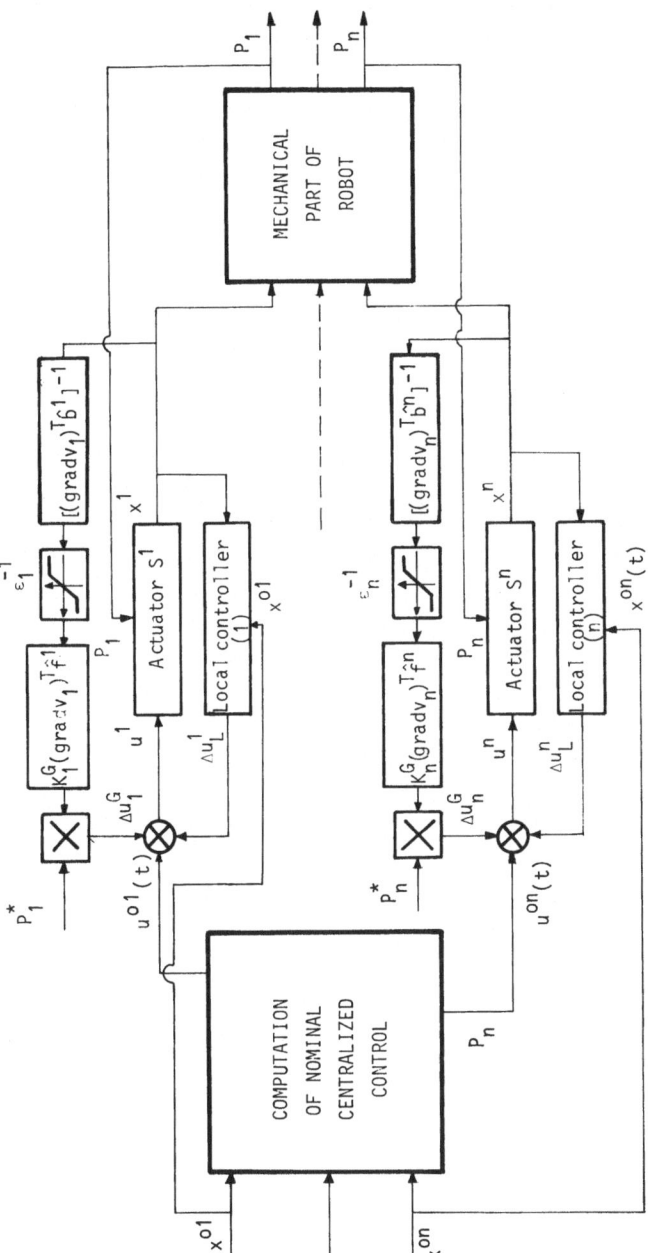

Fig. 5.A.1. Control scheme which includes global control in the
form (5.A.3) - if the centralized nominal control
is applied

The only difference is that now we have to take into account the global control (5.A.3) when we estimate the coupling between the subsystems (joints).

Instead of numbers ξ_{ij}, now we introduce numbers ξ^*_{ij} which fulfill the following inequalities:

$$(\text{grad} v_i)^T \hat{f}^i \Delta \bar{P}_i + (\text{grad} v_i)^T \hat{b}^i \cdot$$

$$\cdot \{-K^G_i [(\text{grad} v_i)^T \hat{b}^i]^{-1} (\text{grad} v_i)^T \hat{f}^i \Delta P^*_i\} \le$$

$$\le \sum_{j=1}^{n} \xi^*_{ij} v_j, \quad \text{for} \quad |(\text{grad} v_i)^T \hat{b}^i| > \varepsilon_i$$

$$(\text{grad} v_i)^T \hat{f}^i \Delta \bar{P}_i + (\text{grad} v_i)^T \hat{b}^i \cdot$$

$$\cdot \{-K^G_i \varepsilon^{-1}_i \text{sign} [(\text{grad} v_i)^T \hat{b}^i] \cdot (\text{grad} v_i)^T \hat{f}^i \Delta P^*_i\} \le$$

$$\le \sum_{j=1}^{n} \xi^*_{ij} v_j, \quad \text{for} \quad |(\text{grad} v_i)^T \hat{b}^i| \le \varepsilon_i \quad (5.A.5)$$

The numbers ξ^*_{ij} estimate not only the coupling $\Delta \bar{P}_i$ between the subsystems, but they also estimate the effects of the global control upon the system stability. The inequalities (5.A.5) must be satisfied for all points in the state space around the nominal trajectory $x^o(t)$ which fulfill (4.A.30). Since the global control is selected to compensate for coupling between the subsystems, then the numbers ξ^*_{ij} must satisfy:

$$\xi^*_{ij} \le \xi_{ij} \quad \text{for} \quad i,j = 1,2,\ldots,n \quad (5.A.6)$$

Now, the analysis of the exponential stability of the model of the state deviation around the nominal trajectory (5.2.11) can be performed by testing the following inequalities:

$$G^* v_o \le 0 \quad (5.A.7)$$

where the elements of the n×n matrix G^* are given by:

$$G^*_{ij} = -\beta_i \delta_{ij} + \xi^*_{ij} \quad (5.A.8)$$

If the conditions (5.A.6) are fulfilled, then, obviously, the tests of the exponential stability (5.A.7) will be "easier" fulfilled than the tests (4.A.65), which are valid when the global control is not applied.

In other words, by the application of the global control, the effects of coupling are reduced and therefore the fulfillment of the exponential stability tests is relaxed. As we have explained in Appendix 4.A, the fulfillments of conditions (4.A.65) and of $\eta > \alpha$ (where η is given by (4.A.67), and α is the demanded degree of the shrinkage of the region around the nominal trajectory to which the state of the system has to belong during the trajectory tracking) guarantee that the robot is practically stable around the trajectory. When we introduce the global control, then the degree of shrinkage of the region, around the nominal trajectory, to which the state of the robot belongs can be estimated by (instead by (4.A.67)):

$$\eta^* = \min_{i=1,2,\ldots,n} \left| -\beta_i v_{io} + \sum_{j=1}^{n} \xi_{ij}^* v_{jo} \right| / v_{io} \qquad (5.A.9)$$

Obviously, the conditions (5.A.6) imply that $\eta^* > \eta$, i.e. if the global control is introduced, then the region, to which the system state belongs, must have greater degree of shrinkage than if just the nominal control and local controllers are applied.

We can see that the global control (5.A.3) can stabilize the robot by compensating for the effects of dynamic moments. The function $\bar{K}_i^G(\Delta x^i)$ in this case is selected in such a way that the global control directly compensates for the effects of coupling upon the system stability. However, to implement the global control (5.A.3) we have to realize on-line computation of the expression $[(\text{gradv}_i)^T \hat{b}^i]^{-1}(\text{gradv}_i)^T \hat{f}^i$ which demands a few floating-point multiplies and adds for each subsystem in which we introduce the global control.[*)]

The simpler solution for the function $K_i^G(\Delta x^i)$ in (5.2.10) is in the form (5.2.13). The stability analysis with the global control (5.2.13) can be performed in analogous way as in the case of (5.A.3). Now, the numbers ξ_{ij} estimating both the coupling and the global control have to satisfy inequalities:

$$(\text{gradv}_i)^T \hat{f}^i \Delta \bar{P}_i + (\text{gradv}_i)^T \hat{b}^i [-K_i^G \bar{f}^i / \bar{b}^i \Delta P_i^*] \leq \sum_{j=1}^{n} \xi_{ij}^* v_j \qquad (5.A.10)$$

for all states in the region defined by (4.A.30). If the second order model of the actuator (subsystem) is considered ($n_i = 2$), then we get:

[*)] Precisely, we have to realize ($2n_i+1$) multiplies, ($2n_i-2$) adds and one floating-point division for each control signal.

$$(\text{grad}v_i)^{T\hat{f}^i}(\Delta\bar{P}_i - K_i^G\Delta P_i^*) \le \sum_{j=1}^{n} \xi_{ij}^* v_j \qquad (5.A.11)$$

However, for $n_i = 3$ the last inequalities do not hold. In that case, the global control (5.2.13) does not compensate directly for the coupling, but we may adopt it as an approximative form (which, in essence, ignores the delay in the actuator). Once we determine the numbers ξ_{ij}^* which satisfy (5.A.10), the procedure for stability analysis is reduced to examination of the tests (5.A.7) and determination of η according to (5.A.9).

We can test stability conditions (5.A.7) for various forms of global control. We can use various approximative dynamic models of the robot for on-line computation of ΔP_i^* and determine ξ_{ij}^* acc. to (5.A.5) and check for which of these models the stability tests (5.A.7) are fulfilled. The same holds if we implement force feedback as the global control. In this way we can iteratively determine in which joint we must apply global control and which approximative model might be used for it. So, we can determine the simplest form of the global control which ensures fulfillment of the conditions (5.A.7), and this procedure has been applied in the software package for computer-aided synthesis of control for manipulation robots (see Section 5.3).

At last, let us mention that if the local nominal control and local servo systems are applied, then we may also introduce the global control which is analogous to the form (5.A.3). The global control (5.2.16) is, now, introduced in the form:

$$\Delta u_i^G = \begin{cases} -K_i^G[(\text{grad}v_i)^{T\hat{b}^i}]^{-1}(\text{grad}v_i)^{T\hat{f}^i}P_i^*(x) & \text{for } |(\text{grad}v_i)^{T}b^i| > \varepsilon_i \\ -K_i^G\varepsilon_i^{-1}\text{sign}[(\text{grad}v_i)^{T\hat{b}^i}](\text{grad}v_i)^{T\hat{f}^i}P_i^*(x) & \text{for } |(\text{grad}v_i)^{T}b^i| \le \varepsilon_i \end{cases}$$

$$(5.A.12)$$

The only difference between global control (5.A.3) and (5.A.12) is in implementation of the function $P_i^*(x)$ instead of the function $\Delta P_i^*(x)$ since in this case the global control has to compensate for the total dynamic moments $\bar{P}_i(x)$. However, if just the local nominal control, the local controllers and the global control (5.A.12) are applied we cannot analyze robot for its asymptotic (or exponential) stability, but we have to analyze practical stability of the system and that is out of the scope of this book.

Appendix 5.B
Centralized Optimal Regulator

We shall consider the problem of synthesis of the centralized robot control for the positioning and tracking of trajectories, which is based on an *optimization* procedure. As with other complex systems, with the robot control synthesis too, a dilemma arises of whether to carry out the so-called *optimal* or *suboptimal* synthesis. In the optimal synthesis the control should be such to *minimize* a certain numerical *criterion*. In the given task we want to ensure the robot is tracking the given trajectories with a defined accuracy (or, it is positioned with certain accuracy), the control is as simple (and cheap) as possible from the standpoint of implementation and maintainance, robust to variations of the parameters, reliable, etc. It is difficult, however, to express all these requirements in terms of precise numerical criteria. On the other hand, when the criteria are defined, the optimal control synthesis is associated with serious numerical problems yielding usually extremly complex and unacceptable control laws [1].

Nevertheless, there have been a number of attempts to realize the robot control synthesis by minimizing a certain criterion. For example, the synthesis has been carried out by minimizing the time necessary to position the robot at a certain point [2], or by minimizing the accelerations at the joints [3]. However, many of the problems in these optimizations require significant simplifications, resulting in the unacceptable solutions.

The *optimal quadratic regulator* represents an *analytic* solution to control, which minimizes the *standard quadratic criterion*, when applied onto a linear system. This approach can also be used for the synthesis of robot control.

Let the trajectories of all the coordinates of the robot state vectors $x^o(t)$ be given (by the higher control level). The task is to ensure the tracking of the trajectories. The robot model in the centralized form is given by (3.2.27)^{*)}:

$$\dot{x} = \hat{a}(x) + \hat{B}(x)N(u) \tag{5.B.1}$$

Let us suppose the *nominal centralized control* $u^o(t)$, satisfying (4.4.8), has been introduced:

$$\dot{x}^o(t) = \hat{a}(x^o(t)) + \hat{B}(x^o(t))u^o(t), \qquad t \in (0, \tau) \tag{5.B.2}$$

The model of the robot state deviation around the nominal trajectory $x^o(t)$ and nominal control $u^o(t)$ is given by (4.4.9):

$$\Delta\dot{x}(t) = \bar{a}(x^o(t), \Delta x(t)) + \hat{B}(x^o(t), \Delta x(t))N(u^o(t), \Delta u) \tag{5.B.3}$$

It should be ensured, that the model of deviation is being stabilized around $\Delta x = 0$ (i.e. around the nominal trajectory $x^o(t)$). The so-called standard quadratic criterion is introduced in the form:

$$J = \int_0^\tau \frac{1}{2}[\Delta x^T(t)Q(t)\Delta x(t) + \Delta u^T(t)\underline{R}(t)\Delta u(t)]dt + \frac{1}{2}\Delta x^T(\tau)Q_T\Delta x(\tau)$$

$$\tag{5.B.4}$$

where $Q(t)$ and Q_T are the positive semidefinite matrices of dimensions $N \times N$, and $\underline{R}(t)$ is a $n \times n$ positive matrix. The quadratic criterion (5.B.4) includes two requirements: the first and the third term express the minimization of states deviations from the nominals (which is required in the task) and the second term represents the minimization of the input signals (i.e., indirectly, the energy) - see the explanation in Appendix 3.A.

If we would endeavour to minimize the criterion (5.B.4), using the nonlinear robot model (5.B.3), we would encounter a lot of problems, for, only a *numeric* solution to control could be found, and not a solution in analytic form - as a function of state coordinates. This means, we would obtain an *open loop* control of the robot (for each initial state, a different numerical solution is obtained, which is extremly difficult to realize). To solve this optimization problem in an analytic form, an *approximate*, *linearized model* of the robot should be considered [4].

---

^{*)} For an explanation of the notation see Chapters 3 and 4.

In Section 4.4. we showed how, starting from the nonlinear model of deviations (5.B.3), we can obtain the linearized time-varying robot model in the form of (4.4.10)

$$\Delta \dot{x}(t) = \tilde{A}_L(x^o(t))\Delta x + \tilde{B}_L(x^o(t))N(u^o(t), \Delta u) \qquad (5.B.5)$$

It is known [4] that the optimal control minimizing the critetion (5.B.4) for the linearized model (5.B.5), is obtained in the form of the centralized regulator:

$$\Delta u(t) = -\underline{R}^{-1}(t)\tilde{B}_L^T(x^o(t))K(t)\Delta x(t) = -D(t)\Delta x(t) \qquad (5.B.6)$$

where K(t) is the positive definite symmetric matrix, N×N, which is the solution of the *Riccati type differential matrix equation*:

$$-\dot{K}(t) = K(t)\tilde{A}_L(x^o(t)) + \tilde{A}_L^T(x^o(t))K(t)+Q(t) -$$

$$- K(t)\tilde{B}_L(x^o(t))\underline{R}^{-1}(t)\tilde{B}_L^T(x^o(t))K(t) \qquad (5.B.7)$$

$$K(\tau) = Q_T$$

Matrix D(t) is the N×N matrix of feedback gains. The solution (5.B.6) holds under the assumption that the matrix pair $\tilde{A}_L$, $\tilde{B}_L$ is controllable on the interval (0, τ), that all the system states coordinates Δx are measured by sensors, and the constraint on the input amplitude $N(u^o, \Delta u)$ is neglected.

It is obvious that the control (5.B.6) is linear with respect to the robot state coordinates Δx(t), i.e. it represents a set of linear feedbacks with respect to the coordinates Δx(t) to the inputs Δu(t) with the gains equal to the elements of matrix D(t). However, these gains are time-varying, which makes the realization much more difficult and requires a large memory capacity to store the time-varying gains, calculated in advance. The on-line solving of differential equation (5.B.7) is hardly attainable.

To simplify the realization of the control law, instead of the linear time-varying model (5.B.5), we use the *linearized time-invariant* robot model which is obtained by "averaging with respect to time" of the model (5.B.5). Thus we obtain an approximate model in the form:

$$\Delta \dot{x}(t) = A_L\Delta x(t) + B_L N(u^o(t), \Delta u(t)) \qquad (5.B.8)$$

where A_L is a constant N×N matrix, while B_L is a constant matrix of dimensions N×n. Instead of the criterion (5.B.4), we introduce:

$$J = \int_0^\infty \frac{1}{2}[\Delta x^T(t)Q\Delta x(t) + \Delta u^T(t)\underline{R}\Delta u(t)]dt \qquad (5.B.9)$$

where Q is a positive semidefinite N×N matrix, and $\underline{R}$ is a positive definite n×n matrix. It should be noted that the criterion is defined on an infinite time interval (around the nominal trajectory), but, since the linearized model (5.B.8) is time-invariant, this makes no problem. It is known that the control minimizing the criterion (5.B.9) for the model (5.B.8) is given as the linear regulator with constant gains:

$$\Delta u(t) = -\underline{R}^{-1}B_L^T K\Delta x(t) = -D\Delta x(t) \qquad (5.B.10)$$

where K is a positive definite N×N matrix which is the solution of the algebraic matrix equation of Riccati type:

$$KA_L + A_L^T K + Q - KB_L\underline{R}^{-1}B_L^T K = 0 \qquad (5.B.11)$$

where D represents the n×N matrix of constant feedback gains. Such linear optimal regulator is represented schematically in Fig. 5.B.1.

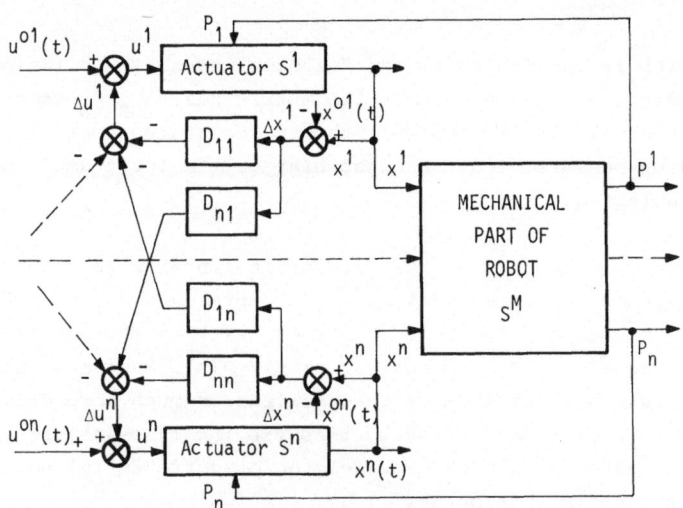

Fig. 5.B.1. Scheme of linear optimal regulator

The linear optimal regulator has the following shortcomings:

(a) The control structure of the optimal regulator is centralized and complex, and requires a large number (N×n) of feedback loops, which complicates the implementation of the control system.

(b) The optimal regulator guarantees the stability of a linearized robot model. As the robot model is, generally, highly nonlinear, it
is questionable if such linear control can also stabilize actual
model. Of course, it is possible to analyze stability of the nonlinear robot model when the linear regulator has been applied.

(c) Robustness of the linear regulator should also be analyzed, because the regulator has been synthesized for the linearized robot
model and for certain fixed values of its parameters.

(d) The "optimality" of the linear regulator is also questionable, both
because of the approximate nature of the model and because of an
arbitrary choice of the weighting matrices Q and $\underline{R}$, on which, to a
great extent, depends behaviour of the system when the linear regulator is applied. Various methods have been developed for selection of the weighting matrices to satisfy certain requirements [1],
but such an approach complicates substantially the synthesis.

In order to overcome the problems arising from the application of the
approximate linearized model (5.B.8), instead of the exact nonlinear
model, it is possible to introduce the additional *nonlinear* control.
Let write the nonlinear robot model (5.B.3) in the form:

$$\Delta \dot{x}(t) = A_L \Delta x(t) + f(\Delta x(t)) + B_L N(u^o(t), \Delta u(t)) \tag{5.B.12}$$

where f is the vector function of the order N, given as:

$$f(\Delta x(t)) = \bar{a}(x^o(t), \Delta x(t)) - A_L \Delta x(t) \tag{5.B.13}$$

Here we assumed that $B_L = B(x^o(t), x(t))$. It can be proved that the
control minimizing the criterion (5.B.9) for the model (5.B.12) is given in the form [5]

$$\Delta u(t) = -\underline{R}^{-1} B_L^T K \Delta x(t) - \underline{R}^{-1} B_L^T (A_L - K B_L \underline{R}^{-1} B_L^T)^{-1} K f(\Delta x(t)) \tag{5.B.14}$$

In other words, the control is in the form:

$$\Delta u(t) = \Delta u_{lin}(t) + \Delta u_N(t) \tag{5.B.15}$$

where Δu_{lin} is the linear, and Δu_N nonlinear part of the control law.

The block-scheme of such control is presented in Fig. 5.B.2. An essential difference between the scheme in Fig. 5.B.1. and the one in Fig. 5.B.2. is in the introducing the nonlinear control component which, in the latter case, plays the role of the global control for the compensation of nonlinear effects in the robot model. Obviously, the robot control (5.B.14) is extremely complex: besides of the complex structure of the optimal regulator (Δu_{lin}) the on-line calculation of the complete robot model is introduced, which brings about all the problems related to the implementation and robustness (see Section 5.2.2).

It should be noticed that we have assumed that the nominal programmed control $u^o(t)$ had been introduced. However, the optimal regulator can also be applied in the case when the nominal control was not being introduced, but it is necessary, then, to examine the practical stability of the robot around the nominal trajectory.

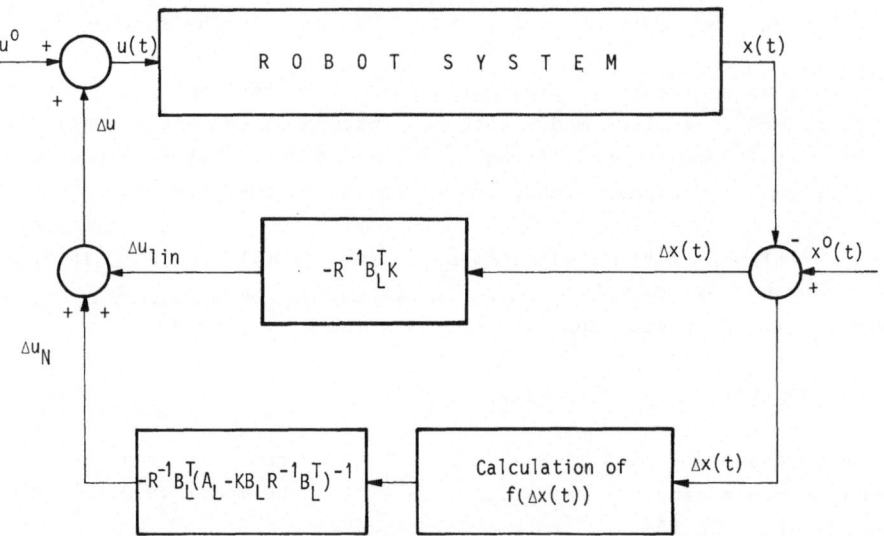

Fig. 5.B.2. Scheme of linear optimal regulator with additional nonlinear control which compensates for the model nonlinearities (acc. [5])

Example 5.B.   For the robot in Fig. 3.2, the linearized time-varying system model around the nominal trajectories, presented in Fig. 4.5., is given in Example 4.3.4. After time-averaging, the linear time-invariant model is obtained whose matrices are:

$$A_L = \begin{bmatrix} 0 & 1 & 0 & 0 & 0 & 0 \\ 0 & -0.132 & 0.254 & 2.77 & 0 & 0 \\ 0 & 0 & 0 & 1. & 0 & 0 \\ 0 & -1.54 & -1.14 & -15.6 & 0 & 0 \\ 0 & 0 & 0 & 0 & 0 & 1 \\ 0 & 0 & 0 & 0 & 0 & -45. \end{bmatrix}$$

$$B_L = \begin{bmatrix} 0 & 0 & 0 \\ 0.58 & -1.75 & 0 \\ 0 & 0 & 0 \\ -1.78 & 10.43 & 0 \\ 0 & 0 & 0 \\ 0 & 0 & 0.361 \end{bmatrix}$$

If the weighting matrices in the criterion (5.B.9) are chosen as

$Q = \text{diag}(0.1, 0.1, 0.1, 0.1, 0.1, 0.1)$,

$\underline{R} = \text{diag}(5., 0.5, 0.05)$

solving the matrix equation of Riccatti type gives:

$$K = \begin{bmatrix} 13180. & 1730. & 2379. & 313. & 0. & 0. \\ 1730. & 706. & 233. & 123. & 0. & 0. \\ 2379. & 233. & 452. & 43. & 0. & 0. \\ 313. & 123. & 43. & 21. & 0. & 0. \\ 0. & 0. & 0. & 0. & 5810. & 135. \\ 0. & 0. & 0. & 0. & 135. & 3. \end{bmatrix}$$

The matrix of feedback gains is:

$$D = \underline{R}^{-1}B_L^T K = \begin{bmatrix} 89. & 38. & 11.5 & 6.6 & 0. & 0. \\ 362. & 51.9 & 77.1 & 10.2 & 0. & 0. \\ 0. & 0. & 0. & 0. & 964.2 & 22.2 \end{bmatrix}$$

Tracking of the nominal trajectories using such optimal regulator with and without nominal centralized control introduced, is shown in Fig. 5.B.3. As can be seen, in the latter case, tracking is poor.

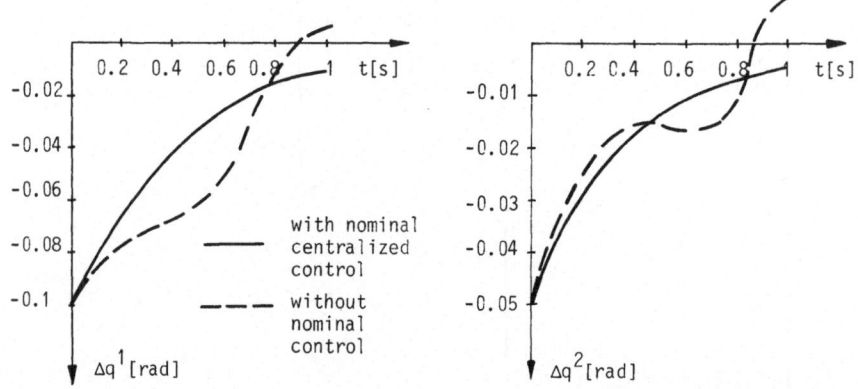

Fig. 5.B.3. Tracking of nominal trajectories using centralized optimal regulator with and without nominal control (for the robot in Fig. 3.2)

References

[1] Vukobratović M., Stokić D., <u>Control of Manipulation Robots: Theory and Application</u>, Series: Scientific Fundamentals of Robotics 2., Monograph, Springer-Verlag, Berlin, 1982.

[2] Kahn M.E., Roth B., "The Near Minimum Time Control of Open Loop Articulated Kinematic Chains", Trans. of the ASME, Journal of Dynamic Systems, Measurement and Control, September, 164-172, 1971.

[3] Young K.K.D., "Control and Optimization of Robot Arm Trajectories", Proc. of IEEE Milwaukee Symp. on Automatic Computation and Control, 175-178, April, 1976.

[4] Popov E.P., Vereschagin A.F., Ivkin A.M., Leskov A.S., Medvedov V. S., "Synthesis of Control System for Robots Using Dynamic Models of Manipulation Mechanisms", Proc. of the Sixth Symp. on Automatic Control in Space, Erevan, USSR, 1974.

[5] Popov, E.P., Vereschagin, A.F., Filaretov, F.V., "Synthesis of Quasi-optimal Nonlinear Control System of Manipulators", (in Russian), Technical Cybernetics, No. 6., pp. 91-101, 1982.

Chapter 6
Variable Parameters and Concept of Adaptive Robot Control

6.1 Introduction

In the proceeding considerations we assumed the robot models and their parameters are being known and determined in advance. Such an assumption holds for the majority of robot parameters, such as masses and inertia moments, link lengths, positions of joints axes, inertia moments of motors, etc. However, some robot parameters change during the work, and they are not always known in advance: these parameters are for example some of the motors parameters, coefficients of viscous and static friction, etc. A common feature of all these parameters is that they change very slowly, so that they can be considered to be quasi-stationary. Very often, their determination is rather difficult. The values of these parameters can be *identified* when the robot starts to work, and they should be checked from time to time and updated appropriately. Most often, these parameters do not affect significantly the functioning of the control system, and they may be considered as being known. A special group form the parameters which can undergo fast and drastic changes during the robot's work, and which cannot be always known in advance. To this group belong, primarily, the parameters of the working object.

In the course of task execution, the robot can move with an empty gripper, grasp the payload and transfer it from one place to another in its workspace (or do some operation on it). In practice, the payload parameters (mass, inertia moments, shape and dimensions) are often known in advance. However, in some cases in the industry, the working object is not defined in advance, but it can have different parameters in dependence of the conditions and other circumstances under which the task is being executed at a given moment. This case appears in modern industry more and more often, especially in those situations when the robot is part of a *flexible technological system*, in which the tasks are frequently changed, so that many of task elements cannot be determined in advance, and the control system has to make decisions on the basis of the information obtained from the sensors, or other subsystems. Thus, the assumption on the known (determined in advance) parameters of the robot system becomes then untenable.

This chapter is devoted to the problems concerning the variation of the robot parameters, and especially of the payload parameters. We shall consider first the problems of control robustness to the variation of payload parameters and then we shall tackle briefly the problem of implementation of adaptive robot control.

6.2 Robustness of Control to Variations of Robot Parameters

If the robot (workpiece) parameters change during the work, the control has to ensure the reliable and smooth functioning of the robot irrespective of the parameters variations. In other words, the robot control should be *robust* to the changes of the workpiece parameters. The theory of control robustness has been fully developed, so that a precise definition of the notion of robustness and different theoretical aspects of this problem will not be the subject of our concern here. Instead, we are going to outline several practical problems related to the variation of workpiece parameters.

It is obvious that the variations in workpiece parameters may substantially influence the behaviour of the robot control system (the quality of trajectories tracking, etc.). If the control has been synthesized under the assumption that the workpiece parameters d have a value d_o (where d is a vector of the workpiece parameters) and the changed parameters are $d_o+\Delta d$ (Δd is the parameter change), it is obvious that the robot's behaviour will not be the same as supposed in the control synthesis. The changes of the workpiece parameters, such as the shape and dimensions, influence the operation of *grasping* the object by the robot gripper. The problem of grasping, however, is beyond the scope of this book. Most often, this problem is solved at higher control levels (strategic and tactical), and this is done on the basis of the information obtained from various sensors (tactile sensors, proximity sensors, cameras, etc.) [1]. The mass of the workpiece, its inertia moments and dimensions (position of the mass centre) influence substantially the robot dynamics, i.e. these parameters are included in the dynamic model of the robot mechanism:

$$P = H(q, d)\ddot{q} + h(q, \dot{q}, d) \qquad\qquad (6.2.1)$$

where, as we have already said, d is a vector of the workpiece parameters of the dimension 4×1, $d = (m_p, J_{px}, J_{py}, J_{pz})^T$, where m_p is the workpiece mass, and J_{px}, J_{py}, J_{pz} are the inertia moments about main inertia axes (variations in the distance of the workpiece mass centre from the gripper might be usually neglected). The variation in the parameters d, from the value d_o to $d_o+\Delta d$ causes a variation in the dynamic behaviour of the robot. If the variation of parameters Δd is known in advance, the control system can be prepared in advance to compensate for this change. However, as we explained it above, this variation is not, generally, known in advance, so that the control system should be robust enough to overcome this variation, i.e. it should ensure that the change of parameters does not cause the robot's malfunctioning.

Let consider briefly the robustness of particular components of control law, synthesized in the preceeding chapters, to the variation of the workpiece parameters.

The local servo system at the i-th joint was synthesized under the assumption that only the i-th joint moves, while all others are kept locked (Chapter 3). In Section 3.3.4. we discussed the influence of the variations of the moment of inertia and gravitational moments on the behaviour of the system. The workpiece parameters influence both the inertia moment of the mechanism about the axis of the i-th joint and gravitational moment of the mechanism. It can be easily shown that the moment of inertia about the i-th joint is a linear function of the workpiece parameters d:

$$H_{ii}(q, d) = H_{ii}^o(q) + d^T \cdot H_{ii}^d(q) \tag{6.2.2}$$

where $H_{ii}^o(d)$ is the value of the moment of inertia of the mechanism about the i-th joint for the case of the no-load gripper, $H_{ii}^d(q) > 0$ is a 4×1 vector whose all the elements are positive. Suppose the velocity gain of the servo system K_v^i has been chosen on the basis of (3.3.39), for the mechanism position q^* and for the parameters values d_o, for which the moment of inertia about the i-th joint $\bar{H}_{ii}(q^*, d_o)$ is:

$$\bar{H}_{ii}(q^*, d_o) = \bar{H}_{ii}^o(q^*) + d_o^T H_{ii}^d(q^*) \tag{6.2.3}$$

If the mechanism is shifted to another position q, and the workpiece parameters become $d_o+\Delta d$, the moment of inertia about the i-th joint will be:

$$H_{ii}(q, d) = H_{ii}^o(q) + (d_o+\Delta d)^T H_{ii}^d(q) \tag{6.2.4}$$

On the basis of (3.3.55), the servo system damping at the i-th joint for the joints position q and the parameters value $d_o + \Delta d$ is:

$$\xi_i = \frac{\sqrt{J_R^i + \bar{H}_{ii}^o(q^*) + d_o^T H_{ii}^d(q^*)}}{\sqrt{J_R^i + H_{ii}^o(q) + d_o^T H_{ii}^d(q) + \Delta d^T H_{ii}^d(q)}} \qquad (6.2.5)$$

where $J_R^i = J_M^i N_v^i N_m^i$.

Now, an analysis can be carried out as in Section 3.3.4. As was shown, in order to ensure the servo system is always (over)critically damped, i.e. that $\xi_i \geq 1$, the chosen velocity gain should be such to correspond to a maximal moment of inertia of the mechanism. Furthermore, it should be ensured that $\bar{H}_{ii}(q^*, d_o) \geq H_{ii}(q, d_o + \Delta d)$, to satisfy $\xi_i \geq 1$. It is easy to show that this requirement can be fulfilled if such mechanism position q^* is chosen, for which

$$\bar{H}_{ii}^o(q^*) = \max_q H_{ii}^o(q), \qquad H_{ii}^d(q^*) = \max_q H_{ii}^d(q) \qquad (6.2.6)$$

and if $d_o = \max d$, so that all the elements of the Δd have to be negative. Therefore, the servo system gains should be synthesized for the largest allowed values of the workpiece parameters.

As we have explained above, it is not necessary the workpiece parameters be known in advance, but for each robot in the stage of its designing it is necessary to determine the range of variations of these parameters, i.e. their maximum allowed values (the so-called, maximal robot load - the largest workpiece mass which can be carried by the robot). If the servo system gains are calculated for the maximum values of the payload parameters, the servo system will be (over)critically damped. However, if the allowed variation of robot parameters is large, the servo system damping will change from the critical (for $d = d_o$) to the highly overcritical (for $d = 0$):

$$\xi_i(d=0) = \frac{\sqrt{J_R^i + \bar{H}_{ii}^o(q^*) + d_o^T H_{ii}^d(q^*)}}{\sqrt{J_R^i + H_{ii}^o(q)}} \qquad (6.2.7)$$

This means, if $J_R^i + d_o^T H_{ii}^d(q^*) + \bar{H}_{ii}^o(q^*) \gg J_R^i + H_{ii}^o(q)$, then $\xi_i(d=0) \gg 1$, i.e. for the no-load gripper, the servo system would be highly overdamped, which indicates the robot's behaviour would be extremely nonuniform. As this nonuniformity is undesirable, such a solution is not satisfactory.

As can be seen, the influence of workpiece parameters depends substantially on the parameters of the robot mechanism itself, $\bar{H}^O_{ii}(q^*)$ and on the moment of inertia of the servomotor rotor J^i_R. If $J^i_R + \bar{H}^O_{ii}(q^*) \gg d^T_O H^d_{ii}(q^*)$, then the damping of the "no-load" robot, cannot be high, i.e. the mass and the moment of inertia of the workpiece exhibit no great effect. In other words, if the moment of inertia of the motor rotor J^i_R and of the mechanism itself are significantly larger then the moment of inertia produced by the workpiece $d^T_O H^d_{ii}(q^*)$, the influence of the workpiece on the servo system's behaviour is not significant. This means that in the case of the relatively large motors and heavy mechanism links, compared to the nominal robot payload, the influence of the payload is not significant. Contrary, if $d^T_O H^d_{ii}(q^*) \gg J^i_R + \bar{H}^O_{ii}(q^*)$, i.e., if the mechanism (and the motor) are relatively light in comparison with the planned workpiece mass, the influence of the workpiece may be great.

A similar analysis may also be carried out for the gravitational moments. The workpiece mass influences the gravitational moment about the i-th joint axis and contributes to the servo system's steady state error. The steady state error can be eliminated in different ways, as we demonstrated it in Section 3.3.4. One of the ways is through the on--line calculation of gravitational moments. However, as the workpiece parameters are not known in advance, compensation of the gravitational moments, caused by the presence of the workpiece, cannot be achieved in this way. How large will be the steady state error which the workpiece causes at the i-th servo system, it depends on the gripper position with respect to the i-th joint axis, as well as on the allowed mass of the workpiece. If the servo system would always be compensated with respect to the maximum allowed mass of the workpiece, then, in the case of no-load gripper, it would cause a steady state error of the opposite sign.

When all the robot joints move simultaneously, the workpiece parameters affect the dynamic forces (moments) (6.2.1) that are loading the servo systems. The workpiece parameters influence all the force components that had been considered in Section 4.2. These forces may substantially change in dependence of the mass and the moments of inertia of the workpiece carried by the robot (or, if the gripper is not loaded). The dynamic control, whose task is to compensate dynamic force, must also take into account the dynamics of the workpiece.

If the nominal centralized control is applied, it should also "include" the working object. If the workpiece parameters (and their variation) are known in advance, i.e. if it is known whether the gripper at a particular moment is carrying the workpiece (and what are the parameters characterizing it), or the gripper is "empty", the nominal programmed control can be calculated from the overall model (6.2.1) (including also the workpiece). However, if the workpiece parameters are not known, the nominal control has to be calculated for the absence of the workpiece, which means that the workpiece dynamics will not be compensated, not even at the nominal level (along the nominal trajectory), which causes a delay in tracking of the trajectory. If, however, the nominal centralized control was synthesized under the assumption of the maximax payload parameters d, it might happen that the given trajectory is "overshooted" which is, as we have explained before, unacceptable. If the mass and moment of inertia of the workpiece are small in comparison with the mass and inertia moments of the mechanism links, the delay in the trajectory tracking may be negligible.

A similar situation also arises when the local nominal control is applied: the control has to be calculated under the assumption of a minimal moment of inertia of the mechanism about the joint axis, which means that the no-load gripper should be assumed in order to avoid the overshoots of the trajectory. However, if such a control is involved, the dynamics of the mechanism and of workpiece become more pronounced, because the nominal control does not compensate for the nominal mechanism's dynamics.

Finally, if all the requirements concerning the accuracy of fast trajectories tracking are such that the global control has to be introduced this control should also compensate the effect of workpiece dynamics. If the global control introduced is in the form of force feedback, then the robot will be robust to the variation of workpiece parameters, because the force sensors measure the total dynamic forces, irrespective of whether they come from the mechanism, or from the workpiece (Section 5.2.1). Such control can compensate for the effect of both the mechanism dynamics and the workpiece dynamics, irrespective of their parameters. As we explained above, the major shortcoming of this solution is the direct connection between the sensors and the inputs to the actuators, which may cause the system to oscillate, as a result of elastic effects directly "transmitted" to the control system.

If dynamic forces are calculated on-line using one of the approximate models (Section 5.2.2), the workpiece parameters may be taken into account, provided they are known. If, however, they are not known, the global control cannot compensate for the change of the workpiece parameters. The same holds for computed torque method (Section 5.4).

As can be seen from all the above, the extent to which the variation in workpiece parameters will influence the robot's behaviour depends mostly on the relative ratio of the allowed variations of the workpiece (the robot's "capacity") and the mechanism (links) parameters. If the maximum workpiece mass assumed is small in comparison to the masses of links, the workpiece effect may be considered as weak, and the local nominal and global control is sufficiently robust to overcome this effect. If, on the other hand, the maximum workpiece mass assumed is of the same order of magnitude as (or larger than) the mass of robot links, then, the workpiece effect becomes significant, and the question arises of whether the control synthesized is sufficiently robust to ensure a uniform and accurate tracking of trajectories (i.e. positioning) when the workpiece is changed. Furthermore, if the links masses are small, or they are of the same order of magnitude, as the mass of the working object, some additional problems usually arise: the links having small masses are relatively thin, so, when loaded with "heavy" objects, the resulting *elastic effects* are greater than in the case of "heavy" and "rigid" links. A "heavy" working object causes elastic bending of the links, which generates a new problem, either from the point of view of the accuracy of positioning and the robot's trajectories tracking, or the appearance of oscillations (especially in the force feedbacks). The need to overcome the effects of these elastic modes by the robot control system makes the synthesis and realization of the control much more complicated [2]. However, the novel "elastic" robots that have appeared recently (suitable from the point of view of material expenditure, energy consumption and reduction of actuators power) have a relatively great allowed load and a more complex control system capable of overcoming both the changes in workpiece and the elastic effects. In all the above considerations, we have assumed the robot links to be rigid; the control of elastic manipulators is beyond the scope of this book.

It should be emphasized that the effect of the unknown workpiece mass
(as well as of other unknown and variable mechanism parameters) is
very important for the direct-drive robot. For this reason, the adaptive control has to be used with these robots, which will be considered in the following section.

Example 6.2. Using the manipulator shown in Fig. 3.2, we shall demon-
strate how the workpiece parameters influence the manipulator's behaviour. We shall assume that a workpiece of maximal mass m_p = 5 [kg]
might be placed at the mass centre of the third link, and the corresponding inertia moments are: J_{px} = 0.01 [kgm^2], J_{py} = 0.01 [kgm^2], and
J_{pz} = 0.02 [kgm^2]. Let assume first the servo systems at all three joints be synthesized without taking into account the workpiece (i.e.,
for m_p = 0). The corresponding servo systems gains are given in Table
4.1. (see Example 4.3.2). The positioning of the first link for the
case of no-load and the case the robot is carrying workpieces of different masses is illustrated in Fig. 6.1. As we can see, the overshoot
of trajectories appears when the robot is loaded by a workpiece, because the system is undercritically damped. For this reason, the servo
systems gains have to be synthesized by taking into account the maximum workpiece mass allowed. The gains synthesized are presented in
Table 6.1. and the corresponding simulation of positioning (Fig. 6.2)
shows the servo system is always overcritically damped. It is also evident that the positioning time in the two cases is not the same.

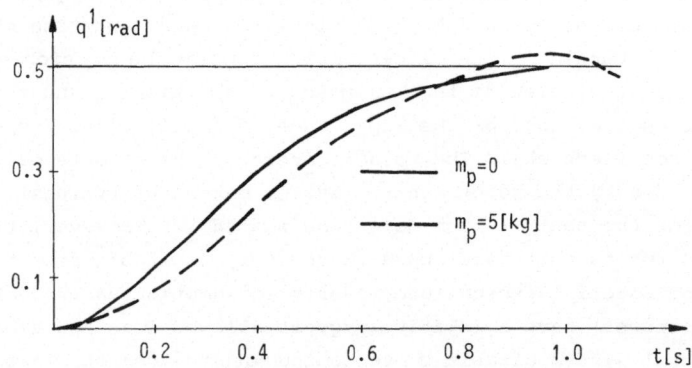

Fig. 6.1. Positioning of the first robot joint for different
workpiece masses (the robot shown in Fig. 3.2. and
the gains in Table 4.1)

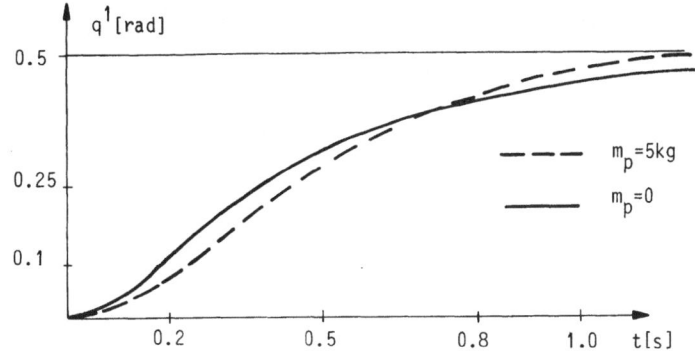

Fig. 6.2. Positioning of the first robot joint for different
workpiece masses (the robot shown in Fig. 3.2. and
the gains in Table 6.1)

JOINT GAINS	1	2	3
K_p $[\frac{V}{rad}]$	132.5	128.6	1714.
K_v $[\frac{V}{rad/s}]$	60.2	13.1	166.

Table 6.1. Servo systems gains for the case of workpiece
mass m_p = 5 [kg] (the robot shown in Fig. 3.2)

If it is necessary to ensure the tracking of trajectories presented in
Fig. 4.5, then the nominal centralized programming control, calculated
under the assumption of the no-load gripper, should be introduced. The
tracking of trajectories using such nominal control for different workpiece masses is illustrated in Fig. 6.3. It can be seen that the greater is the workpiece mass, the greater is the delay. The trajectory
tracking illustrated in Fig. 6.4. corresponds to the centralized control that has been calculated assuming the maximum workpiece mass. It
is evident that the tracking of trajectories by applying such control
for the no-load robot is not satisfactory.

Finally, if the tracking of trajectories shown in Fig. 4.5. is realized by using the local nominal control, local feedbacks and global control in the form of force feedbacks (Example 5.2.1), the tracking of
trajectories is, practically, of the same quality, irrespective of the
workpiece parameters (Fig. 6.5). This illustrates the robustness of
such control to the variation of workpiece parameters.

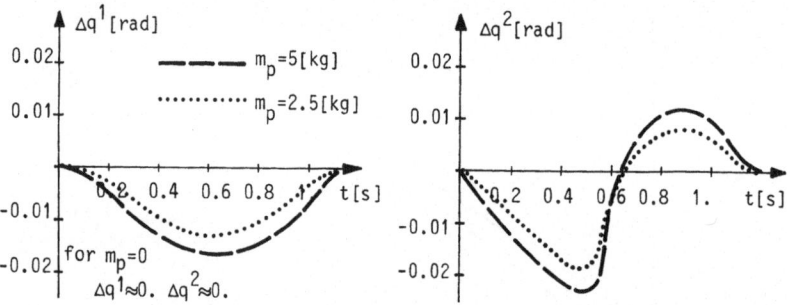

Fig. 6.3. Tracking of trajectories using nominal centralized control and local feedbacks, for different workpiece masses (nominal control calculated for $m_p = 0$)

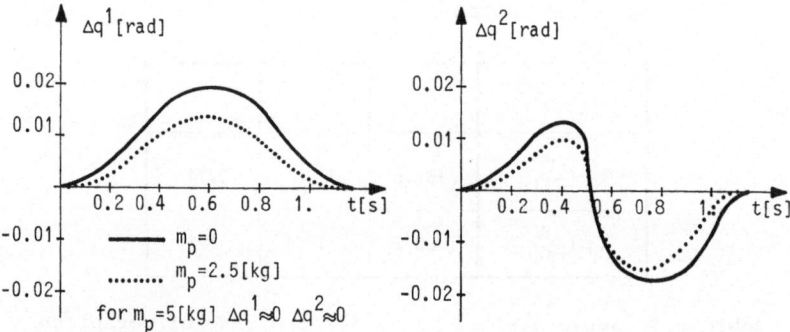

Fig. 6.4. Tracking of trajectories using nominal centralized control and local feedbacks, for different workpiece masses (nominal control calculated for $m_p = 5$ [kg])

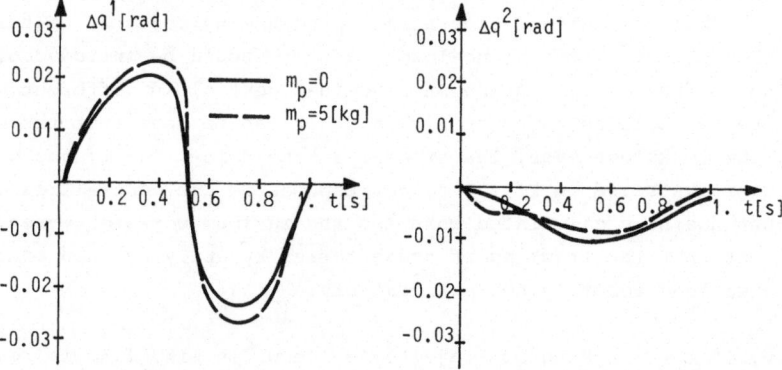

Fig. 6.5. Tracking of trajectories using local nominal control, local feedbacks and global control in the form of force feedbacks

Exercises

6.1. Prove the expression (6.2.2) is correct.

6.2. Explain the statement that the local programmed control should be synthesized under the assumption of the "no-load" robot gripper (in the case when the workpiece parameters are not known). How can the workpiece parameters influence the tracking of trajectories by the local nominal control?

6.3. a) For the second joint of the manipulator presented in Fig. 2.6. and the servo system gains in Example 3.3.4, calculate the steady state error at the second joint, caused by the gravitational moment at the position $q^{o2} = 1.573$ [rad], $q^{o3} = 0.7$ [m]. Use the data on the mechanism and the actuator at the second joint from Example 3.3.4. Assume the workpiece masses at the end of the third link are $m_p = 0$ [kg], $m_p = 1.5$ [kg], $m_p=3$[kg].

   b) If the compensation of gravitational moment is introduced by its on-line calculation (Fig. 3.15, equation (3.3.60)) find out when a greater steady state error is made: (1) when the compensating moment is calculated neglecting the workpiece mass, and the workpiece mass involved is $m_p = 3$ [kg], or, (2) when the compensating moment is calculated for the workpiece mass $m_p = 3$ [kg], and the positioning has been done for the "no-load" robot. Calculate what will be the error in positioning of the robot hand caused by these steady-state errors.

6.4. Repeat the previous exercise a), but assuming now the third link mass is $m_3 = 2$ [kg], and the other data are the same as above. Why the relative increase in steady state error when the robot is carrying the heaviest object, in comparison to the "no-load" robot, is now much larger than in the preceeding case. Assuming the tip of the third link is elastically bent for 0.02 [mm] under the action of a moment of 1 [Nm], calculate the error in the manipulaor tip positioning, caused by the steady state error at the second joint and the elastic deformation of the link, for different masses of the workpiece.

6.5.[*] Explain why the procedure for the analysis of practical robot stability through the analysis of the exponential stability (Section

4.A.2), when the centralized nominal control is applied, cannot
be used when the workpiece parameters has been changed with respect to their "nominal" values (for which the nominal control has
been calculated).

6.3 The Concept of Adaptive Robot Control

Up to now we have considered the various control laws in which all the
control parameters (gains) are constant and which do not vary in dependence of either the robot working regime, or the variation in the
robot's parameters. Such *non-adaptive* control can be, as we have seen,
sufficiently robust, so that the changes in workpiece parameters (as
well as in other variable parameters of the system) do not influence
the robot's behaviour. The control robustness depends on the chosen
law, the information included in that law, the choice of gains, as well
as on the ratio of the variable parameters and the known constant parameters of the system. A direct analysis of the practical robot stability can serve to investigate the robot stability for the variable
workpiece parameters, and thus, to test the robustness of the non-adaptive control to the variation of these parameters. Thus, it is possible to determine the range of variation of the workpiece parameters for
which the non-adaptive control synthesized, can guarantee the robot
stability [3, 4]. Although the workpiece parameters are not known, it
is always possible to estimate the range of variation in workpiece parameters that is to be expected in a particular task. If the analysis
of practical stability shows the non-adaptive control chosen is sufficiently robust to encompass the predicted range of parameters change,
such control can be considered as satisfactory. The majority of commercial robots use exactly this solution: they make use of the control
with fixed gains, which assumes a certain allowable variation in the
workpiece parameters.

However, if the predicted variation in the workpiece parameters would
exceed the capability of the robust control to "overcome" it, the control with variable gains should be introduced. Such control can *adapt*
to the variation in workpiece parameters. For example, if a variable
velocity gain is realized at the local servo systems, it is possible
to achieve the servo systems have always the same damping, and thus
work uniformly, irrespective of the variation in workpiece parameters
(Section 3.3.4). The servo systems gains should change then in depen-

dence of the robot position and workpiece parameters (i.e. in depen-
cence of the moments of inertia of the mechanism and the workpiece
about the joint axis). The *adaptive control* thus realized can guarantee
a satisfactory behaviour of the robot for a wide range of variations
in the workpiece parameters.

Adaptive robot control can be realized in different ways [5, 6]. In
the majority of the approaches, the control parameters and gains are
adjusted in dependence of the instantaneous values of the variable system parameters. Thus in the above example, it is the velocity gains
that should adapt to the change of workpiece parameters. If the workpiece parameters are known (defined at higher control levels, or via
the robot programming language), the executive level should only adjus-
ed the servo systems gains in dependence of the parameters values. In
other words, in the stage of task planning at the strategic level, or
in the course of the robot programming, the workpiece parameters should
be defined and servo systems gains determined in accordance with the
instantaneous values of workpiece parameters. The gains can be calculated on-line according to the procedure described in Section 3.3.2,
and they are determined as a function of moments of inertia of the mechanism and workpiece about the joint axis. Alternatively, the gains
can be calculated in advance for different moments of inertia (workpiece parameters), stored in the computer memory, and, when the information about the workpiece parameters is obtained, those gains values
are taken from the memory that correspond to the actual values of the
payload parameters.

However, if the payload parameters are not known in advance, it is necessary to ensure their *identification* in the course of the robot work.
The on-line identification of these parameters can be realized in different ways. The information about the error of tracking of a given
trajectory may be the basis for the identification of variations in
the system's parameters. This identification is also possible to realize via the functions of sensitivity of the robot model to the parameters variations. However, such an identification usually requires a
large amount of calculation, which is obvious if we bear in mind the
complexity and nonlinearity of the robot model. Because of the need to
determine robot parameters in on-line regime (in the course of the movement), in order to ensure a fast adaptation of control parameters,
the identification can be implemented only with the aid of a fast and
powerful microcomputer. The identification can be considerably accelerated and simplified by introducing force sensors to measure either

moments at robot joints, or the forces at the points of contact of the payload and the gripper (see Section 7.3). On the basis of the information about forces (moments), the parameters of the working object can be determined in a relatively simple way (with much less calculation involved). In this way, the *transition process* which lasts while the identification algorithm is carrying the determination of workpiece parameters and while the adjustment of servo gains is completed, is substantially shortened.

The structure of such adaptive control is illustrated in Fig. 6.6. This approach to adaptive control is called the *indirectly decentralized adaptive control*, because it retains a decentralized structure with respect to the local servo systems [6]. The advantage of such control is in its relative simplicity: it does not require a great deal of calculation and neither substantially complicates the control structure.

Certainly, this is only one of the possible variants of adaptive robot control. Various algorithms have been developed for *centralized* [7, 8] and *decentralized* [9, 10] adaptive control, such as are the *self-tuning* (decentralized or local) *PID controllers*, the approach using the *model referenced adaptive control* etc.

The main problem with all these algorithms for adaptive control is their numerical complexity, which makes their implementation difficult and expensive. For this reason, it is always advisable to examine if the implementation of such control is necessary, or the problem can be solved using the robust non-adaptive control, whose implementation is usually much simpler.

Example 6.3. For the manipulator presented in Fig. 3.2. we have synthesized the non-adaptive control in the form of local nominal control and local servo system feedbacks whose gains are given in Table 4.1 (Example 6.2). The tracking of nominal trajectories (Fig. 4.5) for different workpiece masses is illustrated in Fig. 6.7. As can be seen, the tracking for the workpiece masses m_p = 5 [kg] is much less satisfactory than in the case of the no-load robot. For this reason, the adaptive control is included (Fig. 6.6) in which the identification of workpiece mass is realized on the basis of the force measurements at the points of contact of the gripper and the workpiece. On the basis of the identified mass (and moments of inertia) of the workpiece, the mechanism moments of inertia and velocity servo gains are calculated

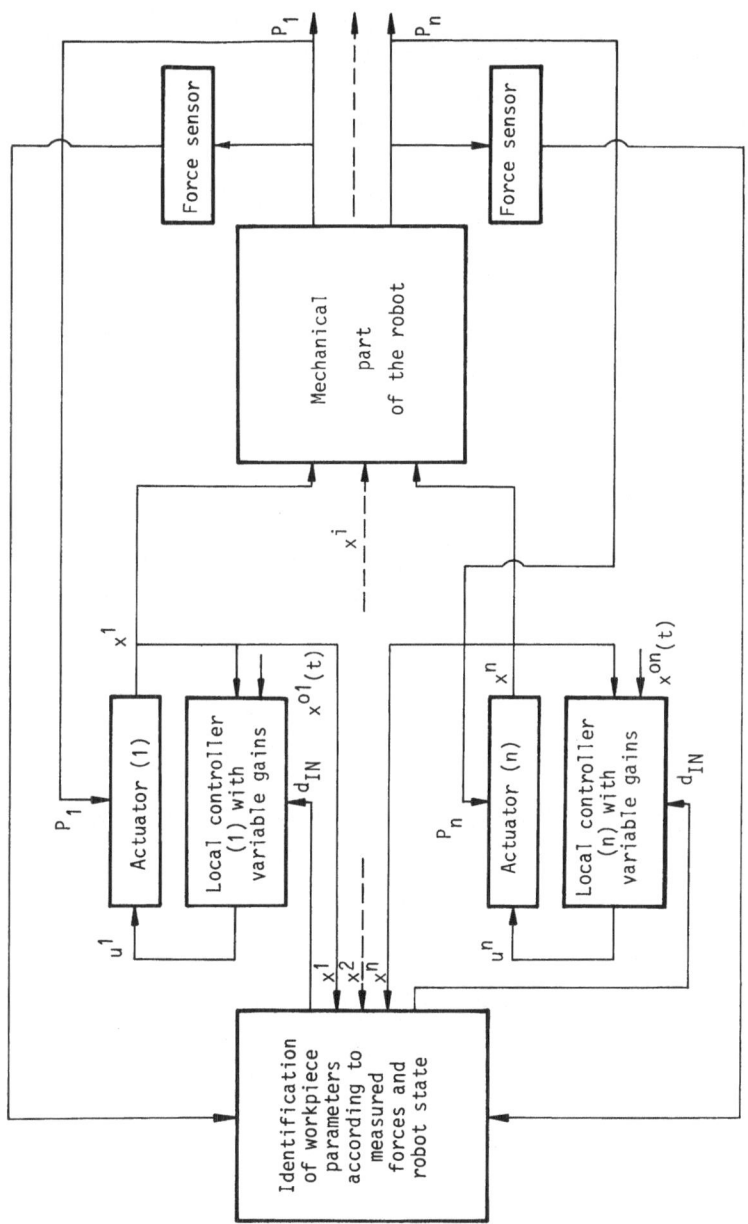

Fig. 6.6. Structure of indirect decentralized adaptive control with identification of workpiece parameters via force measurement and robot states

on-line, which should be such to ensure the damping of all servo systems is permanently critical.

The tracking of trajectories for different workpiece masses is illustrated in Fig. 6.8. It can be seen that during the transition process, while the workpiece parameters are identified and the gains adjusted, the tracking is poor, but it improves from the moment of completion of the transition process. As the transition period is short, the adaptive control ensures a satisfactory tracking of trajectories for a wide range of changes in workpiece parameters.

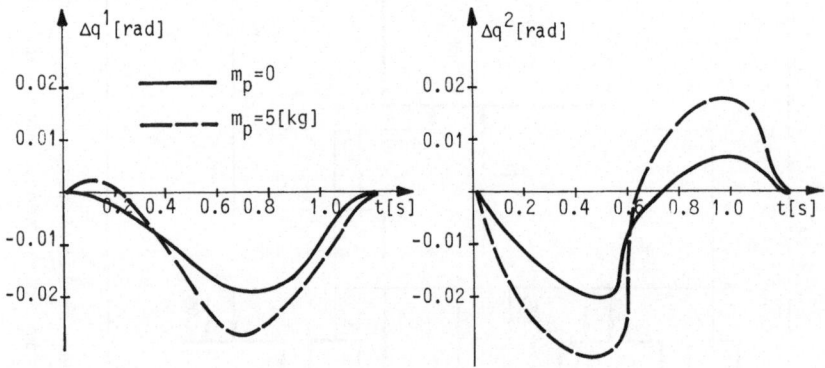

Fig. 6.7. Tracking of nominal trajectories with non-adaptive decentralized control for different workpiece masses

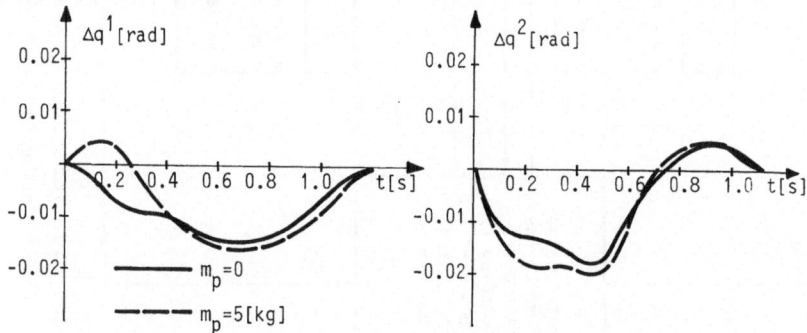

Fig. 6.8. Tracking of nominal trajectories with adaptive decentralized control for different workpiece masses

Exercises

6.6. For the manipulator presented in Fig. 3.2. determine the velocity gains at all three servo systems, so that all of them are critically damped for the position $q^{o1} = 0$, $q^{o2} = 0$, $q^{o3} = 0$, if the workpiece parameters are: a) $m_p = 2.0$ [kg], $J_{px} = J_{py} = J_{pz} = 0.005$[kgm^2]; b) $m_p = 5$ [kg], $J_{px} = J_{py} = J_{pz} = 0.01$ [kgm^2]; c) $m_p = 10$ [kg], $J_{px} = J_{py} = J_{pz} = 0.02$ [kgm^2]. The manipulator data are given in Table 3.1. and Table 3.2., while the resonant frequencies of the structure about the particular joints for $m_p = 0$ are given in Exercise 3.28. and Exercise 3.31.

6.7. Determine the number of operations (adds and multiplies) to be carried out for the given workpiece parameters (m_p, J_{px}, J_{py}, J_{pz}) and the given mechanism position q^{oi} to determine inertia moments of the mechanism about all three joint axes of the manipulator shown in Fig. 3.2., and on the basis of them, calculate the velocity servo gains such that the servo systems are critically damped (see the preceeding exercise). Try to minimize the number of operations.

6.8. If the force feedbacks are used to identify the workpiece parameters for the robot shown in Fig. 3.2., the algorithm for one iteration to calculate the mass and inertia moment of the workpiece requires $n_s = 30$ adds and $n_M = 25$ multiplies. Taking into account the result of the preceeding problem, determine the number of microprocessors to be used in parallel to identify the workpiece parameters, calculate inertia moments of the mechanism about all three joints, and calculate the velocity gains at all three servos (as in the preceeding problem) every 50 [ms], using the microprocessor:

a) INTEL-80-80 (one add operation lasts 0.8 [ms], and one multiply 1.5 [ms]), or

b) INTEL-80-87 (add 35 [μs], multiply 65 [μs]).

Assume the machine time is used only for addition and multiplication (calculation of a sine or a cosine function is equivalent to one multiply plus two adds).

6.9.[*] Write in a high programming language a programme for on-line calculation of velocity gains for all three joints of the robot shown
in Fig. 3.2. as a function of workpiece parameters (m_p, J_{px}, J_{py},
J_{pz}) which should be considered as the input variables to the
programme. Determine the velocity gains such that all three servo
systems are critically damped for all three values of workpiece
parameters and all mechanism positions q^{oi}. The instantaneous positions of the mechanism joints q^{oi} should also be considered as
input data. (For the input data q^{oi} and m_p, J_{px}, J_{py}, J_{pz}, determine the outputs K_v^i, i = 1,2,3). Try to minimize the number of
operations.

6.10.[*] Assume the velocity gains for the servo systems of the robot presented in Fig. 3.2. have been calculated in advance for 6 different values of workpiece mass m_p = 0, 1, 2, 3, 4, 5 [kg] and then
stored. When the identification algorithm has determined the actual workpiece mass, the velocity gain is obtained by the linear
interpolation in between the stored values. Write in a high programming language a programme which will, for the input value of
the workpiece mass m_p, calculate the velocity gains for the servo
systems by linear interpolation, using the stored values. How
much this programme is simpler than the programme in 6.9? As the
velocity gain in this case changes only as a function of workpiece mass, what are the robot's joints positions for which the
velocity gains should be calculated and stored? Explain what are
benefits and what is lost by assuming the velocity gains vary only
in dependence of the workpiece mass, and not of the robot position?

6.11.[*] Explain what are the advantages and what are disadvantages of indirect adaptive control via the force measurement in comparison
to global force feedback (Section 5.2.1). (Suggestion: Take into
account elastic effects of robot links and their "transmission"
to the control system, on the one hand, and the complexity of
calculation, on the other).

References

[1] Vukobratović M., (Ed.), <u>Introduction to Robotics</u>, Springer-Verlag, 1988.

[2] Book J.W., Maizza Neto O., Whitney E.D., "Feedback Control of Two Beam, Two Joint Systems with Distributed Flexibility", Journal of Dynamic Systems, Measurement and Control, Trans. of the ASME, Vol. 97, 424-431, 1975.

[3] Vukobratović M., Stokić D., Kirćanski N., "Towards Non-Adaptive and Adaptive Control of Manipulation Robots", IEEE Trans. on Automatic Control, Vol. AC-29, No 9, 841-844, 1984.

[4] Stokić D., Vukobratović M., "Is Adaptive Control Necessary for Manipulation Robots, and if so, to what Extent?", Proc. of Symp. Third ICAR, Versailles, 1987.

[5] Vukobratović M., Stokić D., Kirćanski N., <u>Non-Adaptive and Adaptive Control of Manipulation Robots</u>, Monograph Series: Scientific Fundamentals of Robotics 5, Springer-Verlag, 1985.

[6] Vukobratović M., Kirćanski N., "An Approach to Adaptive Control of Robotic Manipulators", Automatica, Vol. 21, No 6, 1985.

[7] Dubowsky S., "Application of Model Referenced Adaptive Control to Robotic Manipulators", Journal of Dynamic Systems, Measurement, and Control, Trans. of the ASME, Vol. 101, 1979.

[8] Timofeev A.V., Ekalo Yu.V., "Stability and Stabilization of Programmed Motions of Robots", (in Russian), Automatica and Remote Control, No 20, 1976.

[9] Kiovo A.J., Guo T.H., "Adaptive Linear Controller for Robotic Manipulators", IEEE Trans. on Automatic Control, Vol. 28, No 20, 1983.

[10] Landau Y.D., <u>Adaptive Control</u>, M. Dekker, New York, 1979.

Chapter 7
Control of Constrained Motion of Robot

7.1 Introduction

In all the tasks we have considered up to now, the robot does not come into contact with the objects in the workspace, apart from those that are being transferred. However, this does not hold for the one of the most important industrial application of robots, namely, for the *assembly* of machine parts. In this case, the robot comes into contact with the objects in its environment and experiences actions of the external environmental forces. Similarly, in the processes like cutting, grinding, polishing and forging, the robot gripper has to act upon the given object by certain forces. These external forces acting on the robot gripper make the robot control much more complex. Hence, this chapter will be devoted to the synthesis of control for the robots involved in the realization of the tasks of this type. First our attention will be focused on the assembly process, as one of the most important and most delicate tasks in which the action of external forces is encountered. However, some general approaches to control of constrained motion of robots will be also presented.

If the robot comes into contact with the objects in the workspace, the reaction forces acting upon the robot are the functions of both the moments in the joints and coordinates and velocities of all the joints. When a robot moves in *free space*, it represents an *open kinematic chain*; when it comes into contact with the external objects, the robot becomes a *closed kinematic chain*.

Dynamic models of closed kinematic chains are more complex than the models of open kinematic chains, which makes the synthesis of the robot control more complex.

First, we shall consider briefly the problems encountered in the robotic assembly processes, from the point of view of both modelling and control synthesis. Then, we shall present some of the solutions to these problems that are nowdays successfully used in the industrial robotics, and then we shall consider some schemes for solving constrained motion control of robots in general case.

7.2 An Analysis of Assembly Process by Robots

Assembly of parts is one of the most sophisticated tasks of the industrial robotics. Therefore, the solving of this problem is of greatest
importance, especially if we bear in mind the application of robots in
the *flexible technological lines*. As is known, the industry automation
is achieved by application of either specialized machines-automates,
or the robots for automatic assembly of parts. The application of highly-
-specialized automates is justified only in the mass production of the
one and the same products in large series (let say, at least one mil-
lion copies a year, and for a number of years). In the case of smaller
series, the development and application of the highly specialized machines is not justified any more, because these machines usually cannot be used when some, even minor, changes in the shape and dimension
of the product are to be introduced. As the products of modern industry are manufactured in the relatively small series, which are in addition, frequently changed, the introduction of flexible technological
systems is the preferable choice for the realization of automated production. The central role in these flexible system is played by robots,
which enable an easy reprogramming and switching from the production
of the one product to another. Therefore, it is essential that the process of assembly by robots is effectively solved, not only from the
point of view of the efficiency, accuracy, and speed of work, but also from the point of view of an easy change of the working task and
maximal flexibility (though the robot flexibility is inevitably limited by its mechanical characteristics and adaptability to the variation in task conditions).

Assembly processes may vary substantially in dependence of the type of
elements involved, the mode of their joining, etc, but in all cases,
the following stages can be distinguished: the stage of approaching
the object, the stage of its grasping, the stage of transferring the
object to the spot of assembling, the stage of its joining to another
element, etc. It is clear that the robot in realizing this complex process has to perform three kinds of motion: the so-called *"gross motion"*
related to the robot movement in the obstacle-free space (e.g. the
transfer of the working object from one place to another, etc), *"fine
motion"*, related to the robot motion in an environment containing objects (for example, the stage of grasping the workpiece, the stage of
mating, etc), and *"interface motion"*, representing all transition
kind of motion between gross motion and fine motion (for example, the

stage of approaching the workpiece, or, the stage just before mating the objects, etc). In the preceeding chapters we have considered the gross motion: the robot moves with a high speed along the given trajectories (or, between the given positions), and, as we have seen, the degree of accuracy required for the realization of this motion determined whether dynamic or nondynamic control had to be synthesized (Chapters 4-6). The interface motion, which, according to some analyses [1], takes the greatest amount of the assembly time, can be reduced substantially by ensuring a high precision of the gross motion (in the robot stopping and in approaching the workpiece). In other words, the control synthesis for the interface motion can be carried out in a way quite analogous to that used for the gross motion.

Here we shall focus our attention on the realization of the fine motion in case the robot gripper comes into contact with the objects in its environment. More precisely, we shall focus our attention on the very stage of mating two elements, which represents the most sophisticated stage in an assembly process. The stage of parts mating may be realized in various ways, depending on the nature of the elements involved; one of the possibilities is to use two robots, each of them holding one of the two elements that are to be assembled. In principle, the mating stage can be reduced to the problem of inserting the workpiece (peg) held by the robot into a fixed hole, as shown in Fig. 7.1. It is obvious that the assembly process involving bilateral manipulation (two manipulators-robots) may also be reduced to this case.

The problem of controlling the robot during the assembly of machine elements (insertion of the peg into a fixed hole) is, in principle, the problem of accurate positioning. If all geometrical parameters of the peg and of the hole and their positions in space were ideally known, and if the robot positioning (tracking of the given trajectories) was ideally accurate, then, it would be possible to insert the peg into the hole using only the robot control that has been considered in the previous chapters. However, as none of the conditions could ever be fulfilled, the peg insertion into the hole is inevitably accompanied by the occurrence of contact between the peg and the hole, which results in the appearance of the *reaction forces*. The effects of these forces are of crucial importance for a successful assembly operation. To prevent potential demaging of the assemblying objects, it is obvious that such robot control should be ensured which would minimize these forces. However, it should be noticed that in some tasks the robot itself has to exert certain force on the elements, in order to realize the assembly operation. Hence, the appearance of the reaction forces between the elements at the stage of their mating is practically unavoidable. Therefore, to synthesize the control which would take into account the effects of these forces, it is necessary to carry out a detailed analysis of all the diverse phenomena appearing in this process. First of all, it is necessary to analyze the static and dynamic forces acting via the workpiece upon the robot gripper.

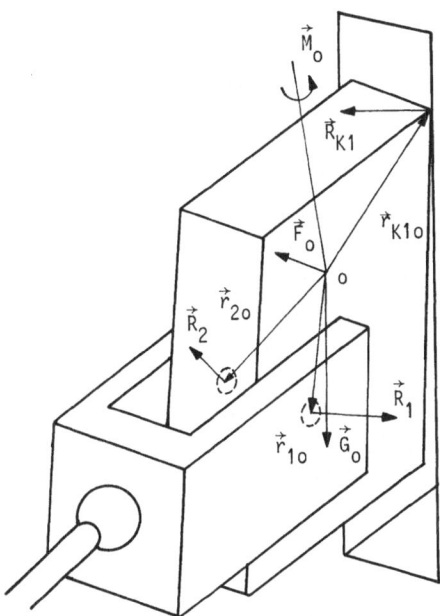

Fig. 7.1. Mating phase of the "peg-in-hole" task

The analysis of the assembly process may be approximately reduced to an analysis of the static forces appearing between the workpiece (robot gripper) and the hole. Of course, a complete and precise notion of all the aspects of this process can be obtained by a dynamic analysis, i.e. by setting out a complete dynamic model of the robot experiencing the reaction force of the hole [2]. Now we shall analyze briefly the forces appearing during the insertion operation and the situations which may arise in the stage of the insertion.

Consider the insertion of a cylindrical object (peg) whose base diameter is $2r_m$ into a cylindrical hole of the diameter $2R_m$. It is straightforward that the quantity $2(R_m-r_m)$ represents the clearance gap between the object and the hole. Obviously, the larger this gap is, the simpler is the insertion process. For this reason, the assembling capability of a robot is judged from the minimal value of the gap between the elements that can be assembled by the robot without causing any damage to the objects, while ensuring fully reliable assembling of the elements. To insert the peg into a hole, it is necessary to ensure the peg is accurately positioned at the mouth of the hole, which is usually achieved on the basis of either precise information about the disposition (arrangement) of the objects in the workspace (known in advance), or the sensory information (visual feedback, and similar). The task of precise initial positioning is of the greatest importance for a successful assembly operation.

Let us analyze briefly the forces which can arise due to the peg-hole interaction when the peg is brought to the mouth of the hole [2, 3].

The purpose of this analysis is to examine the effect of friction and the problem of potential jamming, as well as the possibility of using force feedbacks for controlling the manipulator with the aim of avoiding the undesired effects of hole reaction forces on the insertion. As we have already pointed out, regardless of how well the manipulator has been positioned, contact between the peg and the hole may occur, so that reaction forces result.

In this force analysis none of the aspects of dynamics is neglected. It is assumed that the dynamic coefficient of friction between the peg and the hole surface is the same everywhere and that it equals the static coefficient of friction. This assumption has been introduced for simplicity sake, but may easily be removed. Furthermore, it is assumed that both the peg and hole walls are completely rigid, so that, in our considerations deformation is not considered as this would make the model much more complex.

Obviously, the given peg and hole geometry ensures that, upon insertion, contact occurs simultaneously at most two points. (This does not hold for special cases of coaxiality of the peg and cylindrical hole, when contact may occur at an infinite number of points along the joint generatrix). One of the two contact points has to be between the hole edge and the peg cylindrical surface, and the other on the edge of the peg cylinder base and hole cylindrical surface. It is assumed that the peg has already entered the hole, i.e. no jamming has occured at the hole edge itself.

In analyzing the forces arising during the insertion, four possible situations are considered (under the above assumption that the peg has already passed the hole base) [2].

1. No contact occurs between the hole and the object.

2. One contact point exists.

3. Two contact points exist.

4. Contact is realized along the joint generatrix.

Let us consider each of these situations, respectively.

Case 1, when there is no contact between the hole and peg, represents free manipulator movement without reaction forces from the hole, so this situation does not differ dynamically from the task of transferring the workpiece in free space along a desired trajectory and with a desired orientation. The manipulator dynamics is described by the mathematical model of open chain dynamics.

Case 2 introduces the problem of the unknown reaction force acting upon the manipulator and affecting its dynamics. The problem of determining the reaction force at the hole-object contact points arises. Fig. 7.2. illustrates the case of contact between the hole edge and the cylindrical surface of the peg, while Fig. 7.3. shows another possibility of single contact, namely, contact between the edge of the cylindrical base of the peg and the cylindrical surface of the hole. Both figures show the hole and peg section along the plane determined by the contact point and the symmetry axis of the hole cylinder. The reaction forces at contact points are marked in the figures. Three components of the reaction force at the point K1 or K2 are to be determined in the direction of the coordinate frame axes fixed at the contact point. In fact, the perpendicular component of the reaction force N_{K1} (perpendicular to the peg cylinder generatrix on which the point K1 is situated, i.e. perpendicular to the hole cylinder generatrix on which K2 is situated) is to be determined, as well as the tangential component T_{K1} (or T_{K2}) in the direction of slipping (if it occurs) of the peg along the hole edge (Fig. 7.2) or hole surface (Fig. 7.3).

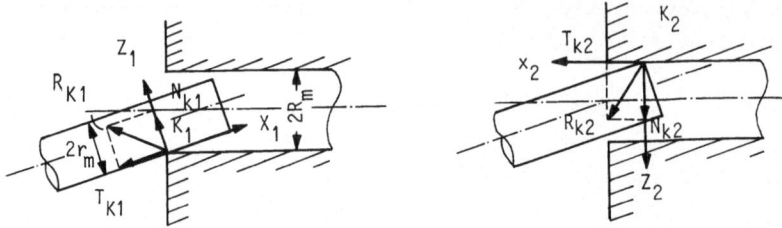

Fig. 7.2. Case of one contact point in assembly process - contact between hole edge and peg surface

Fig. 7.3. Case of one contact point in assembly process - contact between edge of peg and hole cylinder surface

Two cases are possible. Case 2.a), when friction is such that slipping exists in the peg-hole contact, and 2.b), when no slipping occurs, so that the peg has no tangential linear velocity. Assuming that the coefficient of friction μ between the peg material and hole material is known, the following condition should be investigated

$$T_{Ki} < \mu N_{Ki}, \qquad i = 1,2 \qquad\qquad (7.2.1)$$

In case 2.a) the peg has a linear velocity at the contact point in the direction of T_{Ki} and the reaction force has to satisfy the slipping condition

$$T_{Ki} = \mu N_{Ki}, \qquad i = 1,2 \qquad\qquad (7.2.2)$$

In case 2.b) the reaction force at the momentary contact point satisfies the condition (7.2.1). In case 2.a) the contact point changes, and the tangential linear velocity of the object at the point Ki has to be in the direction T_{Ki}. In case 2.b) the peg can have only rotational velocity about the point Ki and the contact point does not change. This imposes certain constraints on the peg acceleration, namely, the peg as a rigid body can have three degrees of freedom, i.e. three rotation accelerations about the contact point Ki, while the three degrees of freedom of translation with respect to the contact point are constrained (linear accelerations in case 2.b) at the point Ki are equal to zero). It is quite obvious that in case 2.a) the peg's linear velocity at the point Ki can have a tangential component only, while the velocity component in the same direction of but in the sense opposite to N_{Ki} must be equal to zero. (If this velocity component has the same direction and sense as N_{Ki}, contact is only fictitious).

In case 2 the forces acting upon the peg are: the inertia force of the peg itself $\vec{F}_O$, the peg gravitational force $\vec{G}_O = m_p \vec{g}$, the forces by which the manipulator (i.e. the gripper) acts upon the peg $\vec{R}_\ell$ (ℓ = 1, 2,...,L, where L denotes the number of contact points between the peg and gripper at which forces $\vec{R}_\ell$ act on the peg), and the reaction force $\vec{R}_{Ki}$ due to contact at the point K_i. The dynamic equilibrium equation of forces acting upon the peg (according to D'Alembert's principle) is

$$\vec{F}_O + \vec{G}_O + \vec{R}_{Ki} + \sum_{\ell=1}^{L} \vec{R}_\ell = 0 \qquad (7.2.3)$$

The moments acting on the peg round the centre of mass are: the moment due to the inertial forces of the object itself $\vec{M}_O$, the moment from to the manipulator via the gripper, i.e. via forces at the gripper-peg contact points, and the moment due to reaction forces at the object--hole contact points. Equilibrium of the moments about the centre of mass of the peg (according to D'Alembert's principle) is

$$\vec{M}_O + \vec{r}_{Kio} \times \vec{R}_{Ki} + \sum_{\ell=1}^{L} \vec{r}_{\ell o} \times \vec{R}_\ell = 0 \qquad (7.2.4)$$

where $\vec{r}_{Kio}$ is the position vector from the centre of mass of the peg 0 to the contact point Ki, $\vec{r}_{\ell 0}$ (ℓ = 1, 2,...,L) is the position vector from the peg's centre of mass 0 to the peg-gripper contact point.

Equations (7.2.3) and (7.2.4) determine the peg dynamics in the case of a single point of contact between the object and hole.

The manipulator dynamics represents the dynamics of a closed kinematic chain.

Case 3, when two points of contact exist, is illustrated in Fig. 7.4. Contact points K1 and K2 are the same as in Case 2. Depending on the character of the reaction forces at contact points, the following cases may occur: Case 3.a), slipping occurs at both contact points; case 3.b) no slipping occurs at any contact point - the peg is jammed; 3.c) it may happen that slipping occurs at one contact point and not at the other - in that case the peg rotates about the point without slipping and loses contact at the other point (if this rotation is permitted by the geometry). However, it may also happen that slipping cannot occur at one contact point (e.g. K2) but can occur at the other (K1), the peg would then rotate about K2, but this is not possible on account of the geometry. In this case, the object becomes jammed.

Which of these three cases will occur it depends on the nature of the reaction forces at the points K1 and K2, i.e. the following conditions should be investigated.

$$T_{K1} < \mu N_{K1}, \qquad T_{K2} < \mu N_{K2} \qquad\qquad (7.2.5)$$

If the condition (7.2.5) is satisfied, there is no slipping at any contact point and case 3.b) occurs. In case 3.a) the reaction forces satisfy the following conditions

$$T_{K1} = \mu N_{K1}, \qquad T_{K2} = \mu N_{K2} \qquad\qquad (7.2.6)$$

In case 3.c) a combination of these conditions occurs, namely,

$$(T_{K1} < \mu N_{K1} \wedge T_{K2} = \mu N_{K2}) \vee (T_{K1} = \mu N_{K1} \wedge T_{K2} < \mu N_{K2}) \qquad (7.2.7)$$

However, in this case, it should be checked whether the peg rotation, which would result on account of the condition (7.2.7) is permitted by the hole geometry. If it is not, jamming which is dynamically undetermined will occur, a peg deformation may result, which would require the introduction of new degrees of freedom (peg and hole flexibility and the like). The dynamics of peg jamming will not be considered here since the elements are assumed to be absolutely rigid bodies and manipulator control should ensure that peg jamming and deformations are avoided. It will suffice to note that no jamming occurs during assembling, i.e. that manipulator control is synthesized in such a way that jamming is avoided. Force feedback must ensure, therefore, that the peg always gets from the manipulator a moment contrary to the external forces moment. Accordingly, case 3.b) should be avoided by applying appropriate control. In case 3.b), since there is no slipping at contact points, both the rotational and linear peg acceleration are constrained to zero, which gives 6 conditions for determining the reaction forces that are to satisfy the condition (7.2.5).

In case 3.a) the peg may have only tangential linear velocity at contact points (if geometry permits them), while linear velocity and acceleration at point Ki in the same direction of but opposite sense to N_{Ki} are forbidden, which, together with the condition (7.2.6) provides the necessary conditions for determining the reaction forces at points Ki.

In case 3.c) conditions (7.2.7) and those constraining peg acceleration to exist in the same direction of but opposite sense to N_{Ki} yield the conditions necessary to determine the reaction forces and peg accelerations; the realizability of these conditions will depend on the geometry.

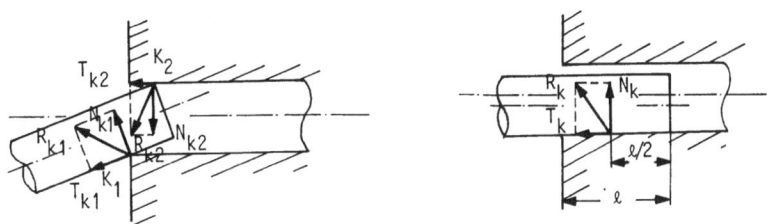

Fig. 7.4. Case of two contact   Fig. 7.5. Case of contact along
          points   joint generatrix

The equations of dynamic equilibrium of forces and moments acting on the peg in this case are of the form

$$\vec{F}_0 + \vec{G}_0 + \vec{R}_{K1} + \vec{R}_{K2} + \sum_{\ell=1}^{L} \vec{R}_\ell = 0 \qquad (7.2.8)$$

$$\vec{M}_0 + \vec{r}_{K1o} \times \vec{R}_{K1} + \vec{r}_{K2o} \times \vec{R}_{K2} + \sum_{\ell=1}^{L} \vec{r}_{\ell o} \times \vec{R}_\ell = 0 \qquad (7.2.9)$$

Case 4 can be reduced to case 2. The reaction force due to the hole acts on the peg along the joint generatrix of the peg's cylindrical surface and hole cylinder (Fig. 7.5). As the reaction force acts uniformly at all contact points along the generatrix, the reaction forces may be represented by one resultant force $\vec{R}_K$ acting at the mid-point of the line of contact between the peg and hole. This special case may be treated like the case with one contact point. Equations (7.2.1) – (7.2.4) hold, so that either slipping or jamming may occur (i.e. the peg cannot slip along the generatrix), but in the last case rotation separating the peg from the hole wall is possible and case 2. occurs (if rotation is permitted by the geometry).

Let us construct the mathematical model of the robot during the insertion process. Starting from the model of the robot dynamics (3.2.2) for the robot moving in free space (open kinematic chain), the mathematical model of the robot dynamics during the insertion of the peg into the hole can be written in the form [3]

$$P = H(q)\ddot{q} + h(q, \dot{q}) + \delta_1 J_{K1} \cdot \vec{R}_{K1} + \delta_2 J_{K2} \vec{R}_{K2} \qquad (7.2.10)$$

where the notation used is the same as in (3.2.2), J_{K1}, J_{K2} are the (n×3) matrices, determined by the position vectors of the peg-hole contact points K1 and K2 with respect to the joints axes, $\vec{R}_{K1}$, $\vec{R}_{K2}$ are the (3×1) vectors of the reaction forces of the hole, while δ_1, δ_2 are Kronecker's symbols which can assume the following values:

$$\delta_1 = \begin{cases} 1 \text{ if there is contact at point K1} \\ 0 \text{ if there is no contact at K1} \end{cases}$$

$$\delta_2 = \begin{cases} 1 \text{ if there is contact at point K2} \\ 0 \text{ if there is no contact at K2} \end{cases}$$

The reaction forces $\vec{R}_{K1}$ and $\vec{R}_{K2}$ are complex functions of the angles, velocities and accelerations of the joints (i.e. of the driving moments P at the mechanism joints). Here, we shall not be concerned with the problems of expressing these forces as explicit functions of q, $\dot{q}$ and $\ddot{q}$ but we shall retain the model in the form of (7.2.10).

However, the model (7.2.10) is not smooth. At the instants of the object-hole contacts (when δ_1 and δ_2 change their values), theoretically "instantaneous jumps" of the system state vector (q, $\dot{q}$) occur as a consequence of the impact of the elements. Modelling of the impact process is extremely complex so that the model of the robot dynamics too is difficult to compose, to simulate it in an exact way, and thus analyze it. It is quite clear that at the moments of the impact, after instantaneous changes in the velocites (and positions) of particular joints, the model of the robot dynamics *changes its structure:* from an open kinematic chain (3.2.2) it becomes a closed kinematic chain (7.2.10).

By combining the model of the mechanical part of the robot (7.2.10) and the models of the actuators at robot joints, the global dynamic model is obtained describing the robot in the assembly process.

Exercises

7.1. For the manipulator presented in Fig. 7.6, determine the matrix J_{K1} in the model (7.2.10) if the peg-hole contact is as shown in the figure. Write the matrix according to the notation used in the figure (the robot has n=4 joints). Show that J_{K1} represents Jacobian matrix for the contact point.

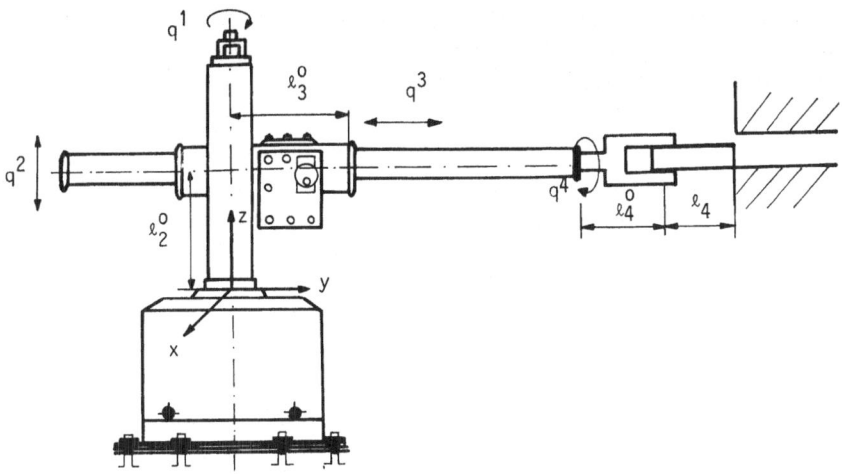

Fig. 7.6. Cylindrical robot (with 4 d.o.f.) in assembly process

7.2. If the models of the actuators at the robot joints are given by
(3.2.4) (second order models), write the global robot model in
centralized form (analogous to (3.2.27)) if the peg insertion
process proceeds with reaction forces acting on the working object. The model of the mechanism dynamics is given by (7.2.10).
(In the total model, the reaction forces $\vec{R}_{K1}$ and $\vec{R}_{K2}$ should be
retained).

7.3 Robot Control in the Stage of Parts Mating

Control of the robot during the assembly process involves solving two
major problems: planning of such trajectories which will provide the
elements are brought to a desired mutual position and ensure their mating, and reducing the (undesired) reaction forces acting between the
assembled elements. These two problems are interrelated: the proper
planning of trajectories diminish the error in mutual positioning of
the elements, and thus, reduce reaction forces. On the other hand, reaction forces provide the basis for planning the trajectories (to mo-
dify the given trajectories) yielding the parts mating.

In order to reduce the effect of reaction forces acting between the
elements it is advisable to measure these forces so that the information obtained can serve as the basis for realization of suitable con-

trol. Thus a visual feedback (camera) or proximity sensors enable planning of the robot trajectories to the very stage of parts mating, i.e.
to the moment just before the insertion of the peg into the hole. However, during the insertion itself, the measurement of reaction forces
provides the best information about the mutual position of the elements
being mated. Because of that, the robotic assembly process is often
solved with the aid of force sensors, introduced to measure the reaction forces acting between the peg and the hole.

As for the disposition of force sensors, different solutions are possible. They can be placed either on the robot itself, or mounted on
the support of the assembled object (at the hole). Obviously, the former solution is more convenient from the point of view of flexibility
and reprogrammability, and it does not require changes in the robot
environment. Sensors can be placed at different points on the robot.
The closer the sensors are to the contact point of the object and hole,
the more accurate information about reaction forces is obtained on the
basis of force measurement. One of the possible solutions is to mount
the force sensors on the gripper, just at the contact points of the
workpiece and the gripper (Fig. 7.7) [4]. The force sensors at contact
points measure the object-hole reaction forces (if contacts exist) and
the dynamic forces of the object itself. Namely, the forces at the contact points of the object and the gripper, $\vec{R}_\ell$, have to satisfy relations (7.2.8) and (7.2.9). In order the measurement of $\vec{R}_\ell$ would supply an
accurate information about the reaction forces $\vec{R}_{K1}$ and $\vec{R}_{K2}$, it is necessary to calculate (on the basis of the known robot angles q, velo-
cities $\dot{q}$ and accelerations $\ddot{q}$) both the inertia force of the object $\vec{F}_o$
and moment $\vec{M}_o$ due to the inertia forces of the object. Obviously, for
relatively low speeds of the gripper motion and small mass and inertia
moment of the object, these forces can be neglected. A shorthcoming of
such a solution to sensors disposition is that, because of the need to
measure relatively small forces, sensors of high precision and sensitivity have to be used. On the other hand, from the point of view of
design, it might be difficult to install sensors at the contact points.

A solution which is used most often is to mount force sensors at the
gripper (wrist) joint (Fig. 7.8). This is usually realized in the form
of the so-called Maltese cross by which the forces and moments at the
wrist joint are measured [5, 6]. The sensors are disposed in all four
cross branches in order to obtain direct information about all three
force components and all three moment components appearing at the gripper joint. However, the measured components of forces and moments do

not represent the direct information about reaction forces acting at
the working object, because, in these forces are also included the dynamic forces of the gripper and working object. Since the gripper mass
is usually much larger than, or equal to, the workpiece mass, the measured forces provide "worse" information about reaction forces if compared with the above solution of force sensors disposition (Fig. 7.7).
On the other hand, to measure force in the wrist joint is technically
much simpler. Besides, as the gripper velocities and accelerations during the assembly process are small, dynamic forces forces and moments
of the gripper and workpiece can be neglected. (It should be borne in
mind that the acceleration of the workpiece and gripper in the direction of the hole reaction forces is equal to zero).

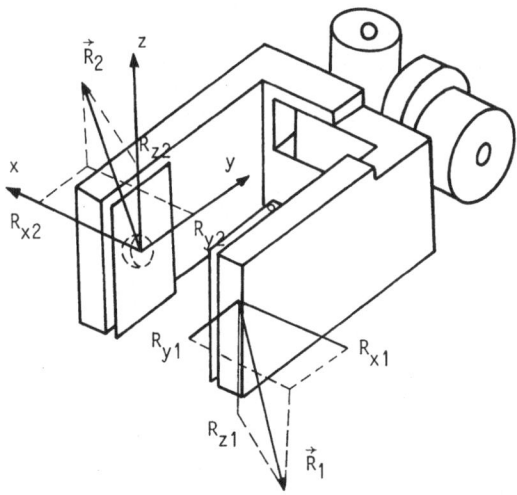

Fig. 7.7. Experimental gripper with force sensors at
the object-gripper contact points [4]

Certainly, there are some other technical solutions for sensors pla-
cing on the robot (for example, force sensors can be placed on the robot base, or in the joints, as considered in Section 5.2, and the like).
It is obvious that the "further" from the workpiece forces sensors are
placed, the "more fouled" information about the reaction forces is obtained, so that the dynamic robot model has to be used to calculate,
on the basis of the measured forces, the actual reaction forces acting
upon the workpiece.

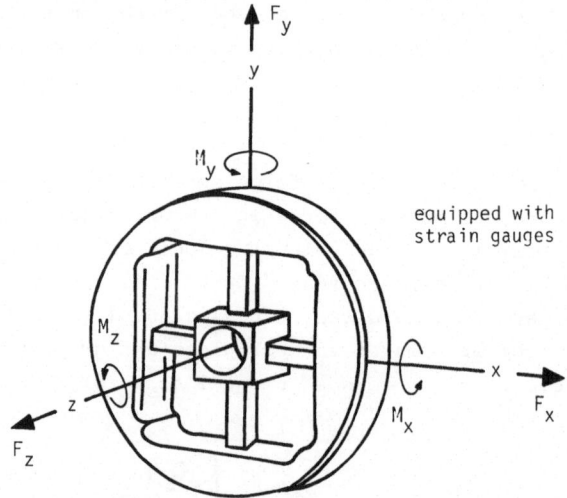

Fig. 7.8. Typical device with force sensors at
the wrist joint [5, 6]

Regardless of where they are situated, force sensors provide information about the reaction forces acting between the objects in the assembly process (i.e. about reaction forces of the hole on the workpiece held by the gripper). After these forces have been measured, it is
possible to establish a feedback loop from the sensor (i.e. from the
calculated momentary values of reaction forces) to the inputs of the
actuators at the joints. In other words, some additional signals (dependent of the reaction forces measured) should be fed to the actuators whose task is to realize the compensating movements of the robot
joints and thus, minimize the reaction forces. Therefore, such compensating control should be realized (via the reaction force feedbacks)
which will minimize the workpiece-hole reaction forces and achieve
compliant motion of the robot and workpiece. Obviously, it is necessary to realize the motion of the gripper and workpiece in the same direction of but in the opoosite sense to the action of the reaction
force.

Force feedbacks can be realized in various ways [7]. Sometimes, it suffices to introduce feedbacks at the gripper joints. The control law
may be either linear or nonlinear. A simple form of force feedback
might be realized when force sensors are installed at the points of
contact of the workpiece and the robot hand.

Let us suppose the sensors at the workpiece-gripper contact points measure the reaction forces acting between the workpiece and gripper, $\vec{R}_\ell$, on the basis of which the reaction forces of the peg-hole $\vec{R}_{Ki}$ can be determined. The torques at robot joints which should be applied in order to minimize reaction forces, and thus, realize the insertion of the peg into the hole may be represented in the following way:

$$\Delta P_i^F = -(\vec{r}_{K1i} \times \vec{R}_{K1} + \vec{r}_{K2i} \times \vec{R}_{K2}) \cdot \vec{e}_i \tag{7.3.1}$$

where $\vec{r}_{k1i}$ and $\vec{r}_{K2i}$ are the position vectors from the axis centre of the i-th joint to the contact point K1 and K2 respectively (see Figs. 7.2 - 7.5) and $\vec{e}_i$ is the unit vector of the i-th joint axis. Let us suppose the inertia forces of the object $\vec{F}_o$, gravitation force $\vec{G}_o$ and the moment due to inertia forces of the object itself $\vec{M}_o$, can be neglected (which is, as we have already said, a realistic assumption in regard to small velocites and accelerations of the workpiece in the stage of parts mating). On the basis of relations (7.2.8) and (7.2.9) for the case of the occurrence of two contact points (i.e. on the basis of (7.2.3) and (7.2.4) for the case of one contact point between the object and hole), expression (7.3.1) can be written (taking into account that $\vec{r}_{K1i} = \vec{r}_{K1o} + \vec{r}_{oi}$, $\vec{r}_{K2i} = \vec{r}_{K2o} + \vec{r}_{oi}$):

$$\Delta P_i^F = (\vec{r}_{oi} \times \sum_{\ell=1}^{L} \vec{R}_\ell + \sum_{\ell=1}^{L} \vec{r}_{\ell o} \times \vec{R}_\ell) \vec{e}_i \tag{7.3.2}$$

where $\vec{r}_{oi}$ is the position vector from the centre of axis of the i-th joint to the mass centre of the workpiece. If we suppose the workpiece parameters are known, expression (7.3.2) can be calculated on the basis of the force sensors information, i.e. from the measured forces $\vec{R}_\ell$. On the basis of the values obtained for the compensating torques ΔP_i^F, it is possible to determine the compensating signals to the actuators inputs which should realize these torques. If we assume the actuators models are of second order (i.e. if the delay in the rotor circuit can be neglected), the compensating signals can be calculated in the form:

$$\Delta u_i^F = K_i^F \bar{f}^i / \bar{b}^i \Delta P_i^F \tag{7.3.3}$$

where $K_i^F > 0$ is the gain in the force feedback loop, and $\bar{f}^i$ and $\bar{b}^i$ are defined by (5.2.12). On the basis of (7.3.2) and (7.3.3) a feedback loop can be established from the force sensors (which measure reaction forces between the object and gripper $\vec{R}_\ell$) to the actuators inputs. The block-scheme of such control is shown in Fig. 7.9. It is not necessary to introduce force feedbacks into all joints, i.e. the compensating moments ΔP_i^F need not to be realized by all the robot joints. The choice of the control structure depends on the mechanical robot structure because the compensating moments at the gripper tip should be realized via the torques at joints. This means that the position vector of the gripper (i.e. of the workpiece) with respect to particular joints $\vec{r}_{oi}$, as well as the directions of the particular joints (i.e. the orts $\vec{e}_i$) determine what are the joints to which force feedbacks should be introduced and what are the joints at which these feedbacks might be neglected.

A special problem in the implementation of the laws (7.3.2) and (7.3.3) represents the choice of gains K_i^F. It should be borne in mind that the control laws (7.3.2) and (7.3.3) include the robot parameters which need not to be known precisely. On the other hand, the force sensors, however precise they may be, introduce always certain noise in the forces measured. Therefore, should the compensating torque be realized exactly according to (7.3.2), it could result in an oscillatory behaviour of the robot. The role of feedback gains K_i^F is to prevent such oscillatory behaviour. However, it is difficult to determine these gains at the stage of robot design (since the modelling of these high frequency modes of the system is very complex) but they are best determined experimentally. Obviously, the appearance of oscillations when force feedbacks are applied in the assembly process may also be prevented in some other ways (e.g. by introducing a damping feedback loops, and the like).

We have considered here a simple scheme of robot control for the assembly process which involves force feedbacks. It should be borne in mind that this control scheme cannot solve the assembly problem in a general way. Moreover, even in the simplest case, this scheme cannot guarantee that peg jamming will be avoided. It is necessary to develop a special strategy (i.e. to plan the gripper motion) so to prevent jamming.

As can be seen, a force feedback can be applied in combination with the position, velocity, and (possibly) current feedbacks. Namely, in the scheme in Fig. 7.9. the feedbacks employed in the position robot control have been retained, to which force feedbacks have been superimposed. The interaction of these two parts of control scheme has not been analyzed. However, a case may arise when the robot positioning feedbacks and force feedbacks are in a "collision". This problem will be considered in the following section.

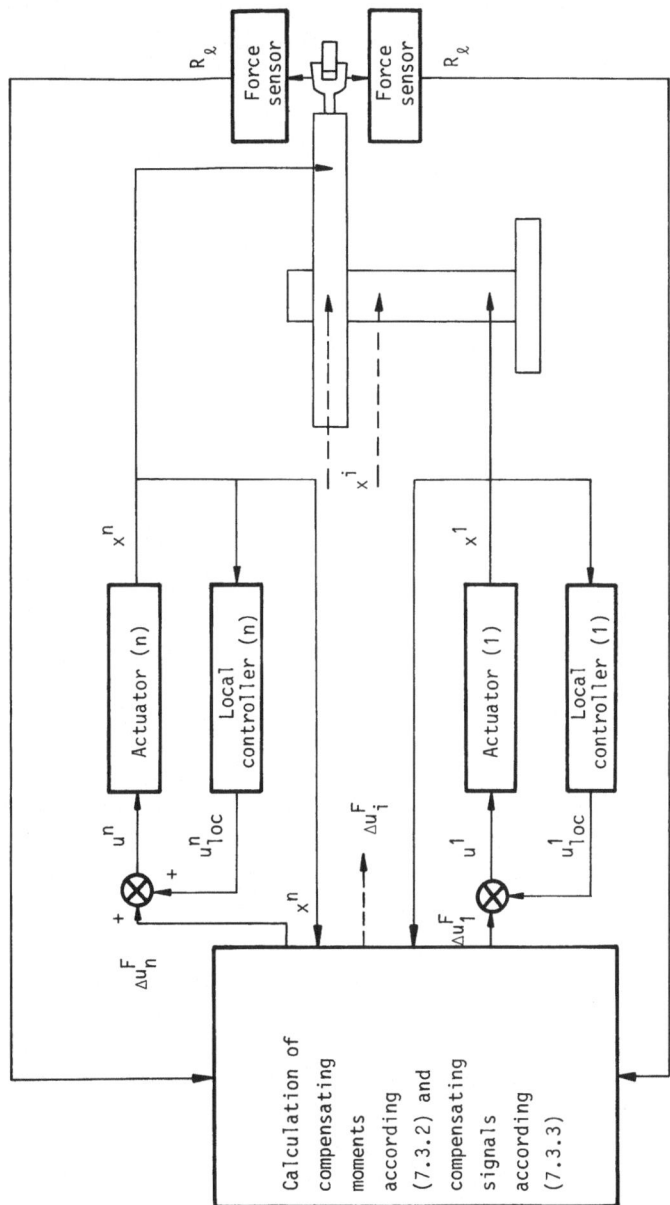

Fig. 7.9. Block-scheme of control with feedbacks by forces measured
by sensors at contact points between the object and
gripper

Several examples of successful robotic assembly using force feedbacks have been realized under laboratory conditions and some of them are described in Ref. [8]. However, the force feedback has not found yet a wider use with the commercially available robots. The reason is in unresolved problems related to the force feedback application [7].

Much better performances in the robotic assembly have been achieved using the specially constructed grippers which enable a passive (with no actuators involved) addaptation of the gripper during the assembly task. Different types of the so-called *passive compliance* have been developed, the best known being the "Remote Centre Compliance" (RCC) developed in the Charles Stark Draper Laboratory [9]. In this solution, a special system of springs is situated between the end-effector and a wrist of the robot which enables easy movements of the workpiece both in the plane perpendicular to the hole axis and in the plane of the rotation about the corresponding axes (Fig. 7.10). In this way the robot (gripper) acquires certain additional degrees of freedom which enable a better *compliance* of the peg position to the hole axis, facillitating thus the peg insertion into the hole. Actually, these passive devices exploit the fact that to achieve easy insertion of the peg into the hole it is necessary to ensure low lateral and rotational stiffness of the gripper. Such low stiffness gripper allow the peg to "*comply with the hole*", i.e. compliance between the peg and the hole is obtained. RCC owes its name to the fact that this device places the *compliance center* at the tip of the peg. A compliance center is a point at mechanical system such that a force applied at that point causes only a translation, while a torque applied around this point causes only rotation around the point. Therefore, if compliance center exists or is artificially created, as RCC does, at the right position on the peg, it allows efficient insertion of the peg into the hole. The advantage of this solution is in that the control system is not complicated either by special feedbacks or additional calculation. In this way, the speed of assembling is increased significantly. However, a disadvantage of these devices is that one passive compliance cannot be applied for different pegs and holes (differing in the dimension, shape, and the like), but usually for each peg-hole combination a proper mechanical adaptor has to be used. As the major characteristic of a robot is its *reprogrammability* (i.e. the ability to switch from one production task to another, that is from one working place to another, what is achieved in a simple and quick way by reprogramming the robot), the application of these passive compliances diminishes robot's capabilities and reduces the robot to the level of a classical machine capable of doing a narrow set of tasks. A frequent solution is that a set of different adaptors is developed for a robot, and when the task is being changed, the adaptor is changed accordingly.

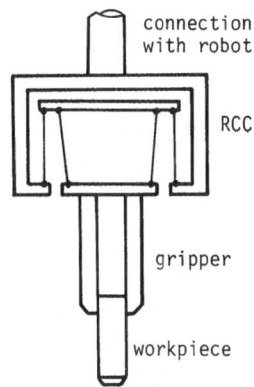

Fig. 7.10. Scheme of a passive compliance (RCC)

Further development of RCC has led to the instrumented RCC (IRCC) whose purpose is also to realize passive adaptation but which includes measurement of contact forces and moments, so that this information can be used in the feedback [10].

Example 7.3. We shall briefly describe an experimental study of the force feedback in the robotic insertion of the workpiece into the hole [4].

For the purpose to study experimentally force feedback loops and their application in robotics a force sensor which could measure all the three components of the force acting between the robot gripper and workpiece has been used. Besides, a special gripper (Fig. 7.11) has been developed and attached to the tip of the minimal configuration of the manipulation robot UMS-2.

Both jaws sides of the experimental gripper were supplied with three-component force sensors. The gripper was designed in such a way that the jaws sides bearing force sensors move parallel with each other.

Each force sensor contained three one-component sensors to measure the forces acting between the payload and gripper. The sensors used were

from the WAZAU company (Berlin), and their measuring range was 20 N.
The force sensors are semiconductors of the strain gauge type of average sensitivity of 1.505 mV/N and have exhibited very good reproduci-
bility. The gripper was covered with rubber to ensure uniform pressure
and friction needed for the efficient gripping of the payload. The
screws with nuts and springs enabled mechanical adjustment of "zero"
in each of the three measuring directions.

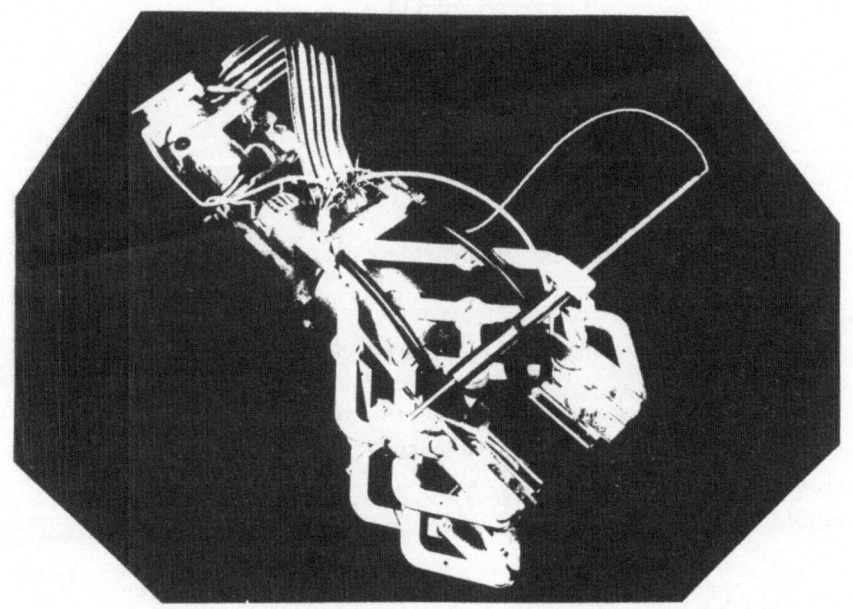

Fig. 7.11. Experimental gripper with force sensors to measure
reaction forces between the gripper and workpiece

The experimental gripper with the force sensors was attached to the
tip of the minimal configuration of the industrial manipulator UMS-2
shown in Fig. 7.6. The robot possesses n=4 d.o.f. (three for the minimal configuration and one for the gripper rotation). The robot control
was realized using a PDP-11-03 microcomputer and either the analogue
or direct digital servo systems.

The force sensors mounted on the gripper enabled direct measurement of
the forces $\vec{R}_{\ell} = (R_{x\ell}, R_{y\ell}, R_{z\ell})^T$. The values of forces obtained from
the force sensors are fed via A/D converter to the microcomputer. The

additional compensating torques which are to be realized by the actuators can be calculated in a direct way as:

$$\Delta P_i^F = \sum_{\ell=1}^{2} \vec{r}_{\ell i} \times \vec{R}_{\ell}$$

In the particular case of the UMS-2 robot, the compensating moments can be calculated in the following way:

$$\Delta P_1^F = (q^3 + \ell_3^0 + \ell_4^0) \cdot [\cos q^4 (R_{y1} - R_{y2}) + \sin q^4 \cdot (R_{z1} + R_{z2})]$$

$$\Delta P_2^F = -\cos q^4 \cdot (R_{z1} + R_{z2}) + \sin q^4 \cdot (R_{y1} - R_{y2}) \tag{7.3.4}$$

$$\Delta P_3^F = R_{x1} + R_{x2}, \qquad \Delta P_4^F = d \cdot (R_{z1} + R_{z2})$$

where d denotes the half-width of the gripper jaws (i.e. the distance from the left to the right force sensor; this distance is measured by means of a corresponding linear potentiometer and, via A/D converter led to the microcomputer).

The aim of the experiment was to investigate the force feedbacks in the process of "fine motion". The manipulator task was to insert a prismatic object into a hole. The clearance gap was relatively large - up to 0.5 mm, so that the insertion could be carried out in a relatively simple way and with no contact occurring between the object and hole. However, to examine the effect of the force feedback, we have positioned the manipulator in such a way that the object and hole come into contact and the resulting reaction forces were measured by the force sensors on the gripper. It was possible to arrange that the reaction force had only one (x) component in the absolute coordinate frame. This component could be compensated by the movement of the first (revolute) manipulator joint. This means that we have established the force feedback to the first joint and the global force gain for the same joint has been varied.

The time-history of the forces appearing at the sensor (i.e. the module of the force vector) during the object insertion are presented in Fig. 7.12. If the global force feedback is opened (by putting the gain $K_1^F = 0$) these forces will reach their maximum. If the control also includes the force feedback (and through it additional driving torques calculated by the microcomputer are applied) these forces become smaller (see the result for $K_1^F = 1$ in Fig. 7.12). However, if the global gain is too high, the robot may start to oscillate and reaction forces may even increase (see the case when $K_1^F=2$). Fig. 7.13. shows the maximum amplitude of the force F_x measured by the force sensor on the gripper (actually, the sensor measures the reaction force R_K acting between the payload and hole, because the payload dynamics forces can be neglected[*]) as a function of the global gain applied in the force feedback K_1^F. It is observed that the maximum reaction force decreases

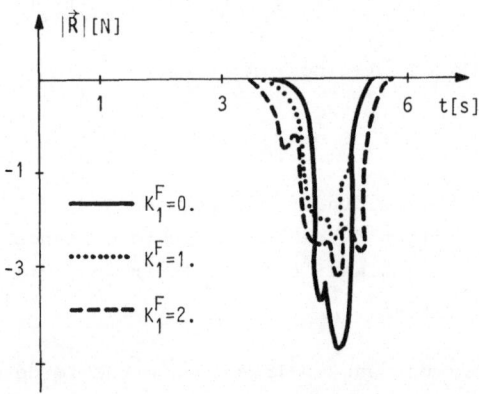

Fig. 7.12. Experimental reaction forces measured in the "peg-in-hole" task for different force feedback gains

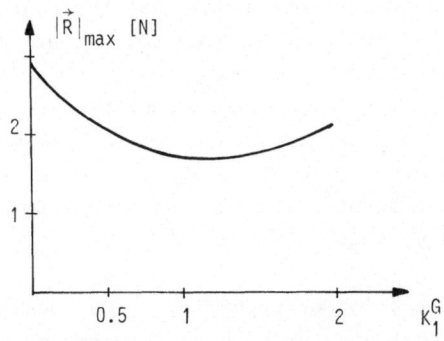

Fig. 7.13. Dependence of maximum reaction force on the force feedback gain in the "peg-in-hole" task

---

[*] We have chosen a relatively light object ($m_p=0.3$ kg) in order to diminish the effect of its dynamics on the insertion process.

for $K_1^F \leq 1$ and increases again for $K_1^F > 1$. This means that force gain should be carefully selected in order to avoid oscillatory motion of the robot when the force feedback gain is too high.

Exercises

7.3. Write the expressions for the compensating torques in the robot joints (7.3.2) for the case the peg and hole are in contact (Figs. 7.2 - 7.5), if the $\vec{R}_\ell$ forces are measured at the points of contact of the workpiece and gripper (as in Fig. 7.11) and if the assumption on neglecting the inertia forces $\vec{F}_O$, gravitational force $\vec{G}_O$ and the moment due to the inertial forces of the workpiece itself $\vec{M}_O$, is not valid (i.e. it is necessary to introduce these forces and the moment into (7.3.2)). Determine these forces and moments via coordinates and velocities of the robot joints for the robot UMS-2 in Fig. 7.6. Was such correction necessary?

7.4. In the experiment described in Example 7.3, the control of the robot shown in Fig. 7.6. is realized by means of a microprocessor which on-line calculates the nominal centralized control according to (4.4.20), local servo system (both the position and velocity) feedbacks according to (4.4.16) and additional compensating moments (on the basis of the forces measured at the gripper-object contact points) according to expressions (7.3.4), (the corresponding signals for the actuators are calculated using (7.3.3)). Write down a dynamic model for the robot UMS-2 in Fig. 7.6 and the models of the actuators (D.C. servomotors - second order models). Determine what is the minimal number of numerical operations (additions and multiplications) needed to calculate the control signals (i.e. the inputs to actuators u^i), on the basis of the given nominal trajectories q^{oi}, $\dot{q}^{oi}$ on the actual coordinates q^i and joint velocities $\dot{q}^i$ ($i = 1, 2, 3, 4$) and on the basis of forces measured by force sensors $(R_{x1}, R_{y1}, R_{z1})^T$ and $(R_{x2}, R_{y2}, R_{z2})^T$, as well as of the gripper jaws d at each sampling period. Assume the nominal joint acceleration $\ddot{q}^{oi}$ is obtained by numerical differentiation of the nominal velocity $\ddot{q}^{oi} \approx (\dot{q}^o(t+\Delta t) - \dot{q}^o(t))/\Delta t$, where Δt is the sampling period. Take that the calculation of a sine or cosine function requires one multiply and two add operations.

7.5. On the basis of the results of the preceeding exercise, determine what is the shortest sampling period that can be attained (i.e.

what is the shortest time needed to calculate input signals for
the actuators), if the control is realized using:

a) PDP-11-03 microprocessor (one add operation lasts 100 [μs],
   and one multiply operation 250 [μs]),

b) a microprocessor INTEL-80-80 (one add 0.8 [ms] one multiply
   1.5 [ms]),

c) a microprocessor INTEL-80-87 (one add 35 [μs], one multiply
   65 [μs]).

Assume the processor time is used only for the operations of addition and multiplication.

7.6. Draw a detailed scheme for the control of the UMS-2 robot according to Example 7.3 and Exercise 7.4, so that all the necessary
A/D and D/A converters, sensors and actuators are marked.

7.7.* Write in a high-level programming language a programme to calculate the control (inputs to the actuators) for the robot UMS-2
according to Example 7.3. and Exercise 7.4. during one sampling
period. Assume the programme input variables are: the nominal joints coordinates $q^{oi}(t)$ and velocites $\dot{q}^{oi}(t)$, actual joints coordinates $q^i(t)$ and velocities $\dot{q}^i(t)$, sensory forces $(R_{x1}(t), R_{y1}(t),$
$R_{z1}(t))^T$ and $(R_{x2}(t), R_{y2}(t), R_{z2}(t))^T$ and the information on the
gripper jaws $d(t)$; whereas the programme output variables are:
the input signals to the actuators $u^i(t)$ at the time instant t
(i.e. at the current sampling period). Assume all the necessary
data about the mechanism, actuators and feedback gains are given
(and represent either input data or constants in the programme).

7.4 Hybrid Position/Force Control of Robots

To the moment, we have considered in this chapter special task in robotics, namely, the robotic assembly of machine parts and elements. As
we have seen, this task can be divided into a series of subtasks (stages), the most sophisticated of them being the stage of parts mating
itself. The task of the robotic assembly belongs to a wider class of
tasks in which the robot comes into contact with the objects in the

workspace, so that these objects "act" upon the robot by reaction forces and thus constrain the robot motion in certain directions. In other
words, in this class of tasks, the robot motion in the particular directions is *constrained* by its contact with one or more surfaces. For
example, when the robot comes into contact with a surface in the workspace, the robot's motion in the direction perpendicular to that surface is constrained, but the robot can act upon that surface by a certain force (Fig. 7.14). It is obvious that the task of part assembling (i.e. of inserting a peg into a hole, Fig. 7.1) belongs to the class
of tasks with partialy constrained motion. As we have already mentioned,
to this class of tasks also belong the tasks of robotic cutting, forging, grinding, deburring and many others.

In this section we shall consider some basic problems concerning the
control in this class of tasks. It is quite obvious that, in these
tasks, it should be ensured the robot position in certain directions
(in which its motion is not constrained) is controlled (its positioning is realized in accord with some definite requirements, i.e. the
robot should track a desired trajectory), while in some other directions (in which the robot motion is constrained) the force or moment applied by the robot is to be controlled. Therefore, in these tasks involving partial constraint to the robot motion, some of the robot's
d.o.f. are controlled by position, and other by force. Because of that
this type of control is named *hybrid position/force control* [11, 12].
The robot control in the stage of parts mating, considered in Section
7.3, belongs to this type of control.

However, we have not considered yet the general relationship between
position control and force control, but we have only described the force feedback control in a special case of inserting the workpiece into
a hole. In this section we shall broaden the notion of position control and force control to encompass the whole class of tasks involving
partially constrained motion of the robot.

When the robot comes into contact with an object in the workspace, then,
in dependence of the specific mechanical and geometrical characteristics of the given contact, various constraints are imposed to the robot gripper motion and to forces the hand can realized. A set of such
constraints which are a natural consequence of the particular relationship between the robot and environment (i.e. of the nature of the contact between them) is called *natural constraints*. For example, with
the robot shown in Fig. 7.14. the robot gripper motion in the direction

perpendicular to the contact surface is naturally constrained, i.e. there is a natural constraint with respect to position. If the friction is neglected, the gripper cannot realize an arbitrary force in the direction tangential to the surface so that there is a natural constraint with respect to force (the force is zero).

To facilitate determination of these constraints to the hand position or to the force by which the hand acts on its surroundings, for each particular task, an appropriate coordinate frame of (or *constraint frame*) is adopted and placed at the position which suits best the particular task. Two characteristic manipulation tasks together with the corresponding constraints are illustrated in Fig. 7.15 [12, 13]. As can be seen, the coordinate frame with respect to which restrictions are defined may be either fixed or movable.

The constraints in Fig. 7.15.a) have been defined under the assumption that the friction between the gripper and crank handle ensures reliable grasping and that the handle can rotate with respect to the crank arm. The robot gripper position constraints are defined by the gripper velocity components with respect to the chosen coordinate frame, which is often more convenient than to define position constraints in a direct way. Similarly, the force constraints are specified via prescribing the values to components of the vector force and moment at the gripper with respect to the same coordinate frame. It is obvious that the gripper position constraint comprises the constraints with respect to position and/or orientation of the gripper, while force constraints include the constraints with respect to forces and/or moments acting upon the gripper.

The constraints in Fig. 7.15.b) have been derived under the assumption that the force in the direction y is limited to zero, since the slot in the screw head allows the screw driver to slip in this direction. All these natural constraints are determined by the nature of the contact itself, not by the desired manipulator motion.

When the desired robot motion and/or the forces to be realized by the robot gripper are prescribed, the so-called *artificial constraints* are thus defined. To define a task to be realized by the robot, is to define the artificial constraints to the gripper motion and to the forces to be realized by the gripper. In Fig. 7.15. apart from the natural constraints, we have also presented the artificial constraints

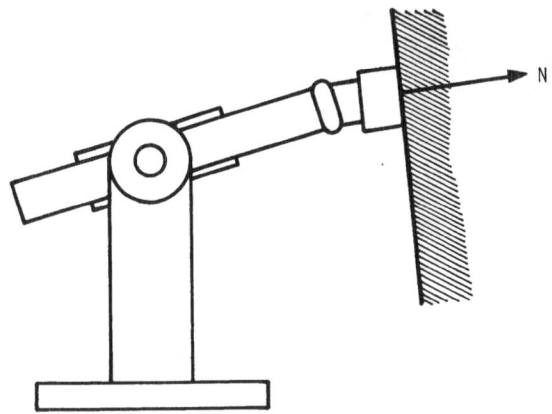

Fig. 7.14. Robot contact with a surface: constraints to robot motion
in the direction perpendicular to the surface

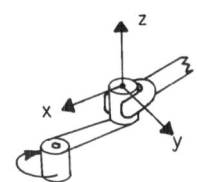

Natural constraints		Artificial constraints	
$x_c = 0.$	$F_z = 0.$	$z_c = 0.$	$F_x = 0.$
$y_c = 0.$			$F_y = 0.$
$\omega_x = 0.$	$M_y = 0.$	$\omega_y = \omega_y^o$	$M_x = 0.$
$\omega_z = 0.$			$M_z = 0.$

(a) Turning crack

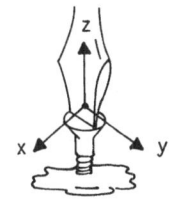

Artificial constraints		Natural constraints	
$y_c = 0.$	$F_x = 0.$	$x_c = 0.$	$F_y = 0.$
$z_c = p\omega_z^o$			$F_z = 0.$
$\omega_z = \omega_z^o$	$M_x = 0.$	$\omega_x = 0.$	$M_z = 0.$
	$M_y = 0.$	$\omega_y = 0.$	

(b) Turning screwdriver

Fig. 7.15. Natural and artificial constraints on robot for
two tasks [12]

which are determined by the production task given by the operator or
by a higher control level (i.e. the desired robot motion and forces it
has to realize). For example, in the task illustrated in Fig. 7..15.b)
it is required the gripper moves about axis z with an angular velocity
ω_z^o, but it does not move in the direction y, while in the direction z
it has to move with speed $p\omega_z^o$ and not to exert any force on the screw
head in the direction x and neither produces moments about axes x and
y (here, p denotes the step of the screw spiral groove); all these requirements directly define the artificial constraints. It is obvious
that when there is a natural constraint on the gripper motion in a certain direction (i.e. with respect to a certain d.o.f.) in the coordinate frame of constraints, then it is necessary to define the artificial constraint with respect to force in that direction, and vice versa. In every task, each gripper d.o.f. in the constraint frame should
be controlled in such a way to satisfy either a position constraint,
or a force constraint.

Therefore, the hybrid robot control should ensure the position control
in the direction (in the constraint coordinate frame) in which there
is a natural force constraint, and to ensure control of the gripper
force in the direction in which there is a natural constraint to the
gripper position. Such control should be realized that it might be applied for different combinations of these constraints with respect to
an arbitrary (suitable) constraint coordinate frame [13].

We have already pointed out that the constraint coordinate frame can
be defined in different ways, which depends on the nature of the particular task. This frame may be fixed to the tip of the robot gripper
or to either a mobile or fixed part of the objects in the workspace
(e.g. on a mobile tool, at the hole into which the peg is being inserted, etc.). Actually, the selection of this frame depends on specific
production task. Therefore this coordinate frame is usually reffered
as *task-specific frame*. However, the constraints defined with respect
to this coordinate frame may easily be transformed into the constraints
on the gripper position and constraints on the force acting on the gripper with respect to the coordinate frame used to define the external
robot coordinates s, and vector forces at the gripper F (Cartesian
frame - see Section 5.5). For simplicity sake we shall assume that the
position constraints have been defined with respect to the coordinate
frame for which the external coordinates s and external forces F were
defined, so that the natural and artificial constraints are defined

with respect to the coordinates of the vector s and vector F. Actually, it should be borne in mind that it is necessary, using the corresponding transformation matrices, to transform the position constraints and force constraints from the task-specific coordinate frame into the coordinate frame of the external coordinates s and forces F.

Let us suppose that in the i-th direction, in which a natural constraint occurs on the i-th component of vector F, is defined a desired trajectory (position) of the corresponding i-th component of vector s which defines the artificial constraint on the gripper position in the i-th direction $s_i^o(t)$. Of course, it is assumed that the component of the external (Cartesian) coordinates vector s_i is defined with respect to the same direction as the i-th component of F_i, i.e. s = $(x_c, y_c, z_c, \psi, \theta, \phi)^T$ and F = $(F_x, F_y, F_z, M_z, M_y, M_x)^T$, where x_c, y_c, and z_c are the coordinates of a fixed point at the gripper with respect to the corresponding coordinate frame and ψ, θ, ϕ are the Euler angles of the gripper (see Section 2.2), whereas F_x, F_y and F_z are the projections of the external resultant forces acting on the gripper, onto the axes of the same coordinate frame, and M_z, M_y, M_x, are the components of the external moments at the gripper with respect to the axes the Euler angles are being defined. Similarly, we can suppose that in the j-th direction, in which there is a natural constraint on the j-th component of vector s, is given a desired "trajectory" (value) of the corresponding component of vector F which defines the artificial constraint on the hand force in the j-th direction, $F_j^o(t)$.

The task of hybrid control is to realize simultaneously the prescribed nominal trajectories (positions) of the gripper coordinates $s_i^o(t)$ and prescribed nominal trajectories of the force components of the gripper $F_j^o(t)$. Hybrid control is based upon the position and force feedbacks, i.e. upon the assumption that the control system possesses the information about the actual position of the robot (hand) and the information on the actual forces acting on the hand. Thus, the control is realized on the basis of the error in position and error in the realized force.

For simplicity sake, matrix S has been introduced with the purpose of choosing the gripper's d.o.f. that are position controlled or force controlled [12, 13]. *Selection matrix* S is of diagonal form, the i-th element on the diagonal has the value 1 if the i-th gripper d.o.f. is force controlled (i.e. if the artificial constriant is prescribed with respect to the i-th component of the vector F), and the value 0, if the i-th gripper d.o.f. is position controlled (i.e. if $s_i^o(t)$ has been prescribed). It is obvious that the dimensions of matrix S are m×m (where m is the order of the external coordinates vector s and of the force vector F; $m \leq 6$ - see Section 2.2). For different tasks and different conditions of the robot-environment contacts the elements of matrix S change during the task execution (because the natural constraints change).

On the basis of the information about actual joint positions q(t) obtained from the position sensors at joints, actual values of the external (Cartesian) coordinates s(t) are calculated using expression (2.2.1). Thus, the control system calculates the error with respect to the hand coordinates (i.e. the difference between the actual coordinates and the prescribed nominal trajectories $s^o(t)$):

$$\Delta s(t) = s(t) - s^o(t) = f(q(t)) - s^o(t) \qquad (7.4.1)$$

However, as we have said above, only certain components of the vector s are "constrained" by the nominal trajectories (positions) $s^o(t)$, and these are those components which have a natural constraint on the force component. Because of that, only the "reduced" vector of error in the gripper position $\Delta s_e(t)$ is relevant to the control, and it is:

$$\Delta s_e(t) = (I_m - S) \cdot [s(t) - s^o(t)] \qquad (7.4.2)$$

By measuring forces $\vec{R}_\ell$ (by means of the force sensors mounted on the robot or on the objects in the workspace - see Section 7.3) the information is obtained about the forces by which the environment acts upon the robot. From the measured $\vec{R}_\ell$, using the corresponding transformation matrices, the actual instantaneous value of the vector F(t) is calculated as

$$F(t) = \sum_{\ell=1}^{L} T_{F\ell} \vec{R}_\ell \qquad (7.4.3)$$

where L is the number of sensors (force transducers), $T_{F\ell}$ is the (m×3) transformation matrix of the information from the ℓ-th sensor to the vector F. In this way, the error in the vector F components in the control system can be obtained as the difference between the actual forces acting on the gripper F(t) and the desired nominal trajectories of the force (and moment) components $F^o(t)$:

$$\Delta F(t) = F(t) - F^O(t) = \sum_{\ell=1}^{L} T_{F\ell} \vec{R}_{\ell} - F^O(t) \tag{7.4.4}$$

As with the position s, for force control too (for the specific task and contact situation) only certain components of F are relevant, i.e. it is relevant the error vector with respect to the gripper force ΔF_e, defined as

$$\Delta F_e(t) = S \cdot [F(t) - F^O(t)] \tag{7.4.5}$$

Thus, for the specific contact situation, i.e. for a defined S, the error in the control system with respect to the gripper position (7.4.2) and the error with respect to the force on the gripper (7.4.5) are obtained. These errors are "divided" into the hand d.o.f. controlled by position (orientation) and the hand d.o.f. controlled by force (moment). On the basis of these errors the controller should generate input signals to the actuators at the robot joints. In this way, different control laws, connecting the signals to the actuators inputs and the errors with respect to the position Δs_e and the errors with respect to force ΔF_e, can be applied.

Fig. 7.16. represents a general scheme of hybrid control [12]. The control consists of two complementary sets of feedbacks loops (the upper with respect to position, the lower ones with respect to force) each of them including its own control law and the both controlling one and the same object - the robot. The control laws have to include the transformation of the external (hand) coordinates into the joint torques (i.e. input signals to the actuators). It is obvious that both sets of the feedbacks control each robot joint in a "cooperative" manner, though each of the hand d.o.f. is controlled only by one of the feedback sets. The control can be realized in external coordinates, and this is done according to the solutions presented in Section 5.5. (Fig. 5.11). If the decentralized control in external (Cartesian) coordinates is adopted, the joints torques can be calculated (in the static case) from the relations (using (5.5.2) and (5.5.3)):

$$\Delta P = J^T(q) \Delta F = J^T(q) \{K_F S[F(t) - F^O(t)] - K_{pe} \cdot (I_m - S) \cdot$$

$$\cdot [s(t) - s^O(t)] - K_{ve}(I_m - S) \dot{s}\} \tag{7.4.6}$$

where K_F denotes a matrix of dimensions m×m which can be adopted in the form $K_F = \text{diag}(K_F^i)$, K_{pe} and K_{ve} are the matrices of dimensions m×m,

$K_{pe} = \text{diag}(K_{pe}^i)$, $K_{ve} = \text{diag}(K_{ve}^i)$. Here K_F^i denotes the force feedback gain, whereas K_{pe}^i and K_{ve}^i are the position and velocity feedback gains in Cartesian coordinates (see Section 5.5). The hybrid control thus obtained is decentralized with respect to external coordinates. This control law is completely realized at the level of hand coordinates; the error calculated in terms of hand coordinates and hand forces is multiplied first with the feedback gains, and then, the transformation is carried out into the joints torques to be realized by the corresponding actuators. (The input signals to the actuators that realize the calculated torques ΔP (7.4.6), can be determined from (5.4.11)).

Another possible solution is to calculate the position and force errors in the hand coordinates and to transform them into the errors of internal (joints) coordinates and of joints torques. First, it is calculated

$$\Delta q_e(t) = J^{-1}(q) \Delta s_e(t) \tag{7.4.7}$$

$$\Delta P_e(t) = J^T(q) \Delta F_e \tag{7.4.8}$$

and then, the control is realized in the joint coordinates (i.e. joints torques are calculated):

$$\Delta P(t) = \Delta P_p(t) + \Delta P_F(t) \tag{7.4.9}$$

Here, $\Delta P_p(t)$ is the vector of correctional torques due to the position control which can be realized via the PID control law:

$$\Delta P_p(t) = K_{PP} \cdot \Delta q_e + K_{Pv} \Delta \dot{q}_e + K_{PI} \int \Delta q_e dt \tag{7.4.10}$$

where K_{PP}, K_{Pv}, K_{PI} are the respective n×n matrices of the position, velocity and integral feedback gains (in joint position control).

In (7.4.9) $\Delta P_F(t)$ denotes the vector of correctional moments due to the force control. This control can be realized as the PI control law:

$$\Delta P_F(t) = K_{FP} \Delta P_e(t) + K_{FI} \int \Delta P_e(t) dt \tag{7.4.11}$$

where K_{FP} and K_{FI} are the (n×n) matrices of the proportional and integral gains in the force feedbacks loops. If the matrices K_{PP}, K_{Pv}, K_{PI}, K_{FP} and K_{FI} are adopted in diagonal form, the control is decen-

tralized with respect to both position and force (decentralization at
the joints level). However, it is obvious that the control about each
joint "participates" in both the position control and force control.
A block-scheme of such a control is shown in Fig. 7.17 [12].

It is evident that such control represents a combination of the control in Cartesian coordinates and the control in joint coordinates:
the difference between the given position $s^o(t)$ and the actual position $s(t)$, as well as the difference between the given force $F^o(t)$ and
the actual force at the gripper $F(t)$ are calculated at the level of
Cartesian coordinates and the feedback gains are introduced at the
joints level. Hybrid control can also be realized in the joints coordinates: the given position $s^o(t)$ and force $F^o(t)$ are transformed from
the Cartesian coordinates (into $q^o(t)$ and $P^o(t)$) and then, the control
is applied which realizes the desired $q^o(t)$ and $P_F^o(t)$. To realize the
given joints positions (trajectories), different control laws can be
applied, which have been considered in the previous chapters. The given
forces (moments) $P_F^o(t)$ are realized by means of force feedbacks, by
transforming the $\vec{R}_\ell$ forces measured by the force sensors into the forces at (moments about) the robot joints (see Section 7.3).

However, irrespective of the concrete law adopted for the realization
of the position control and force control, the essence of hybrid control is in the cooperative realization (by all the joints actuators)
of both the given positions $(I_m-S)s^o(t)$ and the given force at the
gripper $SF^o(t)$, and a consistent position/force control is ensured by
choosing the appropriate S values, i.e. the artificial constraints on
the positions and forces are assigned (in a complementary way).

In should be pointed out that the hybrid control is still at the stage
of laboratory investigations and only some partial solutions of such
control have been used with commercial robots, for some special purposes (e.g. for assembly). As we have already said in the preceeding
chapter, these solutions involve either force feedbacks or some passive compliances.

One of the basic problems in hybrid control is the automatic determination of the natural and artificial constraints on the positions and
forces in different tasks. As we have already mentioned, the natural
and artificial constraints are dependent of the contact situation and
specific task conditions, and these may change in the course of the

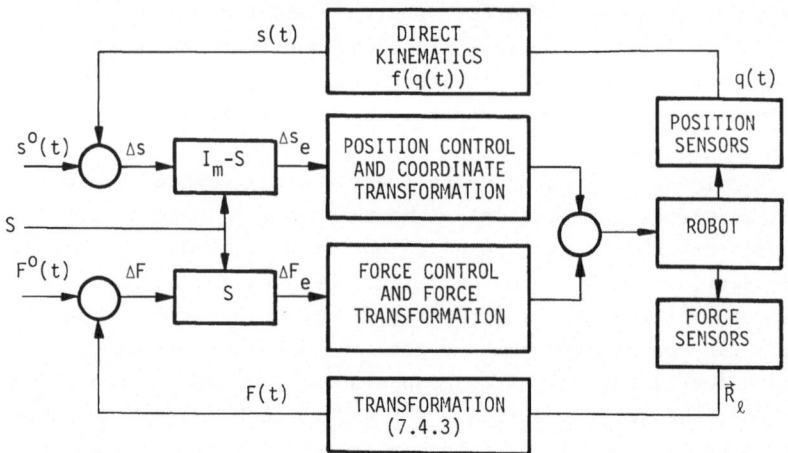

Fig. 7.16. General scheme of hybrid control [12]

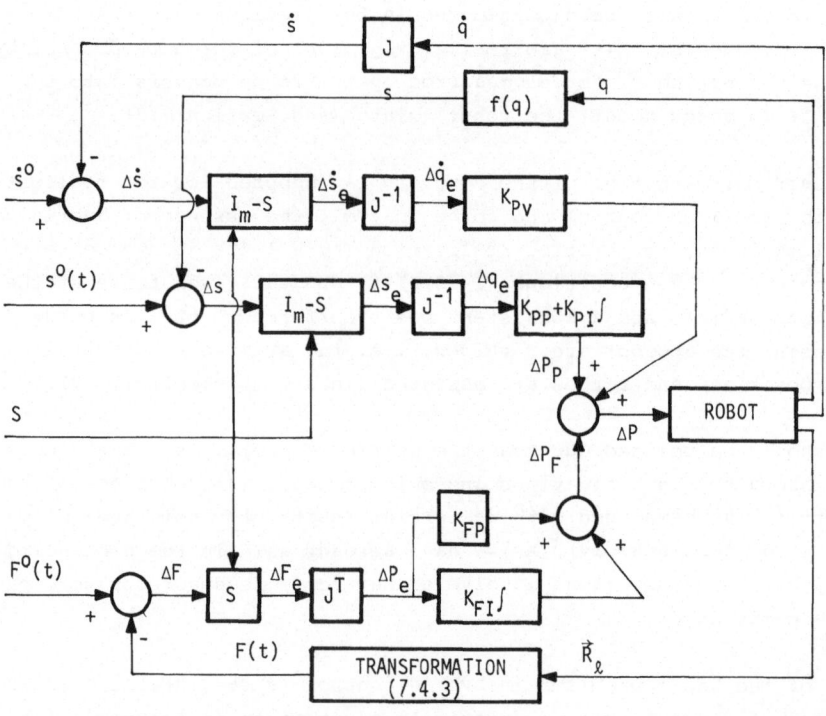

Fig. 7.17. Block-scheme of hybrid control [12]

task execution (depending on whether, and in what a way, the robot comes into contact with the objects in its environment). When a task has been assigned either by the operator or by the higher control level, these constraints should be defined for all the situations which may appear during the task execution. Defining constraints may be a very complex job, so that the task assignement may be really painstaking and may require complex programming (in a high robotic language). Because of that, it has been endeavoured to develop the *automatic planners* [14] which would automatically determine the set of constraints for all the possible situations that may arise in a particular task. Thus the user (the operator of higher control level) would assign in a global way, a desired production task result (e.g. do assembly of two elements, and the like) and the automatic planner would generate a set of artificial constraints which would ensure the task realization. The strategic control level should define the set of matrices S for every possible situation, as well as the set of artificial constraints with respect to positions and forces (in dependence of the given situation in the task execution). Moreover, the automatic planner should also predict (with the aid of sensors and the like) the way in which the robot controller can recognize the situational changes (the robot-object relationship in the workspace) and thus determine what are the artificial constraints (strategy) which should be assigned at a given moment. Methods of automatic planning are still at the experimental stage, because the general solutions are extremely complicated, which is understandable if we have in mind all the variety and great complexity of possible situations [14].

On the other hand, the hybrid control requires that the same actuators realize both the position control and force control. The problem of choosing a control law and feedback gains has not been generally solved. The choice of the force feedback gains is hindered by the lack of a reliable model of the robot-environment contact, as well as by the insufficiently known relationship between the position control and force control. It appears difficult to realize the robust hybrid control which would (using a unique control law and unique feedback gains) satisfy different contact situations, i.e. for the different natural and artificial constraints [15]. One of the fundamental problems is how to ensure a stable robot work with the force feedbacks for different strategies.

In the control laws considered above, (7.4.6) and (7.4.7) - (7.4.11), only some of the necessary terms have been introduced. To improve

the position control, some additional elements can be introduced into
these control laws. For example, it is possible to introduce into the
position control either a feedforward (local or centralized nominal
control), global control, or to apply computed torque method (Section
5.4). A feedforward element [12], and the like, can be also introduced
into the force control. Such hybrid control which takes into account
the robot dynamics is called *dynamic hybrid control* [15, 16].

However, instead to consider dynamic control of robot applying the model of the robot's dynamics expressed in joint coordinates, as we did
in Sections 5.2. and 5.4, we may apply the model expressed in external
(hand) coordinates. Since in hybrid control we intend to control directly the position of the hand and forces applied by the hand upon the
environmental objects, it is natural to express the complete robot dynamics in the hand coordinates [16]. Actually, as the natural and artificial constraints upon the gripper positions and forces are defined
with respect to suitable constraint frame which depends on the specific task (task frame), it is convinient to express the robot dynamics
in such *task-space frame*. It can be shown [16] that the dynamic model
of the robot, when expressed in such task-space coordinates, is obtained in the form similar to that one expressed in the joint coordinates
(3.2.2). Such dynamic model expressed in task-space coordinates might
be used for dynamic hybrid control, i.e. for the control scheme which
controls both position of the gripper and the forces applied by the
gripper and accounts for the entire dynamics of the robotic system. By
such dynamic control, all forces in the system (inertial, centrifugal,
Coriolis, gravity) are compensated for. However, the computation of
the "inertia matrix" and the vectors corresponding to Coriolis, centrifugal and gravity moments in the dynamic model expressed in task-
-space coordinates (or hand coordinates) is even more complex than computation of the matrix H and the vector h of the model (3.2.2). It has
been shown [17] that the current microprocessor might achieve adequate
sampling rate even if the computation of the entire dynamic model in
the hand coordinates is included in the control law. However, the main
problems with this approach are its sensitivity to variation of the
contact situation and to parameters variation, and the problem of selection of appropriate feedback gains for position part of the control
scheme and for the force part of the control scheme [18, 19]. The dynamic hybrid control (similarly to the "computed torque method" considered in Section 5.4) appears to be sensitive to *unmodelled high frequency modes* of the system (flexibilities and nonlinearities in the

manipulator links and joints, in actuators and specially in actuators transmission systems, contact situation between the robot hand and environment, flexibility of force sensors, etc. [17]). This is not suprising, since dynamic hybrid control might be regarded as an extension of "computed torque method" for position/force control. This sensitivity is even more present in dynamics expressed in hand coordinates when force control has to be included, since force sensors and contact with the environment contribute to the model uncertainies.

The interaction between the position control part and force control part has to be clarified. It is the question how a robotic system primarily designed for position control might be used for both position and force control. Therefore, a lot of both theoretical and practical problems have to be solved before the hybrid position/force control will be applied within industrial robotics [20].

It should be noted that, in the above considerations, we have tacitly adopted several approximations. For example, we have neglected the friction forces which complicate the assignement of the artificial constraints with respect to positions and forces [14]. Besides, we have neglected the effect of robot dynamics in the relations (7.4.3) and (7.4.8).

E x e r c i s e s

7.8. Determine natural constraints on the position and force for the gripper which is not in contact with any surface in the workspace (open kinematic chain).

7.9. Determine natural constraints on the positions and forces of the gripper for the contact situations shown in Figs. 7.2 - 7.5. Choose the appropriate coordinate frames to define the constraints, while neglecting the friction between the peg and hole. Determine the artificial constraints which would allow the peg insertion into the hole. How can the control system "recognize" the contact situations presented? How does the friction influence the realization of this task?

7.10. If the force sensors are mounted on the gripper at the points of contact of the gripper and workpiece as in Fig. 7.7, determine the transformation matrices T_{Fe} from (7.4.3), if the vector of

force (and moment) F(t) at the gripper has been defined with respect to the coordinate frame whose origin is at the workpiece mass
centre and the axes are along the main inertia axes of the workpiece. If the workpiece dynamics is taken into account, how will
relation (7.4.3) change?

7.11. Repeat the preceeding task but assuming the force sensors have
been situated at the gripper joints as in Fig. 7.8, and the gripper possesses a spheric joint as in Fig. 2.4. Consider how the
dynamics of the gripper and of the workpiece can influence the
relationship between the forces (and moments) at the gripper F(t)
and the forces $\vec{R}_\ell$ measured at the gripper joint.

7.12. Draw a scheme of the hybrid position/force control, if the control is realized in the joint coordinates: the given position
$s^O(t)$ and the given force $F^O(t)$ are transformed from the Cartesian into the joint coordinates ($q^O(t)$ and $P_F^O(t)$) and the nominals
$q^O(t)$ and $P_F^O(t)$ are realized by the local controllers about joints
with respect to the position and force (the local position controllers are as in Fig. 4.1).

7.13. In the preceeding problem, define the hierarchical levels of the
robot control: strategic, tactical and executive and determine
their functions.

7.14. Draw a scheme of the hybrid position/force control in the case the
position control is realized via the computed torque method (in
joint coordinates), whereas the force control is realized by a
decentralized force controller (in the Cartesian force coordinates).

7.15. Starting from the model of the robot dynamics written in external
(Cartesian) coordinates (see Exercises 5.18 - 5.20) [16] draw a
control scheme of the dynamic hybrid control which includes compensation of the total robot dynamics by "computed torque method"
in Cartesian coordinates. Assume that the position control is realized by simple PD controllers in Cartesian coordinates (5.5.3)
and the force control is realized by PD controller of Cartesian
forces. This means that the commanded Cartesian force F_D is computed analogously to (7.4.6), but the inertia matrix $\lambda(s)$ and
centrifugal moments $\mu(s)$ and gravity moments $p(s)$ - all in Carte-

sian space - have to be introduced (see Exercise 5.20):

$$F_D = K_F S[F_M(t) - F^O(t)] + \lambda(s)\{(I_m - S)[\ddot{s}^O(t) - K_{pe}(s(t) - s^O(t)) -$$

$$- K_{ve}(\dot{s}(t) - \dot{s}^O(t))]\} + \mu(s, \dot{s}) + p(s)$$

where S is the force selection matrix and $F_M(t)$ is measured force vector. Then, the joint torques have to be determined (according to
(5.5.2)) which will realize this commanded Cartesian force F_D. Try to
optimize computation of joint torques taking into account relation between the inertia matrix $\lambda(s)$, Centrifugal term $\mu(s, \dot{s})$ and gravity
term $p(s)$ in Cartesian space and these terms in joint space (see Exercise 5.20).

7.5 Stiffness and Impendance Control of Robots

As we explained in the previous section, hybrid position/force control,
although promising to solve efficiently the problems related to control
of contrained motion of robots, still cannot be effectively applied in
practice. On the other hand, we have mentioned that the passive compliance devices (such as RCC) are very efficient for control of some
specific tasks involving control of constrained motion of robots. These devices actually introduce *compliance* between the robot and the environment and enable *compliance motion* of robots. The basic idea of an
alternative approach to control of constrained motion of robot, so-called *stiffness control*, is to introduce *active compliance* between the
robot and the environment (in the directions where it is needed) rather than to achieve compliance in a passive way [21].

In order to accurately control the position (or trajectory) of the robot gripper it is required to apply very stiff servo controllers around
the joints of the robot. As explained in Chapter 3, the high stiffness
of the controller ensures that the servo will reject all disturbances
which might act upon it (from the dynamic forces of the robot itself,
of from the external forces acting upon the robot, etc.). As it has
been shown the stiffness of the servo is determined by the selected
feedback gain. Although position control requires high stiffness of
the servo, it is limited by the condition that the servo must not excite unmodelled structural modes of the system. Therefore, a finite
stiffness of the servo can be ensured, leading to finite stiffness of
the robot hand.

On the other hand, the force control requires that the system stiffness be as low as possible. Ideally the zero stiffness of the servo will allow to achieve desired force regardless of position disturbances. This means that in order to realize force control we should decrease the stiffness of the servo. If the stiffness of the joints servo systems is decreased the robot gripper will also behave as having low stiffness which will enable appropriate force control. It is well-known from some practical experiments (see Sect. 7.3) that the robot is much more successful in inserting a peg into a hole if the stiffness of the servos is decreased. (However, such solution leads to a poor positioning of the robot when it has to be used as positioning device).

If we have to realize such control tasks which require that the robot hand position has to be controlled in some directions and the force applied by the hand must be controlled in other directions, then we need the robot hand to exibit maximal stiffness in some directions (those in which its position should be controlled) and minimal stiffness in other (in those in which force applied by the hand has to be controlled). The idea of the stiffness control is to try to control the gripper stiffness in various directions depending on the specific task. Since the gripper stiffness depends on the joint stiffness, we can compute the stiffness of the joint servos necessary to achieve a desired gripper stiffness. Therefore, we have to adjust stiffness of the servos in the robot joints (by adjusting the feedback gains) in order to obtain desired stiffness of the gripper.

Let us consider one possible solution to this problem [21]. Let us try to make that the hand of the robot behaves as a general spring with six degrees of freedom. The small displacements of the hand δs (spring) should cause the force F of the hand which can be expressed by (we shall write the expression in hand coordinates, but actually task-space coordinates should be used to express the hand stiffness characteristics):

$$F = -K_{pe}\delta s \qquad\qquad (7.5.1)$$

where K_{pe} is a diagonal 6×6 matrix representing the general stiffness matrix of the robot hand. Remember that δs includes both linear displacements and rotational displacements of the hand, while F is the 6×1 vector which includes both hand forces and hand moments. Therefore

K_{pe} consists of three linear stiffness and three rotational stiffness. Now, we want the hand to exibit different stiffness in various directions depending on the specific task: the stiffness has to be low in those directions in which the force have to be controlled, and they must be high in the directions where the hand position has to be ensured. Therefore, we have to select the matrix K_{pe} according to the specific task being executed by the robot.

We have to determine the torques in the joints that would produce the force (7.5.1) at the robot hand. We recall the static relation between the hand force F and the joint torques P (5.5.3). Therefore, we have to produce the joints torques

$$P = J^T F \tag{7.5.2}$$

in order to achieve the force F (7.5.1). On the other hand, the relation between the small displacements of the hand δs and the small displacements of the joints δq is given by (according to (2.2.5)):

$$\delta s = J(q) \delta q \tag{7.5.3}$$

Combining (7.5.1)-(7.5.3) we obtain:

$$P = -J^T(q) K_{pe} J(q) \delta q \tag{7.5.4}$$

Thus, we obtain the relation between the small joint displacements and the joint torques which ensures that the robot hand behaves as a spring with six degrees of freedom (7.5.1) and with desired stiffness in various directions. Actually, we have to apply such control law which will ensure the relation between joint torques P and the joints displacements δq given by (7.5.4), and by this the desired behaviour of the robot hand (7.5.1) will be achieved.

In [21] the following control law is proposed:

$$P = -J^T(q) K_{pe} J(q) \Delta q - K_v \Delta \dot{q} \tag{7.5.5}$$

where now Δq is the vector of deviation of the joints angles from the desired position q^o, i.e. $\Delta q = q - q^o$, and K_v is a diagonal matrix $K_v = \text{diag}(K_v^i)$. By selecting stiffness matrix K_{pe} according to a specific task and by applying control law (7.5.5), we might achieve that the

robot hand behaves as desired in a specific task: in some directions
it will have high stiffness, while in the other the stiffness will be
low. Thus, by Jacobian matrix $J(q)$ we have mapped the hand stiffness
into the joint stiffness. The role of the damping factor $K_v \Delta \dot{q}$ is to
keep the damping ratio of the system to be (over)critical (and therefore velocity feedback gains K_v^i might be also adjusted). It should be
noted the *stiffness matrix* $J^T K_{pe} J$ between the joint torques and joints
displacement is not diagonal, which means that the coordination of the
joints must be ensured in order to achieve desired stiffness of the
hand.

The stiffness control might be realized also via Cartesian control
(5.5.2), (5.5.3) in the following way [15] (gravity moments might be
directly compensated for):

$$P = -J^T(q) [K_{pe} \Delta s + K_{ve} \Delta \dot{s}] \qquad (7.5.6)$$

where K_{ve} is now 6×6 diagonal matrix $K_{ve} = \mathrm{diag}(K_{ve}^i)$, K_{ve}^i is the gain
in the feedback loop by the i-th Cartesian (hand) velocity $\dot{s}$, and Δs
is the 6×1 vector of the deviation of the hand coordinates from the desired position s^o, $\Delta s = s - s^o$. Here, we directly specify the desired
stiffness K_{pe} and desired damping K_{ve} of the robot hand. The desired
stiffness is specified according to the specific task requirements, as
explained above, and the velocity feedback gains are selected to keep
the damping of the system in each direction to be (over)critical. This
means that in directions in which the stiffness is decreased, the velocity gain K_{ve}^i also must be decreased, and vice versa.

It should be noted that in the stiffness control we have not to specify just the desired stiffness K_{pe} of the robot hand (and damping K_{ve}),
but the control system also has to define desired position s^o (or, q^o).
If pure position control is considered this is the actual desired goal
position of the robot hand. However, in constrained motion control in
directions in which the forces applied by the robot gripper have to be
controlled, the desired position x^o has to be specified in such a way
that the desired force F^o is achieved. The determination of the s^o which
corresponds to desired F^o is not simple since the contact between the
robot and environment (i.e. the actual stiffness of the environment)
considerably affects the relation between s^o and F^o. Namely, the environment itself exibits certain stiffness: under the action of hand forces upon the environmental surface, a deformation of this surface ap-

pears so that the reaction force F_E of the environment (equal to applied hand force F) is given by:

$$F_E = K_E(s_E-s)$$

where K_E is an *effective stiffness* of the *environment* and s_E is contact point between the robot and environment. Therefore, the stiffness of the environment plays main role regarding efficiency of this control.

If the assembly process is considered we have to ensure desired movement of the robot hand in some directions (e.g. in direction in which a peg is inserted into a hole) and in others we have to ensure that the reaction forces are zero. Therefore, by specifying s^O in an assembly process we specify artificial compliance center in similar way that the passive compliance (RCC) does, but in this case the compliance center might be reprogrammed.

It also should be noted that stiffness control does not require explicit application of force sensors (and therefore it belongs to a class of so-called *implicit force control*). However, in practice, application of force sensors to measure forces at the robot hand are required in order to maintain the desired forces. If we assume that we measure the actual forces at the robot hand F, than we can add a term in the control law (7.5.6) which will take into account the error between the desired force F^O and actual force F:

$$P = J^T(q)[K_{pe}\Delta s + K_{ve}\Delta\dot{s} + K_F(F^O-F)] \qquad (7.5.7)$$

where K_F is the 6×6 diagonal matrix of the force feedback gains (the force feedback is implemented in Cartesian space coordinates).

It is quite obvious that the stiffness control of robots is quite similar to simple position control which we have considered in Chapters 3. and 4. Therefore, it is quite understandable that it shows similar characteristics. The stiffness control, besides being simple, is robust to parameter variations and uncertainies. Therefore, a few industrial robots already use some kind of stiffness control for some specific tasks which include constrained motion of the robot. However, the stiffness control is efficient only at low speeds. If the high speeds in the constrained motion control is required, the dynamics of the robot system has to be taken into account.

One possible extension of the stiffness control is by so-called *impendance control* [22]. The impendance control besides controlling the robot hand stiffness, attempts to take into account its dynamic characteristics. Actually, the impendance control attempts to make the robot hand behaves as a system of appropriate inertia characteristics i.e. as a system with desired mass. Thus, this approach intends to control mechanical impedance of the hand against environment and therefore it gets its name: impendance control. This approach instead to control directly forces or positions of the robot hand, attempts to control entire *dynamic relation* between *forces* (applied by the hand upon the environment) and the hand *positions*.

Numerous problems concerning impendance control still have to be solved. Stability of impendance controller has been studied, as well experimental tests have been carried on which show that this control also might be sensitive to unmodelled high frequency dynamics in the robot system and to contact situation, i.e. stiffness of the environments [23, 24]. The application of force sensors both in joints and/or at the robot gripper still has to be elaborated.

In should be mentioned that both stiffness and impendance control might be regarded as a form of realization of hybrid position/force control. However, under hybrid position/force control it is usually assumed control scheme which includes explicit force feedback [12], while stiffness and impendance control represent implicit force control (although the explicit force feedback loops as in (7.5.7) might be included).

The control of constrained motion of robots is a challenging research area, whose successful solution will considerably affect further application of robots in industry and increase their efficiency and productivity.

Exercises

7.16. Assume a six degrees of freedom robot which has to apply a force in z-direction (vertical) upon a frictionless horizontal plane, and to move along the plane. Select the stiffness matrix K_{pe} in the stiffness control (7.5.6) for this specific task.

7.17. Assume that the task from the Exercise 7.16. has to be performed by four degrees of freedom robot presented at Fig. 7.6. Determine the Jacobian matrix for this robot and write explicitely both control laws (7.5.5) and (7.5.6) (or (7.5.7). Introduce selection matrix S. Compare this control laws to that one applied in Section 7.3. Determine the number of operations (adds and multiples) that has to be performed (at each sampling interval) by control microprocessors to apply these control laws.

7.18. Consider the contact situations presented in Figs. 7.2 - 7.5. Determine how the stiffness matrix of the robot hand K_{pe} has to be selected in order to ensure insertion of a peg into a hole. (Neglect the friction between a peg and a hole).

7.19.[*] Consider one-degree of freedom robot in a contact with environment. The system is presented in Fig. 7.18. The system consists of a mass m and the environment and a force sensor. The environment and a force sensor are modelled as a spring with stiffness k_E [24]. Assume that the force F acting upon the mass can be directly controlled. Write the dynamic model of this system (friction is neglected). Assume that the force sensor gives direct information on force $F = k_E x$.

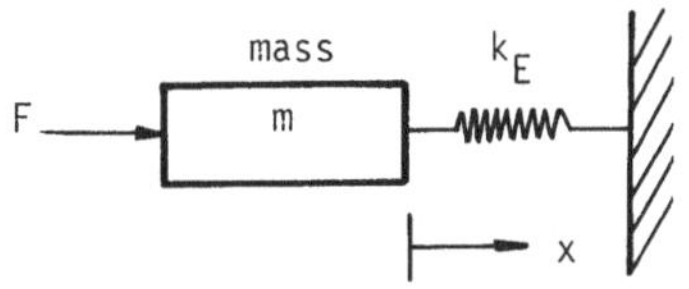

sensor + environment

Fig. 7.18. Simple model of the one-link robot in a contact with the environment [24]

a) Write the stiffness control law for this system (according to (7.5.7)).

b) Write impendance control law for this system [22], assuming that the system has to behave as having apparent mass m_d, i.e. we want that the system behaves as:

$$m_d \ddot{x} + K_V (\dot{x} - \dot{x}^0) + K_p (x - x^0) = F = k_E x$$

where m_d is desired apparent mass, K_p is desired stiffness and K_V is velocity feedback gain.

c) Compare control laws determined under a) and b).

d) The mass is m = 10 [kg], and assumed environment stiffness k_E is 10^3 [N/m]. Determine the velocity feedback gain if we want to make system (over)critically damped and the desired stiffness is K_p = 100 [N/m] in both control laws (assume that the desired mass in the impendance control is m_d = 50 [kg]). What will happen if the actual environment stiffness is $k_E = 10^6$ [N/m]?

e) Consider opposite case. Assume that the environment stiffness k_E is large 10^6 [N/m] and compute the adequate velocity feedback gain. How will the system behave if the actual environment stiffness is low (i.e. $k_E = 10^3$ [N/m])?

7.20.* If in the system considered in the previous exercise, friction force is taken into account, will the system exibit better stability characteristics or not [25]?

References

[1] Nevins J., D. Whitney et al., Exploratory Research in Industrial Modular Assembly, Report R-996, The Charless Starc Draper Laboratory, Cambridge, Mass., 1976.

[2] Simunović S., "Force Information in Assembly Processes", Proc. V International Symposium on Industrial Robots, Chicago, 1975.

[3] Stokić M.D., Vukobratović K.M., "Simulation and Control Synthesis of Manipulator in Assembling Technical Parts", Journal of Dynamic Systems, Measurement and Control, 1979.

[4] Stokić M.D., Vukobratović K.M., Hristić S.D., "Implementation of Force Feedback in Manipulation Robots", International Journal of Robotic Research, No. 4, 1985.

[5] Scheinman V., "Design of a Computer Controlled Manipulator", M.S. Th., Mech. Engineering Dept., Stanford University, 1969.

[6] Nevins L., Whitney D., "The Force Vector Assembler Concept", Proc. First International Conference on Robots and Manipulators, Udine, Italy, September, 1973.

[7] Whitney E.D., "Historical Perspective and State of the Art in Robot Force Control", The International Journal of Robotics Research, Vol. 6., No. 1., 1987.

[8] Nevins J., Whitney D., "Assembly Research", Automatica, Vol. 16, pp. 595-613, 1980.

[9] Drake H., "The Use of Compliance in a Robot Assembly System", IFAC Symp. on Information and Control Problems in Manufacturing Technology, Tokyo, 1977.

[10] DeFazio T., Seltzer D., Whitney D., "The IRCC Instrumented Remote Centre Compliance", Industrial Robot, 1984.

[11] Mason M., "Compliance and Force Control for Computer Controlled Manipulators", MIT Artificial Intelligence Laboratory Memo 515, April, 1979.

[12] Raibert M., Craig J., "Hybrid Position/Force Control of Manipulators", Trans. of the ASME, Journal of Dynamic Systems, Measurement and Control, Vol. 103, No. 2, Juni, 1981.

[13] Craig J., Introduction to Robotics: Mechanics and Control, Addison-Wesley Publishing Company, 1986.

[14] Lozano-Perez T., Mason M., Taylor R., "Automatic Synthesis of Fine Motion Strategies for Robots", Proc. 1st International Symposium of Robotics Research, Bretton Woods, New Hampshire, August, 1983.

[15] Asada H., Slotine J., Robot Analysis and Control, John Wiley and Sons, New York, 1986.

[16] Khatib O., "A Unified Approach for Motion and Force Control of Robot Manipulators: The Operational Space Formulation", IEEE Journal of Robotics and Automation, Vol. RA-3, No. 1, 1987.

[17] Khatib O., Burdick J., "Motion and Force Control of Robot Manipulators", Proc. of IEEE International Conference on Robotics and Automation, 1986.

[18] Eppinger S.D., Seering W.P., "Understanding Bandwidth Limitations in Robot Force Control", Proc. of IEEE International Conference on Robotics and Automation, pp. 904-909, 1987.

[19] Yabuta T., Chona A., Beni G., "On the Asymptotic Stability of the Hybrid Position/Force Control Scheme for Robot Manipulators", Proc. of IEEE International Conference on Robotics and Automation, pp. 338-343, 1988.

[20] Paul R.P., "Problems and Research Associated with the Hybrid Control of Force and Displacement", IEEE International Conference on Robotics and Automation, 1987.

[21] Salisbury J.K., "Active Stiffness Control of a Manipulator in a Cartesian Coordinates", Proc. of IEEE Conference on Decision and Control, Albuquerque, 1980.

[22] Hogan N., "Impedance Control: An Approach to Manipulation: Part I Theory, Part II - Implementation, Part III - Applications", Trans. of the ASME, Journal of Dynamic Systems, Measurement, and Control, Vol. 107, pp. 1-24, 1985.

[23] Kazerooni H., Houpt P.K., Sheridan T.B., "Robust Compliant Motion for Manipulators: Part I - The Fundamental Concepts of Compliant Motion, Part II - Design Methods", IEEE Journal of Robotics and Automation, RA-2, pp. 83-105, 1986.

[24] An C.H., Hollerbach J.M., "Dynamic Stability Issues in Force Control of Manipulators", Proc. of IEEE International Conference on Robotics and Automation, pp. 890-896, 1987.

[25] Wedel D.L., Saridis G.N., "An Experiment in Hybrid Position/Force Control of a Six Dof Revolute Manipulator", Proc. of IEEE International Conference on Robotics and Automation, pp. 1638-1642, 1988.

Appendix
Software Package for Synthesis of Robot Control

In this Appendix listings of subroutines of a very reduced version of the software package for computer-aided synthesis of control of robotic systems are given. The package enables synthesis of the decentralized control for the robot of arbitrary structure (with up to 6 degrees of freedom - up to 6 simple joints). The package includes the following functions:

- Read data on a robot for which the user wants to synthesize control (structure and geometry of the robot, kinematic and dynamic parameters of the robot, data on actuators); these data have to be defined by the user in corresponding files-see description of input data files.

- Form dynamic model of the robot which consists of the model of mechanical part of the robot and of the models of actuators (the model of mechanical part is formed by programmes which listings are given in the book: Dynamics of Manipulation Robots: Modelling, Analysis and Examples, by M. Vukobratović).

- Determine the maximum moments of inertia of the mechanism around the axes of joints and form the models of the subsystems (actuator + moment of inertia of mechanism).

- Compute the local servo systems feedback gains for individual joints of the robot and actuators, using one of two available methods (according to the user choice):

  - by imposing of resonant structural frequence around the given joint and by requiring that the damping ratio must be equal to 1, or

  - by optimal quadratic regulator for individual joint (the user has to specify the weighting matrices and the prescribed (desired) degree of exponential stability, and the user might, if he wants so, introduce an tntegral feedback loop (PID regulator)).

- Determine the transfer functions of the open and closed-loop subsystems (actuator + joint).

- Compute trajectories of the joints for the specified initial and final positions with both triangular and trapezoid velocity profile (with specified movement time and with specified time of acceleration/deceleration).

- Compute nominal driving torques in the joints for the specified trajectories and compute nominal centralized or local programmed control.

- Form linearized model of the robot in the specified points on the nominal trajectory, determine eigen-values of the open-loop and closed-loop matrices of the linearized model (with local servo feedback loops) and check stability in accordance to the user requirements, and re-compute the local control if the conditions of stability of the linearized model are not fulfilled.

- Simulate the robot dynamics around the specified nominal trajectory for the given initial conditions and various control laws (according to the user wishes):

  - with the selected local servo systems,

  - with or without the nominal centralized control or with the local nominal control,

  - with or without global control, and for each joint the user might specify:

    - the global control in the form of force feedback, or,

    - the global control by on-line computation of driving torques which uses approximative models of robot which has to be selected by the user (as an approximative model computation of only inertia matrix of the robot might be introduced, or only gravity moments, or centrifugal and Coriolis moments might be calculated, or the user might include computation of complete dynamic model of the robot as the global control),

- selected form of the global gain (5.2.13) or (5.A.3) (the latter
form is allowed only if the servo is synthesized as local optimal quadratic regulator).

In this Appendix the description of the package structure is given,
and the listings of all subroutines, with brief description of each
subroutine. The guide for forming of the task is also given. The programmes are written for VAX FORTRAN V.4.1 and they have been tested
at the VAX-750 computer under operating system VMS 4.7, but they
can be easely used at other computers which are supplied with FORTRAN
compiler.

The user has to prepare corresponding data files for his selected robot, according to guideline given in the text to follow.

In the Appendix is also given an example of application of the package
for control computation for a robot with three joints presented in Fig.
3.2. The data on the robot are given in Tables 3.3.1. and 3.3.2. The
example includes synthesis of the local servos according to the imposed resonant frequencies of structure, computation of the trajectories
of the joints with the triangular velocity profile, analysis of stability of the linearized model of the robot in two points at the nominal
trajectory, and simulation of tracking of the nominal trajectory with
the nominal centralized control and the local servo systems. The listing of all demanded input data files for this specific robot are given (these data files must be provided by the user), and listing of
the interaction between the user and the package also is given. The
package forms several output data files which include simulation results and results of control synthesis. These data files are also described. Here, only the listing of the output data file which contains
computed local servo gain for the specific example is given, but the
other output data files are not given due to lack of space.

Note: Note that in the presented package many options are not introduced in order to condense the programmes and due to lack of
space. The user can simply extend many functions and easely
introduce various options which will increase the applicability
of this package (for example, the user might introduce computation of the local feedback gains by pole-placement method - see
Section 3.3.3, introduce specification and generation of the
nominal trajectories in external (Cartesian) coordinates - see
Chapter 2, introduce simulation of only individual (decoupled)

subsystems etc.). In lot of subroutines which read data from the input data files and which communicate with the user, the necessary verifications and testing of the data, specified by the user, and also protection functions to prevent nonregular handling of the system are ommitted, and this might produce certain problems in the application of this package. We recommand the user to extend this package by himself in the above described sence, and to add necessary protections (some of them are already included in several subroutines in order to instruct the user how this can be realized wherever it is necessary). We have also to note that in writting these programmes the attempt has been to simplify them as much as possible, in order to make them simple for an user having no experience in software programming. However, in a few subroutines we have used some more sophisticated programming procedures for educational purposes.

To use this package user needs programmes for modelling of mechanical part of the robot. As mentioned before, the programme support for modelling of robot dynamics is given in the listed reference. However, the user may write his own programme for modelling dynamics of his particular robot. He has to write programme which for given joint angles q and velocities $\dot{q}$ computes the inertia matrix H(q) and the vector h(q, $\dot{q}$) (see Exercise 3.5). The programme should be called MODEL and it should include the following two commons: COMMON/UG/ SI(6), SIDOT(6), and COMMON/DINAM/ HH(6,6), H1(6), where SI(6) is vector containing values of joint angles, SIDOT(6) corresponds to joint velocities, HH(6,6) is inertia matrix, H1(6) is vector h.

DESCRIPTION OF SUBROUTINES

Task KG includes the following subroutines:

KG - the main routine which calls all other subroutines

SMINIC - reads the initial values of variable, data on auxiliary parameters and so on.

INPUT_KIN - reads data on the kinematics of the selected robot

MODEL - the main subroutine for computation of the inertia matrix H and the vector of centrif. Coriolis and gravity moments h - this subroutine and all subroutines called are given in ref. M. Vukobratovic: Applied Dynamics of Robots, Springer-Verlag,Berlin, 1989.

COLIAS - the subroutine for the mechanims assembling in the special case

MODCHK - testing of succesfull assembling of the mechanism according to the user's data

INPUT_DIN - reads the data on the dynamic parameters of the robot

SM_MAX_IN - computes the maximal and minimal values of the moments of inertia of the mechanism around the joints axes

INPUT_ACT - reads data on the actuators in the robot joints and forms the models of actuators and joints

INPUTL - reads data on the actuators

SMOPRG - synthesizes the local optimal regulators with or without the integral feedback loop

SMOREG - solves the algebraic equation of Riccati type

SMEIG3 - auxiliary subroutine for determination of the eigen-values and eigen- vectors of a given quadratic matrix

SMLOC_E - synthesizes the local control feedback gains (synthesis in S - domain)

TRAJEK - computes the nominal trajectories (in the joints coordinates)

SMNOMP - computes the nominal moments and the nominal (centralized or local) programmed control

LINANA - subroutine for analysis of the linearized (total) model of the robot via determination of the eigen-values of the system matrix with or without the local controllers

SMLINM - forms matrices of the open-loop linearized model of the robot

SMLGAD - forms matrices of the closed-loop linearized model of the robot

SMEIGN - auxiliary subroutine for computation of the eigen-values and eigen-vectors of a given quadratic matrix

SMSIMD - simulates the total dynamic model of the robot

SMOPTN - subroutine for imposing the control law which will be simulated

SMINGL - for imposing the form and the gain of the global control in simulation

SMCONT - computes the local control during the simulation

SMCONG - computes the selected global gains during the simulation

LIAP1S - computes subsystem Liapunov's function
GRAD1S - computes gradient of the subsystem
Liapunov's function
SMMINV - the auxiliary subroutine for matrix
inversion
The auxiliary SSP subroutines which are used in the task:
GMPRD - for matrix multiplication
MINV - invert a given quadratic matrix
EIGENP,COMPVE,REALVE,HESQR,SCALE - subroutines
for computation of eigen-values and eigen-
vectors of a given quadratic matrix

The above listed subroutines uses the INCLUDE-files with
the COMMON-areas:
MODELM.MOD - COMMON-s for the subroutines related
to the MODEL of the mechanism
dynamics - see above given reference
CONFIG.MOD -   "
IN1:SMCOM.COM - COMMON-s with variables whose
description is given in the
corresponding subroutines
The subroutine SMLOC_E uses the auxiliary data-file
IN1:CONF.DAT for the sake of better communication
with the user.
Note: for description of IN1 see the text to follow

```
-----------------------------------------------------------------
                HIERARCHICAL STRUCTURE OF THE TASK
-----------------------------------------------------------------


LEVEL:     I    II        III        IV          V
           KG
                SMINIC
                INPUT_KIN
                            MODEL
                            COLIAS
                            MODCHK
                INPUT_DIN
                            SM_MAX_IN
                                       MODEL
                INPUT_ACT
                            INPUTL
                SMOPRG
                            SMOREG
                                       SMEIG3
                                                   EIGENP(SCALE...)
                                       GMPRD
                                       MINV
                                       EIGENP
                                                   SCALE
                                                   COMPVE
                                                   REALVE
                                                   HESQR
                SMLOC_E
                TRAJEK
                SMNOMP
                            MODEL
                LINANA
                            SMLINM
                                       MODEL
                                       MINV
                            SMLGAD
                            SMEIGN
                                       EIGENP(SCALE...)
                SMSIMD
                            SMOPTN
                                       SMINGL
                            MODEL
                            SMCONT
                            SMCONG
                                       LIAP1S
                                                   GMPRD
                                       GRAD1S
                                                   GMPRD
                            SMMINV
                                       MINV


-----------------------------------------------------------------
```

INSTRUCTION HOW TO FORM THE TASK:
1. SET AT THE DIRECTORY WHERE ALL SUBROUTINES ARE
   LOCATED AND ASSIGN THE FOLLOWING LOGICAL NAMES
   ASS [directory name] IN1: - the directory where data-files
                                      are located
   ASS [directory name] IN2: -   "
   ASS [directory name] KRO: - directory where source codes
                                      are located

2. Type the following commands for compilation of the subroutines:
$@KG.CPL [ret]

   Form the library kro:MODEL.OLB which includes all subroutines necessary to form the model of the robot dynamics (the
   subroutines are listed in the above given reference).
   Form the library kro:EIGENP.OLB with all SSP subroutines
   which are listed above.

3. Type the following command for linking of the task:
$@KG.LNK

4. The task is executed by the following command:
$ R kro:KG

Before the running of the task the user has to check whether
all necessary input data-files are formed, containing the
data for the given robot (the data-files have the name of
the specific robot selected by the robot) - see instructions for forming of the data-files.

```
$ ! COMMAND FILE FOR SUBROUTINES COMPILATION
$ for kro:input_kin,input_din,input_act,inputl,colias,modchk
$ for kro:smloc_e
$ for kro:domp
$ for kro:sm_max_in
$ for kro:trajek
$ for kro:sminic
$ for kro:smnomp
$ for kro:linana,smlinm,smlgad,smeign,smoptn
$ for kro:smoprg,smoreg,smeig3,smore1,smeig1,smore2,smeig2
$ for kro:smsimd,smingl,smcong,smminv,smcont,liap1s,grad1s
$ lib/create kro:kg
$ lib kro:kg kro:*.obj
$ delete kro:*.obj;*
$ for kro:kg
```

```
$ ! COMMAND FILE FOR LINKING THE TASK: KG
$ LINK kro:KG,KG/LIB,MODEL/LIB,EIGENP/LIB
```

```
*****************************************************************
*               DESCRIPTION OF INPUT/OUTPUT DATA FILES          *
*****************************************************************
```

The package enables the synthesis of control, linear
analysis and simulation of any robot (with up to 6
degrees of freedom). The user has to prepare the data-
files which contains data on the robot for which he
wants to synthesize the control. The package as its
output generates the output data-file and/or prints
output data at the terminal.

All input/output data-files get their names according
to the following rule:
   ROBOT NAME.EXT
where ROBOT NAME is a string of five characters which
contains the name of the robot selected by the user,
EXT denotes the type of data which is put in the corresponding data-file (see the text to follow). When
the user decides for which robot he wants to synthesize the control, he has to prepare the data-files with
data on his robot and in doing this he has to to obey
te above given rule on the files names.

The package requires the following input data-files:

   ROBOT NAME.DAT - default values of certain parameters and
                    initial defining of the actuator matrices
                    for the sake of software reliability
   ROBOT NAME.CNF - data on geometry, structure and kinematics
                    of the robot
   ROBOT NAME.DNM - data on dynamic parameters of the robot
   ROBOT NAME.ACT - data on the robot actuators
   ROBOT NAME.TRA - data on specified joint trajectories
   ROBOT NAME.INT - data on required exponential stability
                    degrees of the robot and on time instants
                    at the nominal trajectory in which the
                    user wants to analyze the stability of the
                    linearized model of the robot
As an output the package generates the following data-files
(names are given according to the above mentioned rule):

   ROBOT NAME.LOC - data on synthesized local feedback gains
   ROBOT NAME.ANG - the nominal joint trajectories as functions
                    of time
   ROBOT NAME.DIN - the nominal joint trajectories, the nominal
                    driving torques, the nominal programmed
                    control as functions of time
   ROBOT NAME.SIM - simulation results: deviations of the state
                    coordinates from the nominal, actual con-
                    trol signals, actual driving torques,
                    actual power of actuators and actual
                    energy consumptions as functions of time

```
*******************************************************************
*                 DESCRIPTION OF DATA-FILE  ***.DAT              *
*******************************************************************
```

The data-file ROBOT NAME.DAT contains default data on
actuators and parameters of control laws. The data on
actuators are given in this file only for the sake of
increasing the system reliability, although the actual
data on actuators are set in the data-file ROBOT NAME.
.ACT and then the models of actuators are formed. Therefore, in this data-file the user might copy data on
the actuators matrices for any robot (it is recommended
to overtake data from the given example for the robot
UMSPR). The actual data for the selected actuators for
the specific robot, the user has to impose in the data-
-file ROBOT NAME.ACT (see description of this data-
file).

File ROBOT NAME.DAT has to be formed according to the
following sequence:
  1) N - the maximal number of degrees of freedom of the
         robot (the actual value is given in the file
         ROBOT NAME.CNF) - format I4
  2) NI(I) - the order of the actuator in the i-th
             joint - format I4
  3) A(ni(i),ni(i),i), B(ni(i),i), F(ni(i),i) - the
     matrices of the model of the i-th actuator -
     - format 3E15.5

     data 2)-3) are repeated in the file N-times
     (it is recommended to copy data 1)-3) from the
     file UMSPR.DAT - see Example)

  4)Q(NI(j),NI(j)) - weighting matrix by the state in the
                     standard quadratic criterion for the local
                     optimal regulator - it must be positive
                     semidefinite and of dimensions NI(j)xNI(j)
                     where NI(j)=max(NI(i)) - format 3E15.5.
  5) R(i) - the weighting elements by the control signals
            in the standard quadratic criteria for the
            local optimal regulators - must be N positive
            numbers - format 3E15.5
  6) QI(i) - the weighting elements by the "integral"
             state coordinates in the standard quadratic
             criteria for the local optimal regulators -
             - must be N positive numbers - format 3E15.5
  7) IOPTOR(i),IOPTV(i),IOPTI(i) - indicators whether in the
             i-th local controller is applied or not:
                 - optimal quadratic regulator
                 - velocity feedback
                 - feedback by rotor current (pressure)
             indicatiors are 1 if yes, 0 if not - format I4
  Data under 7) have to repeat N times (for N local
  controllers).

```
*********************************************************************
*               DESCRIPTION OF DATA-FILE ***.CNF                   *
*********************************************************************
```

The data-file includes data on the robot structure. In the text
to follow brief descriptions of the variables for which data have
to be set are given and the corresponding formats are denoted.
The detailed description of the variables can be found in the
previously mentioned reference.

1) Number of degrees of freedom of the robot (format 1X,I1) −
number of simple joints.

2) Indicator of the type of the first joint (0 − rotational, 1−
linear) (format 1X,F15.6)

3) X coordinate of the unit vector of the first joint with respect
to the coordinate frame attached to the first link e11
(format 1X,F15.6)

4) Y coordinate of the unit vector of the first joint with respect
to the coordinate frame attached to the first link e11
(format 1X,F15.6)

5) Z coordinate of the unit vector of the first joint with respect
to the coordinate frame attached to the first link e11
(format 1X,F15.6)

6) X coordinate of the unit vector of the second joint with respect
to the coordinate frame attached to the first link e12
(format 1X,F15.6)

7) Y coordinate of the unit vector of the second joint with respect
to the coordinate frame attached to the first link e12
(format 1X,F15.6)

8) Z coordinate of the unit vector of the second joint with respect
to the coordinate frame attached to the first link e12
(format 1X,F15.6)

9) X coordinate of the vector from the first joint to the mass center of the first link with respect to the coordinate frame
attached to the first link r11 [m] (format 1X,F15.6)

10) Y coordinate of the vector from the first joint to the mass center of the first link with respect to the coordinate frame
attached to the first link r11 [m] (format 1X,F15.6)

11) Z coordinate of the vector from the first joint to the mass center of the first link with respect to the coordinate frame
attached to the first link r11 [m] (format 1X,F15.6)

12) X coordinate of the vector from the second joint to the mass center of the first link with respect to the coordinate frame
attached to the first link r12 [m] (format 1X,F15.6)

13) Y coordinate of the vector from the second joint to the mass center of the first link with respect to the coordinate frame
attached to the first link r12 [m] (format 1X,F15.6)

14) Z coordinate of the vector from the second joint to the mass center of the first link with respect to the coordinate frame attached to the first link r12 [m] (format 1X,F15.6)

If the joint is linear, then at this place in the file ***.CNF the following data on the robot structure have to be set:

15) The special vector for the linear joint "uii" X coordinate (format 1X,F15.6)

16) The special vector for the linear joint "uii" Y coordinate (format 1X,F15.6)

17) The special vector for the linear joint "uii" Z coordinate (format 1X,F15.6)

18) The special vector for the linear joint "ui,i+1" X coordinate (format 1X,F15.6)

19) The special vector for the linear joint "ui,i+1" Y coordinate (format 1X,F15.6)

20) The special vector for the linear joint "ui,i+1" Z coordinate (format 1X,F15.6)

Description of the futher data to be set are identical as these given for the first joint and they repeat n times for all joints of the robot (n — number of joints)

At the very end of the file the data on the unit vector of the axis of the first joint with respect to the absolute coodinate frame are given:

X coordinate of the unit vector of the axis of the first joint with respect to absolute coordinate frame "e0"(format 1X,F15.6)

Y coordinate of the unit vector of the axis of the first joint with respect to absolute coordinate frame "e0"(format 1X,F15.6)

Z coordinate of the unit vector of the axis of the first joint with respect to absolute coordinate frame "e0"(format 1X,F15.6)

```
**********************************************************************
*            DESCRIPTION OF DATA FILE ***.ACT                        *
**********************************************************************
```

Data file ***.ACT includes data on actuators which have been adopted for the specific robot. In the text to follow the description of the variables whose values have to be given in this file, as well as formats of these values are listed.

1) The first line in the file represents the types of the actuators in the robot joints. Each actuator is denoted by two character symbol:
   i) for D.C. electro-motors the symbol is DC
   ii) for hydraulic actauator the symbol is HD
   The format of data on the actuators types is 6(1X,A2).

Depending on the selected type of the actuator, in the next lines of the data-file the following data have to be set (for the actuator in the first joint):

   - for D.C. electro-motors:
2) The second line contains a data on the order of the actuator model; the vaule migh be: 2 or 3 - format 1X,F15.6

3) The third line contains a value of the mechanical constant of the actuator Kem multiplied by the moment reduction ration of the gear reducer [Nm/A] - format 1X,F15.6

4) The electromotor constant Kme multiplied by the speed reduction ratio of the gear reducer [V/rad/s] - format 1X,F15.6

5) The moment of inertia of the rotor [kg*m**2] - format 1X,F15.6

6) The coefficient of the viscous friction multiplied by the speed reduction ratio and the moment reduction ratio of the gear redu-cer [Nm/rad/s] - format 1X,F15.6.

7) The resistance of the rotor curcuit [Om] - format 1X,F15.6

8) The absolute value of the maximal allowed input voltage signal (for positive values of signals) for D.C. motor [V] - format 1X,F15.6.

9) The absolute value of the maximal allowed input voltage signal (for negative values of signals) for D.C. motor [V] - format 1X,F15.6.

10) The speed reduction ratio of the gear reducer-format 1X,F15.6.

11) The moment reduction ratio of the gear reducer-format 1X,F15.6.

12) The inductivity of the rotor curcuit [H] - format 1X,F15.6.

13) The maximal driving torque of the motor [Nm] - format 1X,F15.6.

14) The motor maximal power [W] - format 1X,F15.6.

15) This values is of no use for D.C. motor model and therefore any number might be put - format 1X,F15.6

16) "   "

   - for hydraulic actuator:

2) The coefficient of viscous friction [Nm/m/s]; 1X,F15.6.

3) The piston area [m**2]; 1X,F15.6.

4) The cylinder volume [m**3]; 1X,F15.6.

5) The mass of piston [kg]; 1X,F15.6.

6) The coefficient of the oil compressibility [N/m**2];1X,F15.6.

7) The gradient of the flow-preassure characteristics of the servo-
   -valve [m**3/s/N/m**2]; 1X,F15.6.

8) The coeficient of the proportionality between the current in the
   servo-valve coil and the oil flow [m**3/s/mA]; 1X,F15.6.

9) The absolute value of the maximum allowed input current for the
   servo-valve (for positive values of current) [mA]; 1X,F15.6.

10) The absolute value of the maximum allowed input current for the
   servo-valve (for negative values of current) [mA]; 1X,F15.6.

The next lines of this file contain values of the same parameters of the actuators in the next joints of the robot, and therefore the descriptions of the lines 2-15 repeat for the next sets of values (16-29, 30-43, etc.). The file contains the sets of data on n actuators in the n joints of robots.

```
*******************************************************************
*                 DESCRIPTION OF DATA-FILE   ***.DNM             *
*******************************************************************
```

The data-file ***.DNM includes data on dynamic parameters of the
robot. In the text to follow brief descriptions of the variables
whose values have to be set in this data-file are given as well
as formats in which the values must be set in the file.

1)   The link type (1- link of the cane type, 0 - not cane)
        (format 1X,F15.6)

2)   Link mass [kg] (format 1X,F15.6)

3)   Moment of inertia of link "Jxx" or "Js" depending on the link
        type [kg*m**2] (format 1X,F15.6)

4)   Moment of inertia of link "Jyy" or "Jn" depending on the link
        type [kg*m**2] (format 1X,F15.6)

5)   Moment of inertia of link "Jzz" (if link type 1) or any number
        (if link type 0) [kg*m**2] (format 1X,F15.6)

6)   Maximal allowable angle (displacement) of joint [rad or m]
        (format 1X,F15.6)

7)   Minimal allowable angle (displacement) of joint [rad or m]
        (format 1X,F15.6)

        Next data are repeated for all links of the robot
        according to the same sequence.

```
*******************************************************************
*                 DESCRIPTION OF DATA FILE    ***.TRA           *
*******************************************************************
```

The data file includes data for the given robot which are used in
trajectory generation. The description of the values for which
the data have to be set are presented according to the sequence
in which they are set in the file.

1) Q0(I) - the vector of the initial positions of the joints given
             in: [m]   - if the joint is linear
                 [rad] - if the joint is rotational
             (format 1X,6F10.5)
2) QF(I) - the vector of the terminal positions of the joints given
             in: [m]   -if the joint is linear
                 [rad] -if the joint is translational
             (format 1X,6F10.5)
3) H   _ sampling period at which the values of joints coordinates
             along the trajectory are computed in [s]
             (format 1X,6F10.5)
4) T   - time duration of the nominal movement (between the initial
             and terminal positions) in [s]
             (format 1X,6F10.5)

```
**********************************************************************
*                   DESCRIPTION OF DATA-FILE   ***.INT               *
**********************************************************************
```

The data-file ROBOT NAME.INT contains data on the time instants along the nominal trajectory in which the analysis of the linearized model stability is required. The exponential stability of the robot linearized model around the nominal trajectory is tested. The desired exponential stability degrees have also to be imposed. The user has to prepare this file according to the following structure:

- the number of points along the nominal trajectory of the robot in which the linearized model has to be analyzed - format I4

- time instant at the nominal trajectory in which the analysis of stability of the linearized model is required - format E15.5 (the user must carefully specify the instant which is within the adopted duration of the nominal movement - this duration is specified in the file ***.TRA)
- the exponential stability degree of the linearized model arond the nominal trajectory which the user wants to achieve - format E15.5 (this value must be positive and usually there is no sense to impose values greater than 20. due to numerical problems, save for some special cases)

The last two data have to be specifeid as many times as is the number of the points at the nominal trajectory around which the user wants to analyze the robot stability (this number is specified in the first line of the file).

```
**********************************************************************
*                DESCRIPTION OF DATA-FILE ***.LOC                    *
**********************************************************************
```

The output data-file ROBOT NAME.LOC contains the data
on the computed feedback gains of the local controllers.
The data-file has the following form:

  - position feedback gain of the local controller
    around the first joint [V/rad] or [V/m] if D.C.
    electro-motor is applied, [mA/rad] or [mA/m] if
    hydraulic actuator is applied – format F12.5

  - velocity feedback gain of the local controller
    around the first joint [V/rad/s],[V/m/s] or
    [mA/rad/s], [mA/m/s] – format F12.5

  - gain in the feedback loop by the rotor current
    if D.C. motor is applied [V/A], or by pressure
    if hydraulic actuator is applied in the first
    joint [mA/N/m**2] – format F12.5

  - integral feedback gain in the local controller
    around the first joint [V/rads], [V/ms] or
    [mA/rads],[mA/ms] – format F12.5

The data on the feedback gains in the local
controllers in the other joints are repeated according
to the same sequence – the data are given for all
joints of the specific robot.

```
**********************************************************************
*                DESCRIPTION OF DATA-FILE ***.ANG                    *
**********************************************************************
```

The output data-file ROBOT NAME.ANG contains data on
the nominal trajectory. The file is the output of the
subroutine TRAJEK. The structure of the file is as
follows:

TE – time instant [s] – format E13.5
Q(6) – nominal trajectories of the robot joints
        the values of the joint coordinates in the
        instant TE at the nominal trajectory
        [rad] or [m] – format 6E13.5
        In the line there are N data (N the number of
        the joints of the specific robot)
DQ(6) – nominal trajectories of the joint velocities
        [rad/s] or [m/s] – format 6E13.5
        In the line there are N data.
DDQ(6) – nominal trajectories of the joint accelera-
        tions [rad/s**2] or [m/s**2]-format 6E13.5
        In the line there are N values.

The listed data are repeated (at each time interval of
0.01 [s]) m times, where m is an integer number obtained if the time duration of the nominal movement (defined in the file ***.TRA) is devided by 0.01.

```
*********************************************************************
*                 DESCRIPTION OF DATA-FILE ***.DIN                 *
*********************************************************************
```

The output data-file ROBOT NAME.DIN contains the data on
the nominal trajectories, the nominal driving torques,
the nominal programmed control, the nominal power and
energy. The file is the output of the subroutine SMNOMP.
The structure of the file is as following:

TIME - the time instant [s] - format E13.5
Q(6) - nominal trajectories of the robot joints -
        values of the joint coordinates at the nomi-
        nal trajectory at the time instant TIME
        [rad] or [m] - format 6E13.5
        In the line there are N data (N the number of
        joints of the user's specific robot).
DQ(6) - nominal trajectories of the joint velocities -
        [rad/s] or [m/s] - format 6E13.5
        There are N data in the line.
DDQ(6) - nominal trajectories of the joint accelera-
        tions [rad/s**2] or [m/s**2] - format 6E13.5
        There are N data in the line.
P0(6) - nominal driving torques around the joints of
        the robot (in the time instant TIME [Nm] or
        [N] - format 6E13.5.
        There are N data in the line.
U0(6) - the nominal programmed control for the actua-
        tors of the joints [V] or [mA]-format 6E13.5
        There are N data in the line.
POWER(6) - the required nominal power in the robot
        joints [W] - format 6E13.5.
        There are N data in the line.
ENERG(6) - the demanded nominal energy in the joints
        of the robot [Ws] - format 6E13.5.
        There are N data in the line.

These data are repeated (at each time interval of 0.01 [s])
m times, where m is the integer number which is obtained when
the time duration of the nominal movement (specified in the
file ***.TRA) is devided by 0.01.

```
***************************************************************
*            DESCRIPTION OF DATA-FILE ***.SIM               *
***************************************************************
```

The output data-file ROBOT NAME.SIM contains data on
simulated (actual) trajectory, driving torques, actual
control signals, power and energy consumptions. The
file is an output of the subroutine SMSIMD and represents the results of the simulation of the robot dynamics in tracking of the nominal trajectory with
the selected control law.
The structure of the file is as follows:

TIME   - time instant [s] - format E13.5
Y(18) - the simulated deviations of the actual
           state coordinates from the nominal trajectory
           (from the file ***.ANG)- the sequence of da-
           ta is: the deviation of the joint angle from
           the nominal [rad] or [m], the deviation
           of the joint velocity from the nominal [rad/s]
           or [m/s], deviation of the current (pressure)
           of the actuator from the nominal values (if the
           third order modelis applied) [A] or [bar] -
           the data are repeated for N joints [N is the
           number of the user's specific robot] -
           - format 6E13.5.
P(6)   - simulated (actual) driving torque around the robot
           joints (in the instant TIME) [Nm] or [N] -
           - format 6E13.5. - torques deeloped during the trac-
           king of the nominal trajectory
           In the line there are N data.
U0(6) - simulated (actual) control signals for the
           actuators [V] or [mA] - format 6E13.5.
           In the line tere are N data.
POWER(6) - simulated (required) actual power in each joint of
           the robot [W] - format 6E13.5.
           In the file there are N data.
ENERG(6) - demanded energy in the robot joints [Ws]
           - format 6E13.5.
           In the line there are N data.

These data are repeated (at each time interval of 0.01 [s])
m times, where m is the integer number which is obtained when
the time duration of the nominal movement (specified in the
file ***.TRA) is devided by 0.01.

```
C------------------------------------------------------------------------
C       ROUTINE:        KG
C------------------------------------------------------------------------
C       FUNCTION:       THE MAIN ROUTINE OF THE TASK-
C                       IT CALLS THE FOLLOWING SUBROUTINES:
C                       SMINIC - reads the initial values of variable, data
C                                on auxiliary parameters and so on.
C                       INPUT_KIN - reads data on the kinematics of the
C                                   selected robot
C                       INPUT_DIN - reads the data on the dynamic parameters
C                                   of the robot
C                       INPUT_ACT - reads data on the actuators in the robot
C                                   joints and forms the models of actuators
C                                   and joints
C                       SMOPRG - synthesizes the local optimal regulators with
C                                or without the integral feedback loop
C                       SMLOC_E - synthesizes the local control feedback gains
C                                 (synthesis in S - domain)
C                       TRAJEK  - computes the nominal trajectories
C                                 (in the joints coordinates)
C                       SMNOMP  - computes the nominal moments and the nominal
C                                 (centralized or local) programmed control
C                       LINANA  - subroutine for analysis of the lineari-
C                                 zed (total) model of the robot via determi-
C                                 nation of the eigen-values of the system
C                                 matrix with or without the local control-
C                                 lers
C                       SMSIMD  - simulates the total dynamic model of the
C                                 robot
C------------------------------------------------------------------------
C       INPUT VARIABLES:
C                       FILE(5) - name of the robot
C                       N       - number of joints of user's  robot
C                       ALFAI(6) - prescribed degrees of the exponential
C                                  stability of local subsystems
C                       ICNVLN   - indicator whether the linearized
C                                  model of the system is stable or not
C------------------------------------------------------------------------
C       OUTPUT VARIABLES:
C                       IOPTOR(6) - indicators whether the local optimal
C                                   regulator is applied in the joint
C                       IOPTIN(6) - indicators whether the local integal
C                                   feedback loop is applied in the local
C                                   controler, or not
C------------------------------------------------------------------------
        include 'in2:smcom.com'
C
        COMMON/OPTPR/ ICNVLN
        COMMON/file/ file(5)
        COMMON/rbsl/ i
        CHARACTER*1 file,filep*5,a$
C------------------------------------------------------------------------
C
C       READ NAME OF THE ROBOT (up to 5 characters)
C       All input/output data-file on robot have the
C       given name of the robot (by convention ROBOT NAME.ext,
C       where ext points to the type of data in the data-file;
C       for example ROBOT NAME.CNF contains data on the robot
C       structure and geometry)
C
```

```
          WRITE (6,2)
          READ (5,3) filep
          DO ipom=1,5
             file(ipom)=filep(ipom:ipom)
          END DO
C
C         raed auxiliary data on the robot
C
          CALL sminic
C
C         read data on the robot
C
          CALL input_kin
          CALL input_din
          CALL input_act
C------------------------------------------------------------------
C
C         computation of the local controllers around all n joints
C         of the robot (n - number of joints)
C
          TYPE 111
 9        DO i=1,n
C
C         user's decision if he wants to apply local optimal
C         controller or not, and if he wants to introduce
C         the integral feedback loop in the i-th joint
C
          TYPE 100,i
          READ(5,101)yes
          IF(yes.eq.'Y')then
             ioptor(i)=1
             ioptin(i)=0
             type 104,i
             read(5,101)yes
             if(yes.eq.'Y')ioptin(i)=1
C
C         set the prescribed exponential stability degree for the
C         i-th subsystem (joint)
C
 10          TYPE 105,i
             READ(5,106)alfai(i)
                IF(alfai(i).lt.1..or.alfai(i).gt.20.)go to 10
C
C         if the local optimal regulator is seleceted
C
                CALL smoprg(i)
          else
C
C         if the synthesis in the S-domain is selected
C
             CALL smloc_e
             ioptor(i)=0
          end if
C
          end DO
C
C         data on local feedack gains are written in the
C         output data-file ***.LOC (if the user wants so)
C
          TYPE 107
          READ(5,101)yes
```

```
          IF(yes.eq.'Y') then
           OPEN(unit=2,name=filep//'.LOC',type='NEW')
              DO 11 i=1,n
                 WRITE(2,106)pk1(i)
                 WRITE(2,106)pk2(i)
                 WRITE(2,106)pk3(i)
  11             WRITE(2,106)pk4(i)
          CLOSE(2)
          end IF
C----------------------------------------------------------------------
C
C         computation of the nominal trajectories (in the joints
C         coordinates)
C
          CALL trajek
C
C         computation of the nominal programmed control
C
          CALL smnomp
C----------------------------------------------------------------------
C
C         linear analysis - linearization of the entire model
C         of the robot and computation of the eigen-values of
C         the open-loop and the closed-loop system matrix
C         (when the local controllers are applied)
C
          TYPE 102
          READ(5,101)yes
          IF(yes.ne.'Y')GO TO 18
          CALL linana
C
          IF(icnvln.EQ.0)THEN
                  TYPE *,' The linearized model is stabilized with'
                  TYPE *,' the prescribed exponential stability degree'
             ELSE
                  TYPE *,' The linearized model is not adequatly'
                  TYPE *,' stabilized by the selected local controllers'
                  TYPE 108
                  READ(5,101)yes
                  IF(yes.EQ.'Y')GO TO 9
          end IF
C----------------------------------------------------------------------
C
C         simulation of the entire dynamic nonlinear model of robot
C
  18      CALL smsimd
          TYPE 112
          READ(5,101)yes
          IF(yes.EQ.'Y') GO TO 18
C----------------------------------------------------------------------
C
  2       format ($,1x,'THE ROBOT NAME [max. length 5 characters]:')
  3       format (a5)
  5       format (i1)
 100      format(' Do you want synthesis of local optimal regulator',/,
       1  $,' for the',I3,'th joint [Y/N]?:')
 102      format(//,$,' Do you want linear analysis [Y/N]?:')
 101      format(a1)
 104      format($,' Want to apply integral feedback loop for the',i3,
       1  'th joint [Y/N]?:')
 105      format($,' Imose the exponential stability degree for the',
```

```
      1   i3,'th joint [1.-20.]:')
106       format(f12.5)
107       format($,' Want to write the feedback gains in the file [Y/N]
      1   ?:')
108       format($,' Want to re-synthesize the local controllers
      1   [Y/N]?:')
110       format(//,' Simulation of nonlinear dynamic model of
      1   the robot',//)
111       format(//,' Synthesis of local controllers around robot
      1   joints',//)
112       format(//,$, ' Want another simulation of robot [Y/N]?:')
C
          end




C-----------------------------------------------------------------
C          SUBROUTINE:     SMINIC
C-----------------------------------------------------------------
C          FUNCTION:       READ DATA REQUIRED FOR THE CONTROL
C                          SYNTHESIS - READ DATA FROM THE FILE
C                          ROBOT NAME.DAT
C-----------------------------------------------------------------
          SUBROUTINE SMINIC
C
          INCLUDE 'IN2:SMCOM.COM'
C
          byte FTOT(14)
          byte FILE(5)
          COMMON/FILE/ FILE
C
          DATA FTOT(1)/'I'/,FTOT(2)/'N'/,FTOT(3)/'1'/,FTOT(4)/':'/,
      1       FTOT(10)/'.'/,FTOT(11)/'D'/,FTOT(12)/'A'/,
      2       FTOT(13)/'T'/,FTOT(14)/0/
C-----------------------------------------------------------------
C
          DO I=1,5
          FTOT(I+4)=FILE(I)
          END DO
C
          OPEN(UNIT=2,NAME=FTOT,TYPE='OLD',ERR=3000)
C
          NUK=0
C
          READ(2,100)N
 100      FORMAT(I4)
 101      FORMAT(3E15.5)
C
C         default data on actuators
C
          DO 1 I=1,N
C
          READ(2,100)NI(I)
          READ(2,101)((A(J,K,I),K=1,NI(I)),J=1,NI(I))
          READ(2,101)(B(J,I),J=1,NI(I))
          READ(2,101)(F(J,I),J=1,NI(I))
C
C         computation of the total order of the robotic system
C
```

```
        NUK=NUK+NI(I)
  1     CONTINUE
C
        READ(2,101)((QQ(I,J,1),I=1,NI(1)),J=1,NI(1))
C
        DO 10 II=2,N
        DO 10 I=1,3
        DO 10 J=1,3
  10    QQ(I,J,II)=QQ(I,J,1)
        READ(2,101)(R1(I),I=1,N)
        READ(2,101)(QI(I),I=1,N)
C
        DO 11 I=1,N
        READ(2,100)IOPTOR(I),IOPTV(I),IOPTI(I)
  11    CONTINUE
C
        CALL CLOSE(2)
        RETURN
C
 3000   TYPE 3001,FILE
 3001   FORMAT(' DIAG*** File ',5A1,'.DAT cannot be found')
        RETURN
        END




C-----------------------------------------------------------------
C       SUBROUTINE: INPUT_KIN
C................................................................
C       FUNCTION: THE MAIN SUBROUTINE FOR INPUT DATA ON
C                 KINEMATIC PARAMETERS OF THE USER'S SPECIFIC ROBOT
C................................................................
C       INPUT: N - number of the joints (NJ = N)
C              I0 - the order number of the joint
C       OUTPUT: Kinematic parameters of the user's robot
C................................................................
        SUBROUTINE INPUT_KIN
C
        INCLUDE 'CONFIG.MOD'
        INCLUDE 'MODELM.MOD'
        COMMON /MODSW/KOD
        COMMON/LINSPE/ UD(6,3),UG(6,3)
        COMMON/ORDGLB/ NUK,NJ
        COMMON/FILE/ FILE(5)
        BYTE file
C
        COMMON /INTR/ SIMB(19,6)
        DIMENSION EPOC1(3)
        CHARACTER*2 OPC1
        CHARACTEr*4 POM1
        CHARACTER*9 FILEP,FILEP2,FILEP4
C
        FILEP=CHAR(FILE(1))//CHAR(FILE(2))//CHAR(FILE(3))//
     >       CHAR(FILE(4))//CHAR(FILE(5))//'.CNF'
C................................................................
C       read data from the file ***.CNF
C
        INEW=0          ! Indicator of new file 1:old; 0:new
        DESNI=1         ! initialization of the coordinate frame
C                         DESNI=1 - right, DESNI=0 - left frame
        OPEN (55,FILE=FILEP,STATUS='OLD',ERR=11)
        INEW=1
```

```
        WRITE (6,14) FILEP
14      FORMAT (/,1X,' DIAG*** File:',a10,'  is found',/)
        read (55,'(1X,i1)') n
            nj=n
        DO IPOM=1,n
            DO JPOM=1,19
                READ (55,'(1X,F15.6)') SIMB(JPOM,IPOM)
            END DO
        END DO
        DO IPOM=1,3
            READ (55,'(1X,F15.6)') EPOC(IPOM)
        END DO
        CLOSE (55)
        if (inew.eq.1) GO TO 12
11      WRITE (6,13) FILEP
13      FORMAT (/,1X,'DIAG*** File ',a10,' cannot be found')
        STOP
C
C       variables substitutions
C
12          do ipom=1,n
                ksi2(ipom)=simb(1,ipom)
                do kpom=1,3
                    eu(ipom,kpom)=simb(kpom+1,ipom)
                    eu1(ipom,kpom)=simb(kpom+4,ipom)
                    r0u(ipom,kpom)=simb(kpom+7,ipom)
                    if (ipom.eq.n) then
                        rt(kpom)=simb(kpom+10,ipom)
                    else
                        r0u(ipom+n,kpom)=simb(kpom+10,ipom)
                    end if
                end do
            end do
C
C       examine the values of the variables
C
            do ipom=1,n
C
C           the vector ei must be the unit vector
C
                eum=sqrt(eu(ipom,1)*eu(ipom,1)+
     >          eu(ipom,2)*eu(ipom,2)+eu(ipom,3)*eu(ipom,3))
                if (eum.gt.1.01.OR.eum.lt.0.99) then
                    write (6,'(''+DIAG*** Vector ei'',i2,
     >              '' is not unit vector'')') ipom
                    stop
                end if
            end do
            do ipom=1,n
C
C           the vector ei,i+1 must be unit vector
C           but there is no need to examine the vector
C           ei,i+1 in the terminal joint of the robot
C
                if (ipom.lt.n) then
                    eu1m=sqrt(eu1(ipom,1)*eu1(ipom,1)+
     >              eu1(ipom,2)*eu1(ipom,2)+eu1(ipom,3)*
     >              eu1(ipom,3))
                    if (eu1m.gt.1.01.OR.eu1m.lt.0.99) then
                        write (6,'(''+DIAG*** The vector ei+1'',i2,
     >                  '' is not unit vector'')') ipom
```

```
                        stop
                end if
             end if
          end do
C
C
C        examine whether there is at least one linear joint

          do ipom=1,n
             if (ksi2(ipom).eq.1) ifl0=1
          end do
          if (ifl0.eq.1) then            !IFL0=1;there is at least
                                          !       one linear joint
             do ipom=1,n                  !=0;there is no linear joints
                if (ksi2(ipom).eq.1) then
                   do kpom=1,3
                      ud(ipom,kpom)=simb(kpom+13,ipom)
                      ug(ipom,kpom)=simb(kpom+16,ipom)
                   end do
                else
                   do kpom=1,3
                      ud(ipom,kpom)=0
                      ug(ipom,kpom)=0
                   end do
                end if
             end do
C
C
C        examine the values of input data

             do ipom=1,n
C
C        test whether the vectors uii,ui,i+1 are unit vectors or not
C        the vector uii must be unit vector
C
                if (ksi2(ipom).eq.1) then
                   udm=sqrt(ud(ipom,1)*ud(ipom,1)+
     >                      ud(ipom,2)*ud(ipom,2)+
     >                      ud(ipom,3)*ud(ipom,3))
                   if (udm.gt.1.01.or.udm.lt.0.99) then
                      write (6,'('' DIAG*** Vector uii'',i2,
     >                       ''is not unit vector'')') ipom
                      stop
                   end if
                end if
             end do
C
C        the vector ui,i+1 must be unit vector
C
             do ipom=1,n
                if (ksi2(ipom).eq.1) then
                   ugm=sqrt(ug(ipom,1)*ug(ipom,1)+
     >                      ug(ipom,2)*ug(ipom,2)+
     >                      ug(ipom,3)*ug(ipom,3))
                   if (ugm.gt.1.01.or.ugm.lt.0.99) then
                      write (6,'('' DIAG*** Vector ui,i+1'',i2,
     >                       '' is not unit vector'')') ipom
                      stop
                   end if
                end if
             end do
```

```
              else
                epocm=sqrt(epoc(1)*epoc(1)+epoc(2)*epoc(2)+
     >                     epoc(3)*epoc(3))
                if (epocm.lt..99.or.epocm.gt.1.01) then
                    stop
                end if
              end if
C
C             test collinearity of the joints axes and the link
C             axes and assembling of the kinetic chain
C
              do ipom=1,n
                call colias(ipom)
              end do
              kod=1
              call model
              call modchk(imod)
              if (imod.lt.0) then
                write (6,'(/,'' FATAL*** ASSEMBLING OF THE MECHANISM:
     >          UNSUCCESSFUL'')')
                return
              else
                write (6,'(/,'' DIAG*** Assembling of the mechanism:
     >          SUCCESSFUL'')')
              end if
              return
        END


C-------------------------------------------------------------------
C          SUBROUTINE: COLIAS
C-------------------------------------------------------------------
C          FUNCTION:         GENERATING OF THE VECTORS NECESSARY FOR
C                            THE MACHANISM (KINEMATIC CHAIN) ASSEMBLY
C                            IF THE VECTORS (Rii,ei) or (Ri,i+1,ei+1) ARE
C                            COLINEAR
C-------------------------------------------------------------------
C          INPUT VARIABLES:
C                  KSI2      joint type
C                  EPOC      the unit vector of the axis of the first joint
C                            with respect to absolute coordinate frame
C                  UD        special unit vector for the i-1-st link if
C                            the i-th joint is linear
C                  UG        special unit vector for the t-th link if the
C                            joint is linear
C                  R0U       link vectors (Rii and Ri,i+1)
C                  EU        the unit vector of the axis of the i-th joint
C                  EU1       the unit vector of the axis of the i+1-st
C                            joint
C-------------------------------------------------------------------
C          OUTPUT VARIABLES:
C                  IAS1      indicator whteher the i-th link is special
C                            since (Ei||Rii): IAS1=0 the link is not
C                            "special", IAS1=1 the link is "special"
C                  IAS2      indicator whether the i-th link is "special"
C                            since (Ei,i+1||Ri,i+1):
C                            =0 the link is not "special"
C                            =1 the link is "special"
C                  RAS       the vectors for the mechanism assembling if
C                            the i-th link is "special"
C-------------------------------------------------------------------
```

```
        SUBROUTINE  COLIAS(I)
C
        INCLUDE 'IN2:CONFIG.MOD'
        COMMON/LINSPE/ UD(6,3),UG(6,3)
        REAL IDEL(3)
        DIMENSION IAS(6)
C
        IAS1(I)=0
        IAS(I)=0
        IAS2(I)=0
C
        IF(I.EQ.1.AND.KSI2(I).EQ.1)GO TO 567
        ROPOC(1)=-EPOC(3)
        ROPOC(2)=-EPOC(1)
        ROPOC(3)=-EPOC(2)
        GO TO 4000
567     DO 403 K=1,3
        ROPOC(K)=UD(I,K)
403     CONTINUE
4000    IND=0
        M=I
        IF(KSI2(I).EQ.1)GO TO 444
        DO 15 K=1,3
        IMUL1=0
        IF(ROU(M,K).EQ.0.)IMUL1=1
        IMUL2=0
        IF(EU(I,K).EQ.0.)IMUL2=1
        IF(IMUL1.NE.IMUL2)GO TO 200
        IF(IMUL1.EQ.0.AND.IMUL2.EQ.0)GO TO 56
        IDEL(K)=0.
        GO TO 15
56      IDEL(K)=ROU(M,K)/EU(I,K)
15      CONTINUE
C
300     IF(IDEL(1).EQ.0.)GO TO 101
        IF(IDEL(2).EQ.0.)GO TO 102
        IF(IDEL(3).EQ.0.)GO TO 103
        IF(IDEL(1).EQ.IDEL(2).AND.IDEL(2).EQ.IDEL(3))GO TO 104
        IAS(I)=0
        GO TO 200
104     IAS(I)=1
        GO TO 201
103     IF(IDEL(1).EQ.IDEL(2))GO TO 104
        IAS(I)=0
        GO TO 200
102     IF(IDEL(3).EQ.0.)GO TO 105
        IF(IDEL(1).EQ.IDEL(3))GO TO 104
        IAS(I)=0
        GO TO 200
105     IAS(I)=1
        GO TO 202
101     IF(IDEL(2).EQ.0.)GO TO 106
        IF(IDEL(3).EQ.0.)GO TO 105
        IF(IDEL(2).EQ.IDEL(3))GO TO 104
        IAS(I)=0
        GO TO 200
106     IF(IDEL(3).EQ.0.)GO TO 107
        GO TO 105
107     TYPE 1016
1016    FORMAT(' FATAL***** The unit vectors of the joint axis and of
```

```
     *the link are zero-vectors*******')
         STOP
200      RAS(M,1)=0.
         RAS(M,2)=0.
         RAS(M,3)=0.
         GO TO 204
201      RAS(M,1)=1.
         RAS(M,2)=0.
         RAS(M,3)=0.
         GO TO 204
202      IF(R0U(M,1).EQ.0.)GO TO 599
         RAS(M,1)=0.
         RAS(M,2)=1.
         IF(IND.EQ.1)RAS(M,2)=-1.
         RAS(M,3)=0.
         GO TO 204
599      RAS(M,1)=1.
         IF(IND.EQ.1)RAS(M,1)=-1.
         RAS(M,2)=0.
         RAS(M,3)=0.
         GO TO 204
444      IAS1(I)=1
         DO 446 K=1,3
         RAS(M,K)=UG(I,K)
446      CONTINUE
         GO TO 666
445      IAS2(I)=1
         DO 447 K=1,3
         RAS(M,K)=-UD(I+1,K)
447      CONTINUE
         GO TO 900
204      IF(IND.EQ.1)GO TO 500
         IAS1(I)=IAS(I)
666      M=I+N
         IF(KSI2(I+1).EQ.1)GO TO 445
         IF(I.EQ.N)GO TO 505
         DO 25    K=1,3
         IMUL1=0
         IF(R0U(M,K).EQ.0.)IMUL1=1
         IMUL2=0

         IF(EU1(I,K).EQ.0.)IMUL2=1
         IF(IMUL1.NE.IMUL2)GO TO 600
         IF(IMUL1.EQ.0.AND.IMUL2.EQ.0)GO TO 79
         IDEL(K)=0.
         GO TO 25
79       IDEL(K)=R0U(M,K)/EU1(I,K)
25       CONTINUE
         IND=1
         GO TO 300
500      IAS2(I)=IAS(I)
900      RETURN
505      IAS2(I)=0
         RETURN
600      IAS2(I)=0
         RAS(M,K)=0.
         RAS(M,K)=0.
         RAS(M,K)=0.
         RETURN
         END
```

```
C-------------------------------------------------------------------
C        SUBROUTINE: MODCHK
C-------------------------------------------------------------------
C        FUNCTION:  EXAMINES IF THE MECHANISM (KINEMATIC CHAIN) CAN BE
C                   SUCCESSFULLY ASSEMBLED; PRINTS THE TRANSFORMATION
C                   MATRIX AND CHARACTERISTIC KINEMATIC PARAMETRS
C        -----------------------------------------------------------
C        INPUT VARIABLES:
C                   Q        (3,3) Transformation matrix of the terminal
C                            link (after mechanism assembling) - An
C                   IAX      indication of the terminal link orientation
C                   SI       joint coordinates
C                   E        the unit vectors of the joints axes with
C                            respect to absolute coordinate frame
C                   RP
C                   RT       position vectors with respect to abs. frame
C                   RTT
C        OUTPUT VARIABLES:
C                   ICHK     indicator of assembling of the mechanism
C                            ICHK=0  successful assembling
C                            ICHK=-1 unsuccessfull assembling - data on
C                                    the mechanism are not consistent
C                                    and the user must change data in the
C                                    file ***.CNF
C        -----------------------------------------------------------
         SUBROUTINE MODCHK(ICHK)
C
         INCLUDE 'IN2:CONFIG.MOD'
         INCLUDE 'IN2:MODELM.MOD'
         DIMENSION RP1(6,6,3),RP2(6,6,3),RP3(6,6,3),RP4(6,6,3),
     >   RP5(6,6,3)
C
         ICHK=0
         DO 1 I=1,3
         LL=0
         DO 2 K=1,3
         IF(Q(I,K).NE.0) LL=LL+1
2        CONTINUE
         IF(LL.EQ.0) ICHK=-1
1        CONTINUE
         TYPE *,' CHARACTERISTIC KINEMATIC VARIABLES'
         TYPE *,' ---------------------------------'
         TYPE *,'        Joint coordinates: '
         TYPE 1099,(SI(I),I=1,N)
1099     FORMAT(6(E13.5))
         TYPE *
         TYPE *,' Transformation matrix of the terminal link An:'
         TYPE 100,((Q(I,J),J=1,3),I=1,3)
100      FORMAT(1X,3F10.5)
         TYPE *
         TYPE *,' Joint axes with respect to absolute coord. frame'
         DO 202 I=1,N
         TYPE 333,I,(E(I,K),K=1,3)
333      FORMAT(/,'   E(',I1,')= ',3(F10.5))
202      CONTINUE
         TYPE *
         TYPE *,' Position vectors of the centers of masses and of
     >   joints'
         DO 303 I=1,N
         IF(I.EQ.N) GO TO 505
```

```
        J=I+1
        IF(I.EQ.3) GO TO 997
        IF(I.EQ.4) GO TO 996
        DO 339 K=1,3
        RP4(I,J,K)=RP(I,I,K)+RP(J,J,K)-RP(J,I,K)
339     CONTINUE

        TYPE 444,I,I,(RP(I,I,K),K=1,3),I,J,(RP4(I,J,K),K=1,3)
444     FORMAT(/,' R(',I1,',',I1,')=',X,3F8.3,
     *  ' R(',I1,',',I1,')=',X,3F8.3)
        GO TO 303
997     DO 887 K=1,3
        RP1(I,J,K)=RP(I,I,K)-RP(J,I,K)
887     CONTINUE
        TYPE 444,I,I,(RP(I,I,K),K=1,3),I,J,(RP1(I,J,K),K=1,3)
        GO TO 303
996     DO 774 K=1,3
        RP2(I,I,K)=RTT(I,I,K)
        RP3(I,J,K)=RTT(I,I,K)+RP(J,J,K)-RP(J,I,K)
774     CONTINUE
        TYPE 444,I,I,(RP2(I,I,K),K=1,3),I,J,(RP3(I,J,K),K=1,3)
        GO TO 303
505     DO 677 K=1,3
677     RP5(I,I,K)=RP(I,I,K)+RT0(K)
        TYPE 666,I,I,(RP5(I,I,K),K=1,3),(RT0(K),K=1,3)
666     FORMAT(/,' R(',I1,',',I1,')=',X,3F8.3,' RT0= ',3F8.3)
303     CONTINUE
        RETURN
        END




C----------------------------------------------------------------
C       SUBROUTINE: INPUT_DIN
C................................................................
C       FUNCTION: THE MAIN SUBROUTINE FOR INPUT DYNAMIC
C                 PARAMETRS OF THE ROBOT MECHANISM
C----------------------------------------------------------------
C       INPUT VARIABLES:
C           NJ - number of joints of the user's specific robot
C           IO - the order number of the joint
C----------------------------------------------------------------
C       OUTPUT VARIABLES:
C           DYNAMIC PARAMETERS FOR THE USER'S SPECIFIC ROBOT
C----------------------------------------------------------------
        SUBROUTINE INPUT_DIN
C
        REAL l(6),lz(6),mm(6),js(6),jn(6),jx(6),jy(6),jz(6)
        COMMON/tipl/ nx,ksi1(6),ksi2(6),ksi3(6)
        COMMON/MIN/ MM,JS,JN,JX,JY,JZ
        COMMON/SPMOM/ GM(6,3),G(6)
        COMMON/ORDGLB/ NUK,NJ
        COMMON/FILE/ FILE(5)
        BYTE file
*
        CHARACTER    opc1*1
        DIMENSION SIMB(17,6)
        COMMON/HMAX/ HMAX(6)
        COMMON/HMIN/ HMIN(6)
        COMMON/MOD_MAX_MIN/ QMAX(6),QMIN(6)
        CHARACTER*9 FILEP,FILEP2
```

```
C.................................................................
14         format (/,1X,' DIAG*** File:',a10,'  has been found',/)
13         format (/,1X,' DIAG*** File ',a10,' cannot be found',/)
2          format ('+',a6,':',F15.6)
4573       format (1x,'Want to compute maximal moments of inertia of the
     >   mechanism',/,1x,' (otherwise you have to directly specify
     >   them) [Y/N]:'$)
4575       format (a2)
4572       format (/,1x,' DIAG*** Computation of maximal moments of
     >   inertia',/)
4574       format (1x,' Specify the maximal moment of inertia around the'
     >   ,i2,' joint:',$)
C
C          read data from the data-file   ***.DNM
C
           FILEP=CHAR(FILE(1))//CHAR(FILE(2))//CHAR(FILE(3))//
     >          CHAR(FILE(4))//CHAR(FILE(5))//'.DNM'
           OPEN (55,FILE=FILEP,STATUS='OLD',ERR=11)
           INEW=1
           WRITE (6,14) FILEP
           DO IPOM=1,NJ
              DO JPOM=1,7
                 READ (55,'(1X,F15.6)') SIMB(JPOM,IPOM)
              END DO
           END DO
           close (55)
           if (inew.eq.1) GO TO 12
11         inew=0
           WRITE (6,13) FILEP
           CLOSE (55)
           STOP
C
C          check parameter KSI1: if it is not 0 or 1 it is put to 0
C
12            do ipom=1,nj
                 if (simb(1,ipom).ne.0.and.simb(1,ipom).ne.1) then
                    simb(1,ipom)=0
                 end if
              end do
C
C          substitution of the variables
C
              do ipom=1,nj
                 ksi1(ipom)=simb(1,ipom)
                 mm(ipom)=simb(2,ipom)
                 g(ipom) =mm(ipom)*9.81
                 qmax(ipom)=simb(6,ipom)
                 qmin(ipom)=simb(7,ipom)
                 if (ksi1(ipom).eq.0) then
                    jx(ipom)=simb(3,ipom)
                    jy(ipom)=simb(4,ipom)
                    jz(ipom)=simb(5,ipom)
                 else if (ksi1(ipom).eq.1) then
                    js(ipom)=simb(3,ipom)
                    jn(ipom)=simb(4,ipom)
                 end if
              end do
              write (6,4573)
              read (5,4575) opc1
              if (opc1.eq.'Y') then
                 write (6,4572)
```

```
C
C         computation of the maximal moments of inertia of the mechanism
C         around the joint axes
C
          call sm_max_in

          do ipom=1,nj
          write (6,'(''Max. mom.inert. H'',i1,i1,''max='',f15.6
     >         )')ipom,ipom,hmax(ipom)
          write (6,'(''Min. mom.inert. H'',i1,i1,''min='',f15.6
     >         )')ipom,ipom,hmin(ipom)
               write (6,'('' ....................'')')
          end do
        else
C
C     the user wants to specify the maximal moments of inertia
C
          do ipom=1,nj
          write (6,4574) ipom
          read (5,*) hmax(ipom)
          end do
        end if
        RETURN
      END




C**************************************************************************
C         SUBROUTINE: SM_MAX_IN
C------------------------------------------------------------------------
C         FUNCTION: COMPUTES THE MAXIMAL AND THE MINIMAL VALUES OF
C                   THE MOMENTS OF INERTIA OF THE MECHANISM AROUND
C                   THE ROBOT JOINTS FOR ALL ALLOWABLE VALUES OF
C                   THE JOINTS ANGLES (LINEAR DISPLACEMENTS) - THE
C                   SUBROUTINE CALLS SUBROUTINE MODEL TO COMPUTE
C                   THE MOMENTS OF INERTIA (inerta matrix HH) FOR
C                   VARIOUS JOINTS ANGLES AND SEARCHES FOR THE MA-
C                   XIMAL AND MINIMAL VALUES OF MOMENTS OF INERTIA
C                   AROUND THE ROBOT JOINTS (diagonal elements in
C                   the inertia matrix HH(i,i))
C------------------------------------------------------------------------
C     INPUT VARIABLES:
C     NS - number of joints of the user's specific robot
C     QMI(6) - the vector of the minimal allowable values of the
C              joints angles (displacements)
C     QMA(6) - the vector of the maximal allowable values of the
C              joints angles (displacements)
C     HH(6,6) - inertia matrix of the robot for given joints angles
C------------------------------------------------------------------------
C     OUTPUT VARIABLES:
C     HMAX(6) - vector of the computed maximal values of the
C               moments of inertia of the mechanism around the
C               robot joints
C     HMIN(6) - vector of the computed minimal values of the
C               moments of inertia of the mechanism around the
C               robot joints
C------------------------------------------------------------------------
C     SUBROUTINE REQUIRED:
C     MODEL   - computates the inertia matrix for the given
C               values of the joints angles (computes HH)
C------------------------------------------------------------------------
```

```
         SUBROUTINE SM_MAX_IN
C
         byte file(5)
         common/file/ file
         INCLUDE 'MODELM.MOD'
         COMMON /HMAX/ HMAX(6)
         COMMON /HMIN/ HMIN(6)
         COMMON /MOD_MAX_MIN/ QMA(6),QMI(6)
         DIMENSION DELTA(6)
         COMMON/ORDGLB/ NUK,NS
C
C        initialization
C
         DO I=1,NS
            HMAX(I)=-1.
            HMIN(I)=1.E+06
            SI(I)=QMI(I)
C
C           determination of increment of the joint angles at which
C           the moments of inertia of the mehanism are computed
C
            DELTA(I)=(QMA(I)-QMI(I))/3.
         END DO
C
C        branching according to the number of joints of the robot
C
         GO TO (1,2,3,4,5,6) NS
C
C         if the robot has one joint only
C
1        CALL MODEL
         DO I=1,NS
            IF (HH(I,I).GT.HMAX(I)) HMAX(I)=HH(I,I)
            IF (HH(I,I).LT.HMIN(I)) HMIN(I)=HH(I,I)
         END DO
         RETURN
C
C        if the robot has two joints
C
2        SI(2)=QMI(2)
         DO J2=1,3
            CALL MODEL
            DO I=1,NS
               IF (HH(I,I).GT.HMAX(I)) HMAX(I)=HH(I,I)
               IF (HH(I,I).LT.HMIN(I)) HMIN(I)=HH(I,I)
            END DO
            SI(2)=SI(2)+DELTA(2)
         END DO
         RETURN
C
C        if the robot has three joints
C
3        SI(2)=QMI(2)
         DO J2=1,3
            SI(3)=QMI(3)
            DO J3=1,3
               CALL MODEL
               DO I=1,NS
                  IF (HH(I,I).GT.HMAX(I)) HMAX(I)=HH(I,I)
                  IF (HH(I,I).LT.HMIN(I)) HMIN(I)=HH(I,I)
               END DO
               SI(3)=SI(3)+DELTA(3)
```

```
            END DO
            SI(2)=SI(2)+DELTA(2)
         END DO
         RETURN
C
C        if the robot has four joints
C
4        SI(2)=QMI(2)
         DO J2=1,3
            SI(3)=QMI(3)
            DO J3=1,3
               SI(4)=QMI(4)
               DO J4=1,3
                  CALL MODEL
                  DO I=1,NS
                     IF (HH(I,I).GT.HMAX(I)) HMAX(I)=HH(I,I)
                     IF (HH(I,I).LT.HMIN(I)) HMIN(I)=HH(I,I)
                  END DO
                  SI(4)=SI(4)+DELTA(4)
               END DO
               SI(3)=SI(3)+DELTA(3)
            END DO
            SI(2)=SI(2)+DELTA(2)
         END DO
         RETURN
C
C        if the robot has five joints
C
5        SI(2)=QMI(2)
         DO J2=1,3
            SI(3)=QMI(3)
            DO J3=1,3
               SI(4)=QMI(4)
               DO J4=1,3
                  SI(5)=QMI(5)
                  DO J5=1,3
                     CALL MODEL
                     DO I=1,NS
                        IF (HH(I,I).GT.HMAX(I)) HMAX(I)=HH(I,I)
                        IF (HH(I,I).LT.HMIN(I)) HMIN(I)=HH(I,I)
                     END DO
                     SI(5)=SI(5)+DELTA(5)
                  END DO
                  SI(4)=SI(4)+DELTA(4)
               END DO
               SI(3)=SI(3)+DELTA(3)
            END DO
            SI(2)=SI(2)+DELTA(2)
         END DO
         RETURN
C
C        if the robot has six joints
C
6        SI(2)=QMI(2)
         DO J2=1,3
            SI(3)=QMI(3)
            DO J3=1,3
               SI(4)=QMI(4)
               DO J4=1,3
                  SI(5)=QMI(5)
                  DO J5=1,3
```

```
                    SI(6)=QMI(6)
                    DO J6=1,3
                        CALL MODEL
                            DO I=1,NS
                            IF (HH(I,I).GT.HMAX(I)) HMAX(I)=HH(I,I)
                            IF (HH(I,I).LT.HMIN(I)) HMIN(I)=HH(I,I)
                            END DO
                            SI(6)=SI(6)+DELTA(6)
                        END DO
                        SI(5)=SI(5)+DELTA(5)
                    END DO
                    SI(4)=SI(4)+DELTA(4)
                END DO
                SI(3)=SI(3)+DELTA(3)
            END DO
            SI(2)=SI(2)+DELTA(2)
        END DO
        RETURN
        end
```

```
C-------------------------------------------------------------------
C      SUBROTINE:  INPUT_ACT
C-------------------------------------------------------------------
C      FUNCTION: COMPUTES THE MODELS OF THE SUBSYSTEMS -
C                ACTUATOR MODELS INCLUDING "MAXIMAL" MOMENT
C                OF INERTIA OF THE MECHANISM AROUND THE JOINT
C                AXIS
C-------------------------------------------------------------------
C      INPUT VARIABLES:
C                ACTUATOR PARAMETERS WHICH ARE READ IN THE
C                SUBROUTINE INPUL
C      HMAX(6) - maximal moments of inertia of the mechanism
C                around the joints axes (computed in SM_MAX_IN)
C      OPC1 - user's option: to print the computed matrices or not
C-------------------------------------------------------------------
C      OUTPUT VARIABLES:
C         THE MATRICES OF THE ACTUATOR'S STATE MODEL
C                dx/dt = A * x   +   b * u  +  f * P
C      (here: x is the state vector of the i-th actuator model
C      u is the input, P is the load - driving torque)
C      NS   - number of joints of the user's specific robot
C      NI(6)- vector contaning the selected orders of the actuator
C             models
C      A(3,3,6) - three dimensional matrix which includes A matrices
C                 of the actuators models for all NS actuators
C      B(3,6) - two dimensional matrix including control distribution
C               b vectors of the actuators models
C      F(3,6) - two dimensional matrix including load distribution
C               f vectors of the actuators models
C      a$(6)  - character vector including information on actuators
C               types
C      AHM,BHM,FHM - matrices analogue to A, B, F but which
C                    include the maximal moments of inertia of
C                    the mechanism HMAX - matrices of the subsystems'
C                    state models
```

```
C-----------------------------------------------------------------------
C       SUBROUTINE REQUIRED:
C          INPUTL - reads data on the selected actuators for the robot
C                   and computes the matrices A,B,F of actuators'models
C-----------------------------------------------------------------------
        SUBROUTINE INPUT_ACT
C
        COMMON/SUBSYS/ A(3,3,6),B(3,6),F(3,6)
        COMMON/SUBSHM/ AHM(3,3,6),BHM(3,6),FHM(3,6)
        COMMON/ACTORD/ NI(6),KI(6)
        COMMON/ORDGLB/ NUK,NS
C
        COMMON/TIP_ACT/ A$(6)
        COMMON/HMAX/ HMAX(6)
        DIMENSION RMEL(6)
        CHARACTER*2 A$,OPC(15),B$,OPC1*1
C..................................................................
2       FORMAT (A1)
3       FORMAT (1X,F16.5)
30      FORMAT (1X,'-------------------------------------------------
     *---------------')
1453    FORMAT (1X,A19)
1256    FORMAT (/,1X,A11,I2)
1387    FORMAT (1X,'Do you want to print the matrices of subsystems
     * models',/,' (which include the maximal mom. of inertia)
     * [Y/N]:',$)
C-----------------------------------------------------------------------
C
C       call subroutine to read data on actuators and computes
C       the matrices A,B,F of the actuators' models
C
        CALL INPUTL
*..................................................................
        WRITE (6,1387)
        READ (5,2) OPC1
        DO I=1,NS
           OPC(1)=A$(I)
           RMEL(I)=F(2,I)
C-----------------------------------------------------------------------
C
C       computation of the matrices AHM, BHM, FHM
C
           C=-1./RMEL(i)/(-1./RMEL(i)+Hmax(I))
           FHM(2,i)=F(2,i)*C
              AHM(1,2,I)=1.
              IF (NI(i).EQ.2) THEN
                 AHM(2,2,i)=A(2,2,i)*C
                 BHM(2,i)=B(2,i)*C
                 RMEL(i)=RMEL(i)*C
              ELSE
                 AHM(2,2,i)=A(2,2,i)*C
                 AHM(2,3,i)=A(2,3,i)*C
*
                 AHM(3,2,I)=A(3,2,I)
                 AHM(3,3,I)=A(3,3,I)
                 BHM(3,I)=B(3,I)
                 RMEL(i)=RMEL(i)*C
              END IF
```

```
C-------------------------------------------------------------------
C
C       print matrices
C
          IF (OPC1.EQ.'Y') THEN
             WRITE (6,1256) ' Subsystem:',I
             WRITE (6,1453) 'Subsystem matrix A:'
             WRITE (6,'(<NI(I)>F16.5)') ((AHM(J,K,I),
     >                                   K=1,NI(I)),J=1,NI(I))
             WRITE (6,1453) 'Subsystem vector b:'
             WRITE (6,3) (BHM(J,I),J=1,NI(I))
             WRITE (6,1453) 'Subsystem vector f:'
             WRITE (6,3) (FHM(J,I),J=1,NI(I))
             WRITE (6,30)
          END IF
       END DO
       END




C-------------------------------------------------------------------
C           SUBROUTINE:  INPUTL
C-------------------------------------------------------------------
C           FUNCTION:   READS DATA ON THE ACTUATORS PARAMETERS SPECI-
C                       FIED BY THE USER IN THE FILE ***.ACT AND
C                       COMPUTES THE ACTUATORS MODELS IN THE STATE
C                       SPACE
C-------------------------------------------------------------------
C           INPUT VARIABLES:
C               PARAMETRS OF THE ACTUATORS SPECIFIED IN THE FILE
C               ***.ACT
C               NS - number of joints of the user's specific robot
C               A$ - character vector including information on the
C                    actuators types (DC - D.C. electro-motor, HD -
C                    hydraulic actuator)
C               NI(6) - vector including ordrers of the actuators
C                    models as selected by the user
C-------------------------------------------------------------------
C           OUTPUT VARIABLES:
C               A,B,F - matrices of the actuators' models in the
C                       state space - see explanation in the subroutine
C                       INPUT_ACT
C-------------------------------------------------------------------
C           NOTE: Due to lack of space in this subroutine all necessary
C                 tests of the input data are omitted. The user is
C                 adviced to add himself this protective tests in
C                 order to protect his package from invalid values of
C                 input parameters
C-------------------------------------------------------------------
       SUBROUTINE INPUTL
C
       COMMON/SUBSYS/ A(3,3,6),B(3,6),F(3,6)
       COMMON/SUBSHM/ AHM(3,3,6),BHM(3,6),FHM(3,6)
       COMMON/ACTORD/ NI(6),KI(6)
       COMMON/ORDGLB/ NUK,NS
       BYTE FILE
       COMMON/FILE/FILE(5)
C
       CHARACTER*2 A$
       CHARACTER*4 DC$(15),HD$(15)
       CHARACTER*1 POM,PP,A1*80
```

```
          DIMENSION DC(15,6),HD(15,6)
          COMMON/TIP_ACT/ A$(6)
          COMMON/UMAX/ UMAX(2,6),UMAX1(6)
          CHARACTER*9 FILEP,FILEP2
          DATA A$/'DC','DC','DC','DC','DC','DC'/
C----------------------------------------------------------------------
14        FORMAT (/,1X,' WARNING*** File:',a10,' is found',/)
13        FORMAT (/,1X,' DIAG*** File ',a10,' cannot be found')
21        format (' DIAG*** The specified order of the actuator ',i1' is
     >    not allowed')
C----------------------------------------------------------------------
C
C         read data from the file ***.ACT
C
          FILEP=CHAR(FILE(1))//CHAR(FILE(2))//CHAR(FILE(3))//
     >         CHAR(FILE(4))//CHAR(FILE(5))//'.ACT'
          INEW=0
          OPEN (55,FILE=FILEP,STATUS='OLD',ERR=11)
          INEW=1
          WRITE (6,14) FILEP
          READ (55,'(6(1X,A2))') (A$(IPOM),IPOM=1,NS)
          DO IPOM=1,NS
             IF (A$(IPOM).EQ.'DC') THEN
                DO JPOM=1,15
                   READ (55,'(1X,F15.6)') DC(JPOM,IPOM)
                END DO
             ELSE IF (A$(IPOM).EQ.'HD') THEN
                DO JPOM=1,15
                   READ (55,'(1X,F15.6)') HD(JPOM,IPOM)
                END DO
             END IF
          END DO
          CLOSE (55)
          IF (INEW.EQ.1) GO TO 12
11        INEW=0
          WRITE (6,13) FILEP
          RETURN
C. . . . . . . . . . . . . . . . . . . . . . . . . . . . .
C
C         if the secected actuator is D.C. electro-motor (DC)
C
12                do i22=1,ns
                  if (a$(i22).eq.'DC') then
                      umax(1,i22)=dc(7,i22)
                      umax(2,i22)=dc(8,i22)
                      n=dc(1,i22)
                      ni(i22)=dc(1,i22)
                      a(1,1,i22)=0.
                      a(1,2,i22)=1.
                      a(2,1,i22)=0.
                      a(1,3,i22)=0.
                      b(1,i22)=0.
                      f(1,i22)=0.
                      f(3,i22)=0.
                      f(2,i22)=-1./dc(9,i22)/dc(10,i22)/dc(4,i22)
                      if(n.lt.2.or.n.gt.3) then
                           write (6,21) i22
                           stop
```

```
                  else if (n.eq.2) then
                   a(2,2,i22)=-(dc(2,i22)*dc(3,i22)/dc(6,i22)+
     >                 dc(5,i22))/dc(4,i22)/DC(9,I22)/DC(10,I22)
                   b(2,i22)=dc(2,i22)/dc(4,i22)/dc(9,i22)/
     >                              dc(6,i22)/DC(9,I22)
                   a(2,3,i22)=0.
                  else if (n.eq.3) then
                   a(2,2,i22)=-dc(5,i22)/dc(4,i22)/DC(9,I22)/
     >                              DC(10,I22)
                   a(2,3,i22)=dc(2,i22)/dc(9,i22)/dc(4,i22)/
     >                              DC(10,I22)
                   a(3,1,i22)=0.
                   a(3,2,i22)=-dc(3,i22)*dc(9,i22)/dc(11,i22)/
     >                              DC(9,I22)
                   a(3,3,i22)=-dc(6,i22)/dc(11,i22)
                   b(2,i22)=0.
                   b(3,i22)=1./dc(11,i22)
                  end if
C. . . . . . . . . . . . . . . . . . . . . . . . . . . . . . . . .
C
C          if selected actuator is hydraulic (HD)
C
                 else if (a$(i22).eq.'HD') then
                  umax(1,i22)=hd(7,i22)
                  umax(2,i22)=hd(8,i22)
                  n=hd(1,i22)
                  ni(i22)=hd(1,i22)
                  a(1,1,i22)=0.
                  a(1,2,i22)=1.
                  a(2,1,i22)=0.
                  a(1,3,i22)=0.
                  b(1,i22)=0.
                  f(1,i22)=0.

                  f(3,i22)=0.
                  f(2,i22)=-1./hd(4,i22)
                  if(n.lt.2.or.n.gt.3) then
                       write (6,21) i22
                       stop
                  else if (n.eq.2) then
                   a(2,2,i22)=-hd(5,i22)/hd(4,i22)-.01*hd(2,i22)*
     >                 hd(2,i22)/(hd(4,i22)*hd(3,i22))
                   b(2,i22)=100.*hd(10,i22)*hd(2,i22)/(hd(4,i22)*
     >                 hd(3,i22))
                  else if (n.eq.3) then
                   pomp=400.*hd(9,i22)/hd(6,i22)
                   a(2,2,i22)=-hd(5,i22)/hd(4,i22)
                   a(2,3,i22)=.0001*hd(2,i22)*hd(11,i22)/hd(4,i22)
                   a(3,1,i22)=0.
                   a(3,2,i22)=-hd(2,i22)*pomp/hd(11,i22)
                   a(3,3,i22)=-hd(3,i22)*pomp*.01
                   b(2,i22)=0.
                   b(3,i22)=pomp*hd(10,i22)*10000./hd(11,i22)
                   f(3,i22)=0.
                  end if
                 end if
                end do
C..............................................................
C
C          print actuators models
```

```
            do i0zz=1,ns
            write (6,'(/,'' Actuator model in the joint:'',I2)')I0ZZ
            write (6,*) ' Matrix A:'
            IO=I0ZZ
            write (6,'(1x,<ni(io)>f15.6)') ((a(jp,jl,io),jl=1,ni(io)
        *                                   ),jp=1,ni(io))
            write (6,*) ' Vector b:'
            write (6,'(1x,f15.6)') (b(jp,io),jp=1,ni(io))
            write (6,*) ' Vector f:'
            write (6,'(1x,f15.6),//') (f(jp,io),jp=1,ni(io))
            end do
            RETURN
        END
```

```
C-----------------------------------------------------------------------
C       SUBROUTINE:     SMOPRG
C-----------------------------------------------------------------------
C       FUNCTION:       COMPUTES LOCAL OPTIMAL REGULATOR FOR THE
C                       SUBSYSTEMS WHICH MIGHT BE EITHER OF THE
C                       SECOND OR OF THE THIRD ORDER, WITH OR
C                       WITHOUT INTEGRAL FEEDBACK LOOP
C-----------------------------------------------------------------------
C       INPUT VARIABLES:
C               I - The order number of the local subsystem for
C                   which we want to synthesize the local controller
C               NI(I) - the order of the i-th subsystem
C               ASHM(NI(I),NI(I),I) - matrix of the i-th subsystem
C                                     (which includes the value of
C                                     the maximal moment of inertia
C                                     of the mechanism around the
C                                     axis of the i-th joint)
C               BSHM(NI(I),I)       - vector of the input distribution
C                                     of the i-th subsystem
C               ALFAI(I)          - the presribed exponential stability
C                                   degree for the i-th subsystem
C               IOPTOR(I)  - user's option:
C                                   =0 the user does not want optimal
C                                      regulator around the i-th joint
C                                   =1 the user wants optimal regulator
C               IOPTV(I)   -        =0 without the velocity feedback
C                                   =1 with the velocity feedback
C               IOPTI(I)   -        =0 without the feedback loop by the
C                                      rotor current
C                                      (if D.C. motor is applied)
C                                      (or, by the pressure if hydraulic
C                                      actuator is applied)
C                                   =1 with the feedback loop by current
C               IOPTIN(I)  - option of introduction of the integral
C                                   feedback loop:
C                                   =0 no
C                                   =1 yes
C               QQ(NI(I),NI(I),I) - weighting matrix by the subsystem state
C                                   in the standard quadratic criterion for
C                                   the i-th subsystem
C               R1(I)             - weighting element by the control signal in
C                                   the local standard quadratic criterion
C               QI(I)             - weighting element by the integral coordi-
```

```
C                              nate (integral of the position error in
C                              the PID regulator) in the local standard
C                              quadratic criterion for the i-th subsys.
C-----------------------------------------------------------------
C        OUTPUT VARIABLES:
C                   PK1(I) - position feedback gain for the i-th subsystem
C                   PK2(I) - velocity feedback gain          "
C                   PK3(I) - feedback gain by current         "
C                   PK4(I) - integral feedback gain           "
C-----------------------------------------------------------------
C        SUBROUTINES REQUIRED:
C                   SMOREG    - computes gains of the local linear optimal
C                               regulator (same holds for SMORE1 and
C                               SMORE2 but for different subsystem orders
C                               different matrices dimensions; these two
C                               subroutines might be ommitted)
C-----------------------------------------------------------------
         SUBROUTINE SMOPRG(i)
C
         INTEGER*2 I
C
         COMMON/GAINS/ PK1(6),PK2(6),PK3(6),PKG(6),PK4(6)
         COMMON/SUBSHM/ ASHM(3,3,6),BSHM(3,6)

         COMMON/ACTORD/ NI(6),KI(6)
         COMMON/OPTION/ IOPTOR(6),IOPTV(6),
        1                IOPTI(6),IOPTG(6),IPOM(18),IOPTIN(6)
         COMMON/SUBCRT/ QQ(3,3,6),R1(6),QI(6)
         COMMON/STDGIM/ ALFAI(6)
C
         DIMENSION QS(3,3),AS(3,3),BS(3,1),PK(3)
         DIMENSION QSS(4,4),ASS(4,4),BSS(4,1),PKS(4)
         DIMENSION QS2(2,2),AS2(2,2),BS2(2,1),PKS2(2)
C-----------------------------------------------------------------
C
C        preparation of the auxiliary matrices ASS,AS, BSS,BS,
C        AS2,BS2,QS,QSS,QS2 - for the various orders of the
C        subsystems
C
         DO 1 I1=1,NI(I)
         BSS(I1,1)=BSHM(I1,I)
         BS(I1,1)=BSHM(I1,I)
C
         DO 1 J=1,NI(I)
         ASS(I1,J)=ASHM(I1,J,I)
       1 AS(I1,J)=ASHM(I1,J,I)
C
         NS=NI(I)
C
C        option: is the integral feedback loop introduced or
C        not- if yes the subsystem order is increased by 1
C
         IF(IOPTIN(I).EQ.1)NS=NS+1
         IF(IOPTIN(I).EQ.0)GO TO 3
C
         DO 2 J=1,NI(I)
         QSS(NS,J)=0.
         QSS(J,NS)=0.
         ASS(NS,J)=0.
         ASS(J,NS)=0.
       2 CONTINUE
```

```
C
          BSS(NS,1)=0.
          ASS(NS,1)=1.
          QSS(NS,NS)=QI(I)
          GO TO 50
C
C         preparation of auxiliary matrices if the subsystem order is 2
C
   3      IF(NI(I).NE.2)GO TO 50
C
          DO 51 I1=1,NI(I)
          BS2(I1,1)=BSHM(I1,I)
          DO 51 J=1,NI(I)
  51      AS2(I1,J)=ASHM(I1,J,I)
  50      CONTINUE
C
C         preparation of auxiliary matrix QSS and the prescribed expo-
C         nential stability degree
C
          DO 21 I1=1,NI(I)
          DO 21 J=1,NI(I)
          QSS(I1,J)=QQ(I1,J,I)
  21      QS(I1,J)=QQ(I1,J,I)
          RS=R1(I)
          ALFA=ALFAI(I)
C
          IF(IOPTIN(I).EQ.1)GO TO 4
          IF(NS.EQ.2)GO TO 4453
C
C         computation of optimal regulator if the subsystem
C         order is 3 and integral feedback loop is not introduced
C
          CALL SMOREG(AS,BS,QS,RS,ALFA,NS,PK,I)
          GO TO 5599
C
C         preparation of the auxiliary matrix QS2
C
4453      DO 54 I1=1,NI(I)
          DO 54 J=1,NI(I)
  54      QS2(I1,J)=QQ(I1,J,I)
C
C         computation of optimal regulator if the subsystem order
C         is 2 and integral feedback loop is not introduced
C
          CALL SMORE2(AS2,BS2,QS2,RS,ALFA,NS,PKS2,I)
C
          DO 55 J=1,NI(I)
  55      PKS(J)=PKS2(J)
          PKS(3)=0.
          PKS(4)=0.
          GO TO 5
C
5599      DO 10 J=1,NI(I)
  10      PKS(J)=PK(J)
          PKS(4)=0.
          GO TO 5
C
C         regulators with the introduced integral feedback loops
C
   4      IF(NS.EQ.4)GO TO 56
```

```
C
C          if the subsystem order is 2 prepare the auxiliary matrices
C
           DO 57 J=1,NI(I)
           AS(J,NS)=0.
   57      AS(NS,J)=0.
           BS(NS,1)=0.
           AS(NS,1)=1.
           QS(NS,NS)=QI(I)
C
C          computation of optimal regulator if the subsystem order
C          is 2 and integral feedaback loop is introduced
C
           CALL SMOREG(AS,BS,QS,RS,ALFA,NS,PK,I)
           DO 58 J=1,NS
   58      PKS(J)=PK(J)
           PKS(3)=0.
           PKS(4)=PK(3)
           GO TO 5
C
C          computation of optimal regulator if the susbsystem order is
C          3 and integral feedback loop is introduced
C
   56      CALL SMORE1(ASS,BSS,QSS,RS,ALFA,NS,PKS,I)
C--------------------------------------------------------------------
C
C          print the computed feedback gains
C
    5      PK1(I)=PKS(1)
           PK2(I)=PKS(2)*IOPTV(I)
           PK3(I)=PKS(3)*IOPTI(I)
           PK4(I)=PKS(4)*IOPTIN(I)
C
           WRITE(5,2000)I
 2000      FORMAT(3X,'Local servo feedback gains in the ',I3,'-th joint',
        1  /)
           WRITE(5,2001)PK1(I),PK2(I),PK3(I),PK4(I)
 2001      FORMAT(3X,'Position feedback gain',G12.5,/,
        1  3X,'Velocity feedack gain',G12.5,/,
        2  3X,'Gain in feedback loop by current/pressure',G12.5,/,
        3  3X,'Integral feedback gain',G12.5,/)
C
           IF(IOPTOR(I).EQ.1)WRITE(5,2002)ALFA
 2002      FORMAT(3X,'Exponential stability degree of local subsystem'
        1  ,G10.5,/)
           RETURN
           END
```

```
C---------------------------------------------------------------------
C      SUBROUTINE:        SMOREG
C---------------------------------------------------------------------
C      FUNCTION:          COMPUTATION OF THE LINEAR OPTIMAL REGULATOR
C                         FOR THE GIVEN LINEAR SYSTEM
C                         (THIS IS WELL-KNOWN PROCEDURE FOR SOLVING OF
C                         THE ALGEBRAIC EQUATION OF THE RICCATI TYPE -
C                         see ref. e.g. Vukobratovic M,
C                         Stokic D., Control of Manipulation
C                         Robots:Theory and Application, Springer-
C                         -Verlag,1982.
C---------------------------------------------------------------------
C      INPUT VARIABLES:
C                 KK - The order number of the subsystem for which the
C                      local optimal regulator is synthesized
C                 N   - the order of the subsystem
C                 A(N,N) - the subsystem matrix
C                 B(N)   - the input distribution matrix of the
C                          subsystem
C                 Q(N,N) - weighting matrix by the state vector in the
C                          local standard quadratic criterion
C                 R      - the weighting matrix by the control signal
C                 ALFA   - the prescribed exponential stability degree
C---------------------------------------------------------------------
C      OUTPUT VARIABLES:
C                 PKK(N) - optimal regulator feedback gains
C                 RK(4,4,KK) - the matrix - solution of the algebraic
C                              equation of Riccati type for the i-th
C                              subsystem
C---------------------------------------------------------------------
C      SUBROUTINES REQUIRED:
C                 EIGENP  - computes the eigen-values of the given
C                           quadratic matrix (SSP-subroutine)
C                 GMPRD   - matrix multiplication (SSP-subroutines)
C                 MINV    - matrix inversion (SSP-subroutines)
C                 SMEIG3  - auxiliary subroutine for computation of the
C                           eigen-values of the given quadratic matrix
C---------------------------------------------------------------------
       SUBROUTINE SMOREG(A,B,Q,R,ALFA,N,PKK,KK)
C
       COMMON/SUBLIA/ RK(4,4,6)
C
       DOUBLE PRECISION AA,EVRA,EVRIA,VECRA,VECIA
       DIMENSION A(N,N),B(N,1),BT(1,3),Q(N,N),PKK(1,N)
       DIMENSION C(1,3),INDIC(6),CPOM(3,3),POM(6,3),LL(3),MM(3)
       DIMENSION PP(3,3),PPP(3,3),P(3,3)
       DIMENSION REALEV(3)
       DIMENSION AA(6,6),EVRA(6),EVRIA(6),VECRA(6,6),VECIA(6,6)
       DOUBLE PRECISION DA(3,3)
C---------------------------------------------------------------------
C
C      The order of the subsystem input is equal to 1 (the input
C      is scalar) - in general case the input order is M
C
       M=1
C
C      transpose the matrix B and compute B*(R**-1)*BT
C
       DO 1 I=1,N
```

```
          DO 1 J=1,M
          BT(J,I)=B(I,J)
  1       C(J,I)=BT(J,I)/R
C
          CALL GMPRD(B,C,CPOM,N,M,N)
C
C         form expanded matrix of the susbsystem
C
          DO 2 I=1,N
          DO 2 J=1,N
          AA(I,J)=A(I,J)
          DA(I,J)=A(I,J)
          AA(I+N,J)=-Q(I,J)
          AA(I,J+N)=-CPOM(I,J)
          AA(I+N,J+N)=-A(J,I)
  2       CONTINUE
C
C         eigen-values of the open-loop subsystem matrix
C
          TYPE 500,KK
500       FORMAT(/,3X,' Eigen-values of the open-loop matrix of the',/,
     1    3X,I1,'.th subsystem (actuator+joint)')
          CALL SMEIG3(N,N,DA,REALEV)
C
          DO 3 I=1,N
          II=I+N
          AA(II,II)=AA(II,II)-ALFA
  3       AA(I,I)=AA(I,I)+ALFA
C
C         computation of the eigen-values and the eigen-vectors of the
C         expanded matrix of the subsystem
C
          NN=N+N
C
          CALL EIGENP(NN,NN,AA,EVRA,EVRIA,VECRA,VECIA,INDIC)
C
C
C         form of the solution of the algebraic equation of Riccati type
C         according to the procedure presented in above reference
C
          M4=0
          I=1
C
 19       IF(I.GT.NN)GO TO 20
          IF(EVRA(I).GT.0)GO TO 7
          M4=M4+1
C
          IF(EVRIA(I).NE.0)GO TO 17
          DO 5 J=1,NN
          POM(J,M4)=VECRA(J,I)
  5       CONTINUE
          GO TO 7
C
 17       I=I+1
          DO 16 J=1,NN
          POM(J,M4)=VECRA(J,I)
 16       POM(J,M4+1)=VECIA(J,I)
C
          M4=M4+1
  7       CONTINUE
```

```
C
          I=I+1
          GO TO 19
  20      CONTINUE
C
          DO 21 I=1,N
          DO 21 J=1,N
          PP(I,J)=POM(I,J)
  21      P(I,J)=POM(I+N,J)
C
C         inverse of the matrix PP
C
          CALL MINV(PP,N,DET,LL,MM)

C
C         calculate (PP**-1)*P
C
          CALL GMPRD(P,PP,PPP,N,N,N)
C
C         memorize the solution PPP of the Riccati's algebraic equation
C         in the matrix RK - necessary for computation of the subsystem
C         Liapunov's function which is required for computation of the
C         gain of global control (see subroutines LIAP1S and GRAD1S)
C
          DO 657 I=1,N
          DO 657 J=1,N
  657     RK(I,J,KK)=PPP(I,J)
C
C         computation of the feedback gains of the optimal regulator
C
          CALL GMPRD(C,PPP,PKK,M,N,N)
C
          WRITE(3,1008)
 1008     FORMAT(/,3X,' Feedback gains of the optimal regulator',/)
          WRITE(3,200)((PKK(I,J),J=1,N),I=1,M)
  200     FORMAT(3X,3E15.5)
C
C         computation of the closed-loop subsystem matrix
C
          CALL GMPRD(CPOM,PPP,P,N,N,N)
C
          DO 12 I=1,N
          DO 12 J=1,N
  12      DA(I,J)=A(I,J)-P(I,J)
C
C         computation of the eigen-values of the closed-loop
C         subsystem matrix
C
          WRITE(5,1009)
 1009     FORMAT(/,3X,'Eigen-values of the closed-loop subsystem
     1    matrix',/)
C
          CALL SMEIG3(N,N,DA,REALEV)
          RETURN
          END
```

```
C-----------------------------------------------------------------------
C         SUBROUTINE:              SMEIG3
C-----------------------------------------------------------------------
C         FUNCTION:                AUXILIARY SUBROUTINE TO COMPUTE EIGEN-
C                                  -VALUES OF THE GIVEN QUADRATIC MATRIX
C-----------------------------------------------------------------------
C         INPUT VARIABLES:
C                  NN      - Matrix order
C                  NM      -     "
C                  DA(NN,NN) - Quadratic matrix
C-----------------------------------------------------------------------
C         OUTPUT VARIABLES:
C                  REALEV(NN) - Vector of the real parts of the eigen-
C                               values
C-----------------------------------------------------------------------
C         SUBROUTINES REQUIRED:
C                  EIGENP - SSP-subroutine to compute the eigen-values
C                             of the given quadratic matrix
C                  REALVE,COMPVE,SCALE,HESQR - Auxiliary subroutines
C                                              called by EIGENP
C-----------------------------------------------------------------------
          SUBROUTINE SMEIG3(NN,NM,DA,REALEV)
C
          DOUBLE PRECISION DA(NN,NN),EVR(3),EVI(3),VECR(3,3)
          DOUBLE PRECISION VECI(3,3)
          DIMENSION REALEV(NN)
          DIMENSION INDIC(3)
C-----------------------------------------------------------------------
C
C         call subroutine EIGENP to compute the eigen-values of
C         the input matrix DA
C
          CALL EIGENP(NN,NM,DA,EVR,EVI,VECR,VECI,INDIC)
C
          DO 10 I=1,NN
   10     REALEV(I)=EVR(I)
C
          WRITE(5,102)
          DO 6 I=1,NN
    6     WRITE(5,103)I,EVR(I),EVI(I),INDIC(I)
  102     FORMAT(//,3X,'Eigen-values ',/,3X,'No.    Real part
         1    Im. part       indic',/)
  103     FORMAT(3X,I4,2E15.5,I4)
          RETURN
          END
```

```
C---------------------------------------------------------------
C          SUBROUTINE:      SMLOC_E
C---------------------------------------------------------------
C          FUNCTION:        COMPUTES LOCAL FEEDBACK GAINS FOR THE ISO-
C                           LATED SUBSYSTEMS (ACTUATOR + JOINT ASSUMING
C                           THAT ALL THE OTHER JOINTS ARE KEPT LOCKED);
C                           CONTROL IS SYNTHESIZED TO KEEP THE DAMPING
C                           RATIO CLOSE TO 1 AND TO KEEP CHARACTERISTIC
C                           FREQUENCY OF THE SERVO (NATURAL UNDAMPED
C                           FREQUENCY) BELLOW ONE HALF OF THE RESONANT
C                           STRUCTURAL FREQUENCY OF THE MECHANISM
C                           (SPECIFIED BY THE USER)
C---------------------------------------------------------------
C      INPUT VARIABLES:
C          I    - the order number of the subsystem for which the servo
C                 gains are to be synthesized
C          NI(i) - the order of the i-th subsystem
C          AHM, BHM - matrices of the subsystem's model (see subrou-
C                     tine INPUT_ACT)
C          the user's options:
C          U0 - resonant structural frequency of the mechanism around
C               the i-th joint [rad/s]
C          OPC - the user wants feedback loop by current/press. (Y/N)
C          ZETA - damping factor of the non-dominant pair of poles
C          OMEGA_N - natural undapmed frequency of the non-dominant
C                    pair of poles - for the third order subsystem
C                    only
C---------------------------------------------------------------
C      OUTPUT VARIABLES:
C          PK1(i) - position feedback gain [V/rad],[V/m],[mA/rad],
C                   [mA/m]
C          PK2(i) - velocity feedback gain [V/rad/s] etc.
C          PK3(i) - gain in feedback loop by current/pressure [V/A]
C---------------------------------------------------------------
       SUBROUTINE SMLOC_E
C
       COMMON/SUBSYS/ A(3,3,6),B(3,6),F(3,6)
       COMMON/SUBSHM/ AHM(3,3,6),BHM(3,6),FHM(3,6)
       COMMON/ACTORD/ NI(6),KI(6)
       COMMON/GAINS/ PK1(6),PK2(6),PK3(6),PK4(6),PK5(6),PK6(6)
       BYTE FILE
       COMMON/FILE/FILE(5)
       CHARACTER*9        FILEP
       COMMON/HMAX/       HMAX(6)
       COMMON/RBSL/       I
       COMMON/TIP_ACT/    A$(6)
       CHARACTER*1 OPC
       CHARACTER*30 l$pom*60
*...............................................................
2      FORMAT (A1)
3      FORMAT (1X,F16.5)
4      FORMAT (1X,A27,I1,A2,E12.5)
6      FORMAT (' Want to introduce feedback loop by current/pressure
     > [Y/N]?:',$)
7      FORMAT (1X,'Specify the resonant structural frequency',/,' of
     > the mechanism [rad/s]')
10     FORMAT (1X,F12.5,35X,A30,I1)
25     FORMAT (1X,'W(S)=',F12.5,'/(S**2+',F12.5,'*S)',/)
26     FORMAT (1X,'W(S)=',F18.5,'/(S**3+',F18.5,'*S**2+',F18.5,
     >            '*S )',/)
```

```
30         FORMAT (1X,'------------------------------------------------------
      >-----------------',/)
35         FORMAT (1X,'G(S)=',F12.5,'/(S**2+',F12.5,'*S+',F12.5,')',/)
36         FORMAT (1X,'G(S)=',F18.5,'/(','S**3+',F18.5,'*S**2+',
      >                F18.5,'*S+',F18.5,')',/)
*......................................................................

           WRITE (6,7)
           WRITE (6,37) I
37         FORMAT (1X,'for the servo in ',i2,' -th joint [rad/s]:',$)
           READ (5,*) U0                        !    U0
C
C          the specified frequency is modified to determine the gains for
C          the case of two equal dominant poles of the closed-loop sub-
C          system;this modification does not affect the computed feedback
C          gains but the algorithms for their computation requires it;
C          the frequency is not modified if there is just one dominant
C          pole
C
           U02=U0
           U0=U0*2.5
C
           IF (NI(I).EQ.3) THEN
              WRITE (6,6)
              READ (5,2) OPC                    !    opc - feedback loop by
           END IF                               !              current
           WRITE (6,30)
C
           if(ni(i).eq.2)then
                   ifl2=1
                   ierrind=1
                   go to 100
           end if
C
C          present the containts of the auxiliary file conf.dat
C          in order to enable the user to select the poles of the closed-
C          -loop subsystem (in the case the order of the subsystem is 3)
C
           open (23,file='conf',status='old')
           do ipom=1,10
              read (23,'(a60)') l$pom
              write (6,'(1x,a60)') l$pom
           end do
           close(23)
1          write (6,'(1x,'' If you want to calculate feedback gains'',/
      >              '' that the pole placement be as in Fig.1, then'',/
      >              '' specify 1, otherwise the gains will be computed'',/
      >              '' to place the poles as presented in Fig.2.:'',$)')
           ierrind=1
           read (5,'(i1)',err=1500) ifl2
C----------------------------------------------------------------------
C
C          Computation of the feedback gains
C
 100       IF (ifl2.eq.1) THEN
C
C          the pair of dominant poles is complex (or real and equal) and
C          non-dominant pole is real (Fig.1.)
C
```

```
          dpot=u0/10.
          pot=2.
          IF (ni(i).eq.2) THEN
              pk1(i)=(u0/5.)**2/bhm(2,i)
              pk2(i)=(2.*u0/5.+ahm(2,2,i))/bhm(2,i)
          ELSE
              IF (opc.EQ.'N') THEN
                  alf=-ahm(2,2,I)-ahm(3,3,I)
                  pk3(I)=0
                  bet=alf*2.*u0/5.-3.*(u0/5.)**2
                  gam=alf*(u0/5.)**2-2.*(u0/5.)**3
                  pk1(I)=gam/ahm(2,3,I)/bhm(3,I)
                  pk2(I)=(bet-ahm(2,2,I)*ahm(3,3,I)+ahm(2,3,I)
     >                    *ahm(3,2,I))/ahm(2,3,I)/bhm(3,I)

              ELSE
                  alf=(pot+1.)*u0
                  pk3(I)=(alf+ahm(2,2,I)+ahm(3,3,I))/bhm(3,I)
                  bet=alf*2.*u0/5.-3.*(u0/5.)**2
                  gam=alf*(u0/5.)**2-2.*(u0/5.)**3
                  pk1(I)=gam/ahm(2,3,I)/bhm(3,I)
                  pk2(I)=(bet-ahm(2,2,I)*ahm(3,3,I)+ahm(2,3,I)
     >                    *ahm(3,2,I)+ahm(2,2,I)*pk3(I))/ahm(2,3,I)
     >                    /bhm(3,I)
                  DO WHILE (pk2(i).LT.0.OR.pk3(i).LT.0)
                      pot=pot+dpot
                      alf=(pot+1.)*u0
                      pk3(I)=(alf+ahm(2,2,I)+ahm(3,3,I))/bhm(3,I)
                      bet=alf*2.*u0/5.-3.*(u0/5.)**2
                      gam=alf*(u0/5.)**2-2.*(u0/5.)**3
                      pk1(I)=gam/ahm(2,3,I)/bhm(3,I)
                      pk2(I)=(bet-ahm(2,2,I)*ahm(3,3,I)+ahm(2,3,I)
     >                        *ahm(3,2,I)+ahm(2,2,I)*pk3(I))
     >                        /ahm(2,3,I)/bhm(3,I)
                  END DO
              END IF
              s3=2.*u0/2.-alf
              WRITE (6,'(//'' Non-dominant pole is (S3):'',E10.3)')S3
          END IF
      ELSE
          IF (ni(i).eq.3) THEN
C
C
C         the dominant pole is real and the nondominant pair of poles
C         are complex (or real) - Fig.2.
C         the third order subsystem
C
5             WRITE (6,'('' The damping coefficient (zeta[0.-1.]):'',
     >               /,'' of the non-dominant pair of poles:'',$)')
              ierrind=2
              read (5,*,err=1500) zeta
888           WRITE (6,'('' Natural undamped frequency (omega_n)''
     >               ,/'' of the non-dominant pair of poles:'',$)')
              ierrind=3
              READ (5,*,err=1500) omega_n
              IF (opc.eq.'Y') THEN
                  omegad=omega_n/10.
                  DO WHILE (pk2(i).LE.0.OR.pk3(i).LE.0)
                      omega_n=omega_n+omegad
                      alfa=2.*zeta*omega_n+u02/2.
                      pk3(i)=(alfa+ahm(2,2,i)+ahm(3,3,i))/b(3,i)
                      beta=omega_n*omega_n+zeta*omega_n*u02
                      pk2(i)=(beta+ahm(2,3,i)*ahm(3,2,i)
```

```
     >                              +ahm(2,2,i)*pk3(i)*b(3,i)
     >                              -ahm(2,2,i)*ahm(3,3,i))/ahm(2,3,i)/b(3,i)
                       gama=.5*u02*omega_n*omega_n
                       pk1(i)=gama/ahm(2,3,i)/b(3,i)
                    END DO
                 ELSE
                    omega_n=(-ahm(2,2,i)-ahm(3,3,i)-u02/2.)/2./zeta
                    gama=.5*u02*omega_n*omega_n
                    pk1(i)=gama/ahm(2,3,i)/b(3,i)
                    beta=omega_n*omega_n+zeta*omega_n*u02
                    pk2(i)=(beta+ahm(2,3,i)*ahm(3,2,i)
     >                              +ahm(2,2,i)*pk3(i)*b(3,i)
     >                              -ahm(2,2,i)*ahm(3,3,i))/ahm(2,3,i)/b(3,i)
                 END IF
              ELSE
C
C
C           the second order subsystem
C
              s2=u02*2.

              s2d=u02/10.
              DO WHILE (pk2(i).LE.0.1)
                 s2=s2+s2d
                 pk1(i)=u02/2.*s2/bhm(2,i)
                 pk2(i)=(u02/2.+s2+ahm(2,2,i))/bhm(2,i)
              END DO
           END IF
        END IF
C
C
C       print computed gains and open-loop and clsed-loop subsystem
C       transfer functions
C
        WRITE (6,'(1X,''Local feedback servo gains:'',/)')
        WRITE (6,4) 'Position  servo  gain   KP(',I,')=',PK1(i)
        WRITE (6,4) 'Velocity  servo  gain   KV(',I,')=',PK2(i)
        IF (NI(I).EQ.3) THEN
           WRITE (6,4) 'Curr./pr. servo gain KS(',I,')=',PK3(i)
        END IF
        WRITE (6,30)
        WRITE(6,'(1X,''Open-loop transfer function:'',/)')
        IF (NI(I).EQ.2) THEN
           WRITE (6,25) bhm(2,i),-ahm(2,2,i)
        ELSE
           WRITE (6,26) ahm(2,3,i)*bhm(3,i),-(ahm(2,2,i)+ahm(3,3,i)),
     >                  ahm(2,2,i)*ahm(3,3,i)-ahm(2,3,i)*ahm(3,2,i)
        END IF
        WRITE (6,30)
        WRITE (6,'(1X,''Closed-loop transfer function:'',/)')
        IF (NI(I).EQ.2) THEN
           WRITE (6,35) bhm(2,i)*pk1(i),-ahm(2,2,i)+bhm(2,i)*pk2(i),
     >                  bhm(2,i)*pk1(i)
        ELSE
           bpom2=pk3(i)*bhm(3,i)-ahm(3,3,i)-ahm(2,2,i)
           bpom1=ahm(2,3,i)*(pk2(i)*bhm(3,i)-ahm(3,2,i))-ahm(2,2,i)*
     >           (pk3(i)*bhm(3,i)-ahm(3,3,i))-ahm(2,1,i)
           bpom0=pk1(i)*bhm(3,i)*ahm(2,3,i)-ahm(2,1,i)*(pk3(i)*
     >           bhm(3,i)-ahm(3,3,i))
           WRITE (6,36) bhm(3,i)*ahm(2,3,i)*pk1(i),bpom2,bpom1,bpom0
        END IF
```

```
        WRITE (6,30)
        RETURN
1500    WRITE (6,'('' DIAG*** Input conversion error'')')
        go to (1,5,888) ierrind
        END
```

```
C--------------------------------------------------------------------
C         SUBROUTINE:    TRAJEK
C--------------------------------------------------------------------
C         FUNCTION:      GENERATES NOMINAL TRAJECTORIES OF THE
C                        ROBOT JOINTS - BETWEEN TWO SPECIFIED
C                        TERMINAL POSITIONS WITH TRIANGULAR OR
C                        TRAPEZOID VELOCITY DISTRIBUTION
C--------------------------------------------------------------------
C         INPUT VARIABLES:
C                 FILE(5)   robot name
C                 N         number of joints of the user's specific robot
C                 Q0(I)     initial values of the joint coordinates
C                 QF(I)     values of the terminal point of the joint tra-
C                           jectories (from the input file ***.TRA)
C                 H         sampling interval at which the trajectories
C                           are computed
C                 T         time duration of the nominal movement (traj.)
C                 BETA      parameter defining the trapezoidal profile of
C                           velocity distribution
C               ^ DLMB
C               | (trajectory)
C               |
C               |
C               +         .--------------.
C               |       .                  .
C               |     .                      .
C               |   .                          .
C               | .                              .
C            ---|----+--------------+------+---------------->
C               0    T1             T2     T      TE(time)
C
C                 T1=T*BETA
C                 T2=T*(1.-BETA)
C
C                 PROFIL    character variable for the user's selection
C                           of the desired profile of the velocity dis-
C                           tribution along the nominal trajectory
C                 KSI2(6)   indicator of the joint type (1- linear,
C                           0 - rotational)
C--------------------------------------------------------------------
C         OUTPUT VARIABLES:
C                 TE        time instant
C                 Q(6)      positions of the joints angles at the nominal
C                           trajectory at the moment TE
C                 DQ(6)     velocities            "
C                 DDQ(6)    accelerations         "
C                 these values are written in the output file ***.ANG
C--------------------------------------------------------------------
```

```
          SUBROUTINE TRAJEK
C
          CHARACTER *2 PROFIL,FILEP*9,STAMPA*1,FILET*9
          CHARACTER *4 IND(6)
          BYTE FILE(5)
          COMMON/FILE/FILE
          COMMON/DINT_TR/ H
          COMMON/ORDGLB/ NUK,N
          COMMON/TIPL/NPOM,KSI1(6),KSI2(6),KSI3(6)
          DIMENSION Q0(6),QF(6),Q(6),DQ(6),DDQ(6),DLTQ(6)
          REAL LMB
C----------------------------------------------------------------------
C
C         read data from the input file ***.TRA
C
          FILET=CHAR(FILE(1))//CHAR(FILE(2))//CHAR(FILE(3))//
     *         CHAR(FILE(4))//CHAR(FILE(5))//'.TRA'

          open(unit=7,file=filet,status='old')
          read(7,210)(q0(i),i=1,n)
          read(7,210)(qf(i),i=1,n)
          read(7,210)h,t
200       format(1x,i3)
210       format(1x,6f10.5)
          close(unit=7)
C
22        FORMAT(6(4X,A5,4X))
100       FORMAT(6E13.5)
C
          TE=0.
          M=T/H
C
C         indication of the joint type (TRA/ROT)
C
          DO I=1,N
          IND(I)=' rad '
            IF(KSI2(I).EQ.1)THEN
            IND(I)=' [m] '
            END IF
          END DO
C
          type *
          type *,' Synthesis of the joints nominal trajectories'
          TYPE 40
40        FORMAT(/,$,' Want to print the nominal trajectories [Y/N]?:')
          READ(5,45)STAMPA
45        FORMAT(A1)
C
          TYPE 10
10        FORMAT($,' Select the velocity profile-triangular or trapezoid
     *  [TA/TP]:')
          READ(5,20)PROFIL
20        FORMAT(A2)
C
          FILEP=CHAR(FILE(1))//CHAR(FILE(2))//CHAR(FILE(3))//
     *         CHAR(FILE(4))//CHAR(FILE(5))//'.ANG'
          OPEN(UNIT=4,FILE=FILEP,STATUS='NEW')
C
          IF(PROFIL.EQ.'TA')GO TO 70
C----------------------------------------------------------------------
```

```
C
C          trapezoid velocity profile
C
25         TYPE 30
30         FORMAT(' Specify the trapezoid profile -
     *     by defining parameter BETA',/,$,' (duration of the
     *     acceleration/deacceleration phase as a part of',/,
     *     ' total trajectory duration)  [0.0-0.5]:')
           READ (5,35) BETA
35         FORMAT (F10.5)
           IF((BETA.LT.0.0).OR.(BETA.GE.0.5)) GO TO 25
C
C
           T1=BETA*T
           T2=(1.-BETA)*T
C
C          Computation of the scalar parameters of the velocity
C          profile LMB and DDLMB and CONST
C
           P=1./((BETA*T**2)*(1.-BETA))
           P1=1./(T*(1.-BETA))
           CONST=0.5*BETA/(1.-BETA)
C
50         CONTINUE
           IF(TE.GT.T)GO TO 65
           IF(TE.LT.T1) THEN
                   LMB=0.5*P*TE**2
                   DLMB=P*TE
                   DDLMB=P
           ELSE IF(TE.LT.T2)THEN
                   LMB=P1*TE-CONST
                   DLMB=P1
                   DDLMB=0.
           ELSE IF(TE.LE.T) THEN
                   LMB=-0.5*P*(T-TE)**2+1.
                   DLMB=P*(T-TE)
                   DDLMB=-P
           END IF
C
C          computation of the nominal positions, velocities and
C          accelerations
C
           DO I=1,N
           DLTQ(I)=QF(I)-Q0(I)
           Q(I)=Q0(I)+LMB*DLTQ(I)
           DQ(I)=DLMB*DLTQ(I)
           DDQ(I)=DDLMB*DLTQ(I)
           END DO
C
C          write the trajectories in the file ***.ANG
C
           WRITE(4,100)TE
           WRITE(4,100)(Q(I),I=1,N)
           WRITE(4,100)(DQ(I),I=1,N)
           WRITE(4,100)(DDQ(I),I=1,N)
C
C          print at the screen
C
```

```
            IF(STAMPA.EQ.'N')GO TO 110
            WRITE(6,22)(IND(I),I=1,N)
            WRITE(6,100)TE
            WRITE(6,100)(Q(I),I=1,N)
            WRITE(6,100)(DQ(I),I=1,N)
            WRITE(6,100)(DDQ(I),I=1,N)
C
110         TE=TE+H
            GO TO 50
65          CONTINUE
C
            CLOSE(UNIT=4)
            RETURN
C--------------------------------------------------------------------
C
C           triangular velocity profile
C
70          CONTINUE
            T1=T/2.
C
C           computation of parameters
C
            P=4./(T**2)
150         CONTINUE
            IF(TE.GT.T) GO TO 160
C
            IF(TE.LT.T1)THEN
                    LMB=0.5*P*TE**2
                    DLMB=P*TE
                    DDLMB=P
            ELSE
                    LMB=-0.5*P*(T-TE)**2+1.
                    DLMB=P*(T-TE)
                    DDLMB=-P
            END IF
C
            DO I=1,N
            DLTQ(I)=QF(I)-Q0(I)
            Q(I)=Q0(I)+LMB*DLTQ(I)
            DQ(I)=DLMB*DLTQ(I)
            DDQ(I)=DDLMB*DLTQ(I)
            END DO
C
C           write in the output fle ***.ANG
C
            WRITE(4,100)TE
            WRITE(4,100)(Q(I),I=1,N)
            WRITE(4,100)(DQ(I),I=1,N)
            WRITE(4,100)(DDQ(I),I=1,N)
C
C           print at the screen
C
            IF(STAMPA.EQ.'N')GO TO 90
            WRITE(6,22)(IND(I),I=1,N)
            WRITE(6,100)TE
            WRITE(6,100)(Q(I),I=1,N)
            WRITE(6,100)(DQ(I),I=1,N)
            WRITE(6,100)(DDQ(I),I=1,N)
C
```

```
90        TE=TE+H
          GO TO 150
160       CONTINUE
C

          CLOSE(UNIT=4)
          RETURN
          END
```

```
C-----------------------------------------------------------------
C       SUBROUTINE:     SMNOMP
C-----------------------------------------------------------------
C       FUNCTION:       COMPUTES THE NOMINAL DYNAMICS OF THE ROBOT
C                       - NOMINAL DRIVING TORQUES AND NOMINAL CEN-
C                       TRALIZED OR LOCAL CONTROL
C-----------------------------------------------------------------
C       INPUT VARIABLES:
C               FILE(5) - robot name
C               Q01(6) - nominal angles (displacements) of the joints
C               QT01(6) -nominal velocities of the joints
C               QU0(6) - nominal accelerations of the joints
C                       the data on nominal trajectory are read from
C                       the file ***.ANG
C               NI(I) - the order of the i-th subsystem
C               N      - number of joints of the user's specific robot
C               A,B,F - matrices of the actuators' models (see
C                       explanations in subroutine INPUTL)
C               UMAX(2,6)  - contraints upon the actuator inputs
C               HMIN   - minimal values of the moments of inertia of
C                       the mechanism around the joints axes (see ex-
C                       planation in subroutine SM_MAX_IN)
C               HH(6,6) - inertia matrix of the robot for given joints
C                       angles(SI) and velocities (SIDOT) - computed
C                       in the subroutine MODEL
C               H1(6) - vector of centrifugal, Coriolis and gravity
C                       moments for given SI and SIDOT - subrout.MODEL
C-----------------------------------------------------------------
C       OUTPUT VARIABLES:
C               P0(6)  - nominal driving torques
C               U0(6)  - nominal programmmed control -feedforward
C                       (centralized or local)
C               POWER(6)- nominal power at each actuator
C               SENERG(6) - nominal power consumption of each actuator
C               DX0(I) - nominal currents/pressures
C-----------------------------------------------------------------
C       SUBROUTINES REQUIRED:
C               MODEL - computes the dynamic model of the robot
C                       mechanism (computes inertia matrix H and
C                       centrifugal, Coriolis, gravity moments h
C                       for the given nominal trajectories)
C-----------------------------------------------------------------
C       SUBROUTINE SMNOMP
C
```

```
        DIMENSION XPOM(6)
        DIMENSION QU0(6)
        DIMENSION SENERG(6),POWER(6)
C
        COMMON/NPOM1/ Q00(6),QT00(6),QU00(6),P00(6)
        COMMON/NPOM2/ Q01(6),QT01(6),QU01(6),P01(6)
        COMMON/SUBSYS/ A(3,3,6),B(3,6),F(3,6)
        COMMON/NKOORD/ P0(6),X0(18),U0(6)
        COMMON/ACTORD/ NI(6)
        COMMON/UMAX/ UMAX(2,6)
        COMMON/HMIN/ HMIN(6)
        COMMON/SKOORD/ DX0(18),PX0(6)
        COMMON/ORDGLB/NUK,NN
        COMMON/FILE/FILE(5)
        BYTE FILE
C
        INCLUDE 'IN2:MODELM.MOD'
        COMMON/MODOPC/ MOD,ITIP
        CHARACTER*9 FILEP
C-----------------------------------------------------------------
C
        TYPE *
        TYPE *,'   Computation of the nominal dynamics of the robot'
        TYPE *,'        - nominal driving torques and nominal'
        TYPE *,'                   programmed control -'
        type 399
 399    FORMAT(/,$,' Want centralized or local nominal programmed
     1 control? [C/L]:')
        read(5,401)YES
C
C       set indicator ILOC: ILOC=0 - local nominal control,
C       ILOC=1 - centralized nominal control
C
        ILOC=1
        IF(YES.EQ.'L')ILOC=0
        TYPE 400
 400    FORMAT($,' Want to print the nominal dynamics [Y/N]?:')
        READ(5,401)YES
 401    FORMAT(A1)
C
        MOD=0
        ITIP=0
        N=NN
C
C       openning of the files ***.ANG to read nominal trajectories
C       and ***.DIN to write the computed nominal control and
C       nominal driving torques
C
        FILEP=CHAR(FILE(1))//CHAR(FILE(2))//CHAR(FILE(3))//
     1       CHAR(FILE(4))//CHAR(FILE(5))//'.ANG'
        OPEN(UNIT=1,NAME=FILEP,STATUS='OLD',ERR=1110)
C
        FILEP=CHAR(FILE(1))//CHAR(FILE(2))//CHAR(FILE(3))//
     1       CHAR(FILE(4))//CHAR(FILE(5))//'.DIN'
        OPEN(UNIT=2,NAME=FILEP,STATUS='NEW')
C
C       read nominal trajectories from the file ***.ANG
C       VR2 is the time instant along the nominal trajectory
C
```

```
      READ(1,100)VR2
      READ(1,100)Q01
      READ(1,100)QT01
      READ(1,100)QU0
100   FORMAT(6E13.5)
C
C     the auxilaiary vectors SI and SIDOT - through them the
C     values of the joint angles and velocities are sent to the
C     subroutine MODEL to compute matrix HH and vector H1
C
      DO 1 I=1,N
      SENERG(I)=0.
      SI(I)=Q01(I)
1     SIDOT(I)=QT01(I)
C
C     call subroutine MODEL to compute HH and H1 for given joints
C     angles and velocities (nominal trajectory)
C
      CALL MODEL
C
C     computation of the nominal driving torque PX0 in the initial
C     point of the nominal trajectory and computation of the third
C     actuator state coordinate (if its order is 3)
C
      DO 2 I=1,N
      PX0(I)=H1(I)
      DO 77 J=1,N
      PX0(I)=PX0(I)+HH(I,J)*QU0(J)
77    CONTINUE
C
      IF(NI(I).EQ.2)GO TO 2
      DX0(I)=QU0(I)-A(2,2,I)*QT01(I)
C
C     if the centralized nominal control is computed then the total
C     nominal driving torque is taken into account
C     if the local nominal control is computed then just the
C     estimation of the mechanism moment of inertia around the
C     joint axis is taken into account
C
      IF(ILOC.EQ.1)THEN
          DX0(I)=DX0(I)-F(2,I)*PX0(I)
        ELSE
          DX0(I)=DX0(I)-F(2,I)*HMIN(I)*QU0(I)
      END IF
      DX0(I)=DX0(I)/A(2,3,I)
2     CONTINUE
999   FORMAT(3X,6E12.3)
C-------------------------------------------------------------
C
C     memorize the nominal torques PX0 in P0
C     and the nominal trajectory in Q00,QT00,QU00 (the previous
C     point) and the time instant in VR1
C
3     VR1=VR2
      DO 4 I=1,N
      P0(I)=PX0(I)
      Q00(I)=Q01(I)
      QT00(I)=QT01(I)
      QU00(I)=QU0(I)
4     X0(I)=DX0(I)
```

422

```
C
C          read next point at the nominal trajectory  from the file
C
           READ(1,100,END=977)VR2
           READ(1,100)Q01
           READ(1,100)QT01
           READ(1,100)QU0
C
           POM=VR2-VR1
C
C          compute QU0(6) - nominal accelerations
C          and send nominal joints angles and velocities through the
C          auxiliary vectors SI and SIDOT into subroutine MODEL
C
    5      DO 6 I=1,N
           SI(I)=Q01(I)
           SIDOT(I)=QT01(I)
    6      CONTINUE
C
C          call MODEL to compute HH and H1 for the nominal trajectory
C
           CALL MODEL
C
C          computation of the nominal driving torques, power and energy
C          consumption, nominal state vectors and programmed control
C
           DO 7 I=1,N
           PX0(I)=H1(I)
           DO 88 J=1,N
           PX0(I)=PX0(I)+HH(I,J)*QU0(J)
   88      CONTINUE
C
           POWER(I)=ABS(P0(I)*QT00(I))
           SENERG(I)=SENERG(I)+POWER(I)*POM
C
           IF(NI(I).EQ.2)GO TO 89
C
C          computation of the third state coordinate of the actuator
C          model (if its order is 3)
C
           DX0(I)=QU0(I)-A(2,2,I)*QT01(I)
C
           IF(ILOC.EQ.1)THEN
                   DX0(I)=DX0(I)-F(2,I)*PX0(I)
             ELSE
                   DX0(I)=DX0(I)-F(2,I)*HMIN(I)*QU0(I)
           END IF
C
           DX0(I)=DX0(I)/A(2,3,I)
C
C          computation of the derivative of the third state coordinate
C
           XPOM(I)=(DX0(I)-X0(I))/POM
C
C          computation of the nominal control if the model order is 3
C
           U0(I)=(XPOM(I)-A(3,2,I)*QT00(I)-A(3,3,I)*X0(I))/B(3,I)
           GO TO 90
C
```

```
C           if the order of the actuator state model is 2, directly
C           compute the nominal control
C
 89         U0(I)=QU00(I)-A(2,2,I)*QT00(I)
C
            IF(ILOC.EQ.1)THEN
                  U0(I)=(U0(I)-F(2,I)*P0(I))/B(2,I)
             ELSE
                  U0(I)=(U0(I)-F(2,I)*HMIN(I)*QU00(I))/B(2,I)
            END IF
C
C           amplitude contraint upon the actuators inputs
C
 90         IF(U0(I).LT.UMAX(1,I).AND.U0(I).GT.(-UMAX(2,I)))GO TO 7
            TYPE 101,VR1,I,U0(I)
 7          CONTINUE
C
C           set computed values to 0 if they are less than 0.0001
C
            DO 7722 I=1,N
            IF(ABS(Q00(I)).LT.0.00001)Q00(I)=0.
            IF(ABS(QT00(I)).LT.0.0001)QT00(I)=0.
            IF(ABS(QU00(I)).LT.0.0001)QU00(I)=0.
            IF(ABS(P0(I)).LT.0.0001)P0(I)=0.
            IF(ABS(U0(I)).LT.0.0001)U0(I)=0.
            IF(ABS(POWER(I)).LT.0.0001)POWER(I)=0.
            IF(ABS(SENERG(I)).LT.0.0001)SENERG(I)=0.
 7722       CONTINUE
C
C           write nominal dynamics in the output file ***.DIN
C
            WRITE(2,100)VR1
            WRITE(2,100)(Q00(I),I=1,N)
            WRITE(2,100)(QT00(I),I=1,N)
            WRITE(2,100)(QU00(I),I=1,N)
            WRITE(2,100)(P0(I),I=1,N)
            WRITE(2,100)(U0(I),I=1,N)
            WRITE(2,100)(POWER(I),I=1,N)
            WRITE(2,100)(SENERG(I),I=1,N)
C
C           print nominal dynamics at the terminal screen
C
            IF(YES.NE.'Y')GO TO 3
            TYPE 200,VR1
 200        FORMAT(' Time:',F12.5,' [s]')
            TYPE 201,(I,I=1,N)
 201        FORMAT(' Joint numbers        ',6(4X,I1,5X))
            TYPE 202,(Q00(I),I=1,N)
 202        FORMAT(' Angles [rad] or [m]:',6E10.3)
            TYPE 203,(QT00(I),I=1,N)
 203        FORMAT(' Velocit.[rad/s-m/s]:',6E10.3)
            TYPE 204,(QU00(I),I=1,N)
 204        FORMAT(' Acceler. [rad/s**2]:',6E10.3)
            TYPE 205,(P0(I),I=1,N)
 205        FORMAT(' Driving torques[Nm]:',6E10.3)
            TYPE 206,(U0(I),I=1,N)
 206        FORMAT(' Control sign.[V-mA]:',6E10.3)
            TYPE 207,(POWER(I),I=1,N)
 207        FORMAT(' Power [W]:           ',6E10.3)
```

```
        TYPE 208,(SENERG(I),I=1,N)
 208    FORMAT(' Energy [Ws]:          ',6E10.3)
        GO TO 3
C
C
 101    FORMAT(' DIAG*** Nominal control signal are greater than the',/,
     1  ' allowable amplitude constraints at the actuators inputs',/,
     2  ' (change nominal kinematics, or change actuators)',/,3X,
     3  ' time =', F15.5, ' in the joint No. ',I2,' the required nominal
     4   signal =', G15.5)
C
 977    CONTINUE
C
C       write the terminal point at the trajectory into output file
C
        WRITE(2,100)VR1
        WRITE(2,100)(Q00(I),I=1,N)
        WRITE(2,100)(QT00(I),I=1,N)
        WRITE(2,100)(QU00(I),I=1,N)
        WRITE(2,100)(P0(I),I=1,N)
        WRITE(2,100)(U0(I),I=1,N)
        WRITE(2,100)(POWER(I),I=1,N)
        WRITE(2,100)(SENERG(I),I=1,N)
C
        CLOSE (2)
C
6680    CONTINUE
        CLOSE (1)
        RETURN
1110    TYPE 1111, FILEP
1111    FORMAT (1X,'DIAG*** File ',a9,1x,'  cannot be found')
        CLOSE (1)
        STOP
        END




C-------------------------------------------------------------------
C       SUBROUTINE:     LINANA
C-------------------------------------------------------------------
C       FUNCTION:       THE SUBROUTINE FOR EXPONENTIAL STABILITY
C                       ANALYSIS OF THE LINEARIZED MODEL OF THE
C                       ROBOT; CALLS SUBROUTINES FOR FORMING OF
C                       THE LINEARIZED MODEL OF THE ROBOT AND FOR
C                       COMPUTATION OF THE EIGEN-VALUES OF THE
C                       LINEAR MODEL; THE STABILITY OF THE SYSTEM
C                       IS TESTED IF ONLY LOCAL CONTROLLERS ARE
C                       APPLIED
C-------------------------------------------------------------------
C       INPUT VARIABLES:
C               FILE(5)-name of the user's robot
C               N      - number of joints of the user's robot
C               T0     - time instant at the nominal trajectory
C               Y0(12)- nominal joints angles and velocities at the
C                       time instant T0 (the nominal state vector of
C                       the mechanical part of the system)
C               DX0(6)- Nominal accelerations of the joint coord.
C               P0(6) - Nominal driving torques in the time instant T0
```

```
C                     TIMIN(*) - vector of time instants in which the analy-
C                                sis of the system stability is required -
C                                this is specifified by the user in ***.INT
C                     ITMIN - number of points at the nominal trajectory in
C                             which we have to analyze the robot stability
C                     ALFAIM - desired exponential stability degree of the
C                              linearized model of the robot
C                     IOPTIN(6) - option: integral feedback loop(1-yes,0-no)
C-----------------------------------------------------------------------
C         OUTPUT VARIABLES:
C                     REALEV(NUK) - Real parts of the eigen-values of the
C                                   linearized global system
C                     RIMAG(NUK)  - Complex parts      "
C                     ICNVLN      - Indication whether the linearized model
C                                   of the system is stabilized or not
C-----------------------------------------------------------------------
C         SUBROUTINES REQUIRED:
C                     SMLINM  - Forming of the matrices of linearized mo-
C                               del of the robotic system (open-loop system)
C                     SMLGAD  - Introduce local feedback loops in the mat-
C                               rices of the linearized model of the system
C                     SMEIGN  - Auxiliary subroutine for computation of the
C                               eigen-values of the given quadratic matrix
C-----------------------------------------------------------------------
        SUBROUTINE LINANA
C
        COMMON/ORDGLB/ NUK,N
        COMMON/STDGIM/ ALFAI(6),ALFAIM
        COMMON/ACTORD/NI(6)
        COMMON/OPTION/ IPOM1(42),IOPTIN(6)
        COMMON/OPTPR/ ICNVLN
        COMMON/AUXIL/ T0,P0(6),Y0(12),DX0(6),U0(6)
        COMMON/CTIMIN/ TIMIN(60),ITMIN
C
        BYTE FILE(5)
        COMMON/FILE/ FILE
        COMMON/CAUXIL/ ITERC,REALPS
        COMMON/INTGAI/ NUKI
C
        DOUBLE PRECISION AL(24,24)
        DIMENSION A0(24,24),BL(24,6)
        DIMENSION REALEV(24),RIMAG(24)
C
        DIMENSION HU(6,6)
C
        BYTE FTOT(14),FTOT1(14)
        DATA FTOT(1)/'I'/,FTOT(2)/'N'/,FTOT(3)/'1'/,FTOT(4)/':'/,
     1       FTOT(10)/'.'/,FTOT(11)/'D'/,FTOT(12)/'I'/,FTOT(13)/'N'/,
     2       FTOT(14)/0/
        DATA FTOT1(1)/'I'/,FTOT1(2)/'N'/,FTOT1(3)/'1'/,FTOT1(4)/':'/,
     1       FTOT1(10)/'.'/,FTOT1(11)/'I'/,FTOT1(12)/'N'/,
     2       FTOT1(13)/'T'/,FTOT1(14)/0/
C-----------------------------------------------------------------------
C
        TYPE *
        TYPE *,' Stability analysis '
        TYPE *
C
        DO 199 I=1,5
        FTOT1(I+4)=FILE(I)
```

```
 199       FTOT(I+4)=FILE(I)
C
C         computation of NUKI - the order of the total systems with
C         local integral feedback loops
C
          NUKI=0
          DO 9000 I=1,N
 9000     NUKI=NUKI+NI(I)+IOPTIN(I)
C
C         open file containing nominal dynamics computed in SMNOMP
C
          OPEN(UNIT=1,NAME=FTOT,TYPE='OLD',ERR=3100)
C
C         open file ***.INT  specified by the user which contains the
C         desired stability degrees and the moments at the nominal
C         trajectories "in which" the user wants to analyze the
C         stability of the linearized model
C
          OPEN(UNIT=2,NAME=FTOT1,TYPE='OLD',ERR=3101)
C
          READ(2,102)ITMIN
 102      FORMAT(I4)
          GO TO 3102
C
 3100     TYPE 3103,FILE
 3103     FORMAT(' DIAG*** File ',5A1,'.DIN cannot be found')
          RETURN
C
 3101     TYPE 3104,FILE
 3104     FORMAT(' DIAG*** File ',5A1,'.INT cannot be found')
          RETURN
C-------------------------------------------------------------------
C
C         reset the counter ITERC of points at the nominal
C         trajectory in which analysis has to be performed
C
 3102     ITERC=1
C
C         read data on nominal dynamics from ***.DIN
C
 1        READ(1,100,END=3201)T0
          READ(1,100)(Y0(I),I=1,N)
          READ(1,100)(Y0(I),I=N+1,N+N)
          READ(1,100)(DX0(I),I=1,N)
          READ(1,100)(P0(I),I=1,N)
          READ(1,100)(U0(I),I=1,N)
C
C         auxiliary read of power and energy - not necessary in
C         this routine
C
          READ(1,100)(U0(I),I=1,N)

          READ(1,100)(U0(I),I=1,N)
 100      FORMAT(6E13.5)
          GO TO 3202
C
 3201     CLOSE(UNIT=1)
          GO TO 3203
C
 3202     CONTINUE
```

```
C
C          test if the next point at the nomonal trajectory around which
C          the stability of the system has to be analyzed is reached or
C          not; if it is reached - go to stability analysis
C          otherwise - read the next data at the nominal trajectory
C
           IF(T0.LT.TIMIN(ITERC)-0.0001)GO TO 1
C
C          read data on desired stability degree
C
  3203     READ(2,101)TIMIN(ITERC)
           READ(2,101)ALFAIM
   101     FORMAT(E15.5)
C-------------------------------------------------------------------
C
C          compute matrices of the linearized model of the system in the
C          user's selected point at the nominal trajectory (without local
C          feedback loops)
C          A0(NUKI,NUKI),BL(NUKI,N)- Matrices of the open-loop linearized
C                                    model of the robot
C
   10      CALL SMLINM(A0,BL,HU)
C
C          put A0 into matrix AL to get double-precision computation
C
           NN=NUKI
           NM=NUKI
           DO 7773 KKK=1,24
           DO 7773 JJJ=1,24
  7773     AL(KKK,JJJ)=A0(KKK,JJJ)
C
           WRITE(6,412) TIMIN(ITERC)
   412     FORMAT(/,' Eigen-values of the open-loop matrix of',/,
      1    ' linearized model of the robot in the time instant',F12.5,'
      2    [s]',/)
C
C          computation of eigen-values of the open-loop system matrix
C
           CALL SMEIGN(NN,NM,AL,REALEV,RIMAG)
C
C          introducing of the local feedback gains in the matrix of the
C          open-loop linearized model of the system A0
C          the matrix AL s obtained: the matrix of the closed-loop linea-
C          rized model of the robotic system
C
   11      CALL SMLGAD(A0,AL,BL)
C
C          computation of the eigen-values of the linearized model of
C          the system in the selected point at the nominal  tajectory
C
           WRITE(6,411)TIMIN(ITERC)
   411     FORMAT(/,3X,' Eigen-values of the closed-loop matrix of',
      1    /,'    linearized model of the robot in the time instant
      2    ',F12.5,' [s]',/)
           CALL SMEIGN(NN,NM,AL,REALEV,RIMAG)
C
C          search for the greatest eigen-value of the system matrix
C          (the eigen-value with the greatest value of the
```

```
C       real part)
C
        PALFA=REALEV(1)
        DO 15 I=2,NUKI
  15    IF(REALEV(I).GT.PALFA)PALFA=REALEV(I)
C
        WRITE(6,4422)PALFA
 4422   FORMAT(/,3X,' Achieved stability degree of the robot',F10.5,/)
C
        IF(PALFA.GT.0)TYPE *,' WARNING*** The robot is not stabilized'
C
C       Check if the exponential stability degree of the global system
C       is greater than the desired stability degree
C
        IF((PALFA).GT.(-ALFAIM))GO TO 50
C
C       increment of the counter of points at the nominal trajectory
C
        ITERC=ITERC+1
C
C       check if the system stability is tested in all desired points
C       at the nominal trajectory or not: if yes - terminate analysis
C       otherwise go to read the next point at the nominal trajectory
C
        IF(ITERC.LE.ITMIN)GO TO 1
C
C       message whether the linearized model of the robot is stable
C       in desired way if just local controllers are applied
C
        ICNVLN=0
C
        GO TO 111
C
  50    ICNVLN=1
C
 111    RETURN
        END




C-------------------------------------------------------------------
C       SUBROUTINE:        SMLINM
C-------------------------------------------------------------------
C       FUNCTION:          COMPUTES THE MATRICES OF THE LINEARIZED
C                          MODEL OF THE ROBOT; COMPUTES THE OPEN-
C                          -LOOP MATRICES OF THE TOTAL SYSTEM BY
C                          ADDING THE MATRICES OF THE ACTUATORS
C                          MODELS TO THE MATRICES OF THE LINEARIZED
C                          MODEL OF THE ROBOT MECHANICAL PART
C-------------------------------------------------------------------
C       INPUT VARIABLES:
C               N        - Number of joints of the user's robot
C               NUK      - system order
C               NUKI     - order of the system with local integral feed-
C                          back loops
```

```
C                    NI(N)  - orders of the subsystems (actuators) models
C                    A(NI(I),NI(I),I)-subsystem matrix of the i-th actuator
C                    B(NI(I),N) - input distribution vector of the i-th
C                                 actuator
C                    F(NI(I),N) - load distribution vector of the i-th
C                                 actuator
C                    Y0(2*N)   -nominal trajectory of the joint angles and
C                                velocities (at the given time instant)
C                    P0(N)   - nominal driving torques
C                    DX0(N)  - nominal accelerations of joint angles
C             OPTIONS:
C                    IOPTIN(N)- Integral feedback loop in the local con-
C                                troller (1-yes,0-no)
C             INPUT VARIABLES FROM THE CALLED SUBROUTINES:
C                    H(N,N)  - inertial matrix H of the mechanical part of
C                                the system for the given state
C                    H0(N)   - vector of gravity, centrifugal and Coriolis
C                                moments (forces) h for the given state
C                    DH(N,N,N)- the first derivative of the matrix H by the
C                                joint angles
C                    DH0(N,N) - the first derivative of the vector h by the
C                                joint angles
C                    DH0DOT(N,N) - the first derivative of the vector h by
C                                the joint velocities
C----------------------------------------------------------------------
C       OUTPUT VARIABLES:
C                    A0(NUKI,NUKI) - system matrix of the open-loop
C                                linearized model of the robot
C                    BL(NUKI,N)  -  input distribution matrix of the
C                                linearized model of the robot
C                    MODLIN      - indicator for computation of the linea-
C                                rized model of mechanical part of the
C                                robot in subroutine MODEL (1-compute,
C                                0-not)
C----------------------------------------------------------------------
C       SUBROUTINES REQUIRED:
C                    MODEL       - computes the nonlinear and the linearized
C                                model of the mechanical part of the robot
C                                for the given joint angles and velocities
C                                (computes matrices H,H0,DH,DH0,DH0DOT)
C                    MINV        - SSP subroutine - inverts the given quad-
C                                ratic matrix
C----------------------------------------------------------------------
        SUBROUTINE SMLINM(A0,BL,HU)
C
        COMMON/DHCOM/ DH(6,6,6),DH0(6,6),DH0DOT(6,6)
        COMMON/DINAM/ H(6,6),H0(6)
        COMMON/MODOPC/ MOD,ITIP,MODLIN
        COMMON/SUBSYS/ A(3,3,6),B(3,6),F(3,6)
        COMMON/OPTION/ IPOM1(42),IOPTIN(6)
        COMMON/ACTORD/ NI(6),KI(6)

        COMMON/ORDGLB/ NUK,N
        COMMON/AUXIL/ T0,P0(6),Y0(12),DX0(6),U0(6)
        COMMON/UG/ SI(6),SIDOT(6)
        COMMON/INTGAI/ NUKI
C
        DIMENSION A0(NUKI,NUKI),BL(NUKI,N)
        DIMENSION HP(6,6),HV(6,6),HU(N,N)
        DIMENSION LL(6),MM(6)
        DIMENSION Y(12),DXU(6),Y1(12)
```

```
C-----------------------------------------------------------------------
C
C         computation of the model of the mechanical part: nonlinear
C         and the linearized model matrices for the given nominal state
C
C         put nominal state in the auxiliary vectors SI and SIDOT
C         to be transferred into subroutine MODEL
C
          DO 2 I=1,N
          SI(I)=Y0(I)
    2     SIDOT(I)=Y0(I+N)
C
C         set indicator to 1 - to compute the matrices of the
C         linearized model
C
          MOD=0
          MODLIN=1
C
          CALL MODEL
C
          MODLIN=0
C
C         form matrices: HU (matrix by the accelerations)
C         HP (matrix by the joints angles) and HV (matrix by the joints
C         velocities) in the linearized model of the robot
C
          DO 3 I=1,N
          II=2-NI(1)
C
          DO 3 J=1,N
          II=II+NI(J)
C
          HP(I,J)=DH0(I,J)
C
          HV(I,J)=DH0DOT(I,J)
C
          HU(I,J)=H(I,J)
C
          DO 3 JJ=1,N
    3     HP(I,J)=HP(I,J)+DH(I,JJ,J)*DX0(JJ)
C
C         introducing of the matrix F
C
          DO 4 I=1,N
          DO 4 J=1,N
C
          HU(I,J)=-F(2,I)*HU(I,J)
C
          HP(I,J)=F(2,I)*HP(I,J)
C
    4     HV(I,J)=F(2,I)*HV(I,J)
C
C         adding of the unit matrix to matrix HU
C
          DO 5 I=1,N
          HU(I,I)=HU(I,I)+1.
C
C         set initially matrix BL to zero
C
```

```
          DO 5 J=1,NUKI
    5     BL(J,I)=0.
C
C         invert the matrix HU
C
          NN=N
          CALL MINV(HU,NN,D,LL,MM)
C
C         form matrices A0 and BL
C
          DO 6 I=1,NUKI
          DO 6 J=1,NUKI
C
    6     A0(I,J)=0.
C
          JJ=0
          II=1
          DO 11 I=1,N
C
          III=II+1
C
                          IF(NI(I).EQ.2)GO TO 15
                          BL(II+2,I)=B(3,I)
C
   15     DO 7 J=1,NI(I)
          JJ=JJ+1
C
          A0(II,JJ)=A(1,J,I)
C
             IF(NI(I).EQ.2)GO TO 77
             A0(II+2,JJ)=A(3,J,I)
             IF(IOPTIN(I).EQ.0)GO TO 7
             A0(II+3,JJ)=0.
             GO TO 7
C
   77        IF(IOPTIN(I).EQ.0)GO TO 7
C
          A0(II+2,JJ)=0.
C
    7     CONTINUE
C
          IF(IOPTIN(I).EQ.0)GO TO 78
          JJ=JJ+1
C
          DO 79 J=1,NI(I)
   79     A0(II+J-1,JJ)=0.
C
C         introducing of "integral state coordinates"
C
          A0(II+NI(I),JJ-NI(I))=1.
C
   78     I2=1
C
              DO 10 I1=1,N
              I3=I2+1
C
          A0(III,I2)=HU(I,I1)*A(2,1,I1)
C
          A0(III,I3)=HU(I,I1)*A(2,2,I1)
C
                 IF(NI(I1).EQ.2)GO TO 8
```

```
C
                A0(III,I2+2)=HU(I,I1)*A(2,3,I1)
C
    8           BL(II+1,I1)=HU(I,I1)*B(2,I1)
C
                    DO 9 IJ=1,N
                    A0(III,I2)=A0(III,I2)+HU(I,IJ)*HP(IJ,I1)
    9               A0(III,I3)=A0(III,I3)+HU(I,IJ)*HV(IJ,I1)
C
            I2=I2+NI(I1)
C
        IF(IOPTIN(I1).EQ.0)GO TO 10
        I2=I2+1
C
   10       CONTINUE
C
        II=II+NI(I)
        IF(IOPTIN(I).EQ.1)II=II+1
C
   11       CONTINUE
            RETURN
            END
```

```
C-----------------------------------------------------------------
C       SUBROUTINE:     SMLGAD
C-----------------------------------------------------------------
C       FUNCTION:       ADDS LOCAL FEEDBACK GAINS INTO THE
C                       MATRICES OF THE LINEARIZED MODEL OF THE
C                       ROBOT AND BY THIS FORMS THE MATRIX OF THE
C                       CLOSED-LOOP LINEARIZED MODEL OF THE SYSTEM
C                       (JUST LOCAL SERVO LOOPS ARE ADDED)
C-----------------------------------------------------------------
C       INPUT VARIABLES:
C               NUK     - system order
C               N       - number of joints
C               NUKI    - order of the total system if integral feed-
C                         back loops are introduced
C               A0(NUKI,NUKI) - system matrix of the open-loop linea-
C                         rized model (computed in SMLINM)
C               BL(NUKI,N) - input distribution matrix of the linea-
C                         rized model of the robot (SMLINM)
C               NI(N)   - orders of the actuators models
C               PK1(N)  - position feedback gain (computed in SMLOC_E
C                         or in SMOPRG)
C               PK2(N)  - velocity feedback gain
C               PK3(N)  - current (pressure) feedback gain
C               PK4(N)  - integral feedback gain
C       OPTION:
C               IOPTIN(N) - local controller with integral feedback
C                         loop (1 -yes, 0-no)
C-----------------------------------------------------------------
C       OUTPUT VARIABLES:
C               AL(NUKI,NUKI) - matrix of the closed-loop linearized
C                         model of the robot - local feedback
C                         gains included
C-----------------------------------------------------------------
```

```
      SUBROUTINE SMLGAD(A0,AL,BL)
C
      COMMON/GAINS/ PK1(6),PK2(6),PK3(6),PKG(6),PK4(6)
      COMMON/INTGAI/ NUKI
      COMMON/OPTION/ IPOM(42),IOPTIN(6)
      COMMON/ACTORD/ NI(6)
      COMMON/ORDGLB/ NUK,N
C
      DIMENSION A0(NUKI,NUKI),BL(NUKI,N)
      DOUBLE PRECISION AL(NUKI,NUKI)
C-----------------------------------------------------------------
C
C     add local gains to the open-loop system matrix
C
      DO 1 I=1,NUKI
      DO 1 J=1,NUKI
  1   AL(I,J)=A0(I,J)
C
      II=0
      JJ=1
C
      DO 2 I=1,N
      II=II+NI(I)
C
      AL(II,JJ)=AL(II,JJ)-BL(II,I)*PK1(I)
      AL(II,JJ+1)=AL(II,JJ+1)-BL(II,I)*PK2(I)
C
          IF(NI(I).EQ.2)GO TO 3
C
      AL(II,JJ+2)=AL(II,JJ+2)-BL(II,I)*PK3(I)
      AL(II,JJ+3)=AL(II,JJ+3)-BL(II,I)*PK4(I)
C
      GO TO 4
C
  3   AL(II,JJ+2)=AL(II,JJ+2)-BL(II,I)*PK4(I)
C
  4   JJ=JJ+NI(I)
C
      IF(IOPTIN(I).EQ.1)II=II+1
C
      IF(IOPTIN(I).EQ.1)JJ=JJ+1
C
  2   CONTINUE
      RETURN
      END
```

```
C-----------------------------------------------------------------
C         SUBROUTINE:              SMEIGN
C-----------------------------------------------------------------
C         FUNCTION:                COMPUTATION OF EIGEN-VALUES OF THE
C                                  GIVEN QUADRATIC MATRIX
C-----------------------------------------------------------------
C         INPUT VARIABLES:
C                    NN    - matrix order
C                    NM    -       "
C                    AL(NN,NN) - quadratic matrix
C-----------------------------------------------------------------
C         OUTPUT VARIABLES:
C                    REALEV(NN) - vector of the real parts of the eigen-
C                                 -values of the given matrix
C-----------------------------------------------------------------
C         SUBROUTINES REQUIRED:
C                    EIGENP - SSP subroutine to compute eigen-values of the
C                                 given quadratic matrix
C                    REALVE,COMPVE,HESQR - auxiliary subroutines called by
C                                 subroutine EIGENP
C-----------------------------------------------------------------
          SUBROUTINE SMEIGN(NN,NM,AL,REALEV,RIMAG)
C
          DOUBLE PRECISION AL(NM,1),EVR(24),EVI(24),VECR(24,24)
          DOUBLE PRECISION VECI(24,24)
C
          DIMENSION REALEV(NN),RIMAG(NN)
          DIMENSION INDIC(24)
C-----------------------------------------------------------------
C
C         call subroutine EIGENP to compute eigen-values of AL
C
          CALL EIGENP(NN,NM,AL,EVR,EVI,VECR,VECI,INDIC)
C
          DO 10 I=1,NN
          RIMAG(I)=EVI(I)
   10     REALEV(I)=EVR(I)
C
          WRITE(6,102)
          DO 6 I=1,NN
    6     WRITE(6,103)I,EVR(I),EVI(I),INDIC(I)
  102     FORMAT(3X,' Eigen-values: ',/,3X,'No.     Real part
        1     Complex part   Indic.',/)
  103     FORMAT(3X,I4,2E15.5,I4)
          RETURN
          END
```

```
C-------------------------------------------------------------------------
C       SUBROUTINE:    SMSIMD
C-------------------------------------------------------------------------
C       FUNCTION:     SIMULATES THE ROBOT DYNAMICS WITH VARIOUS
C                     USER SELECTED CONTROL LAWS
C                     THE SIMULATION IS PERFORMED BY NUMERICAL INTEGRA-
C                     TION OF THE NONLINEAR DYNAMIC MODEL OF THE ROBOT;
C                     FOR NUMERICAL INTEGRATION THE SIMPLE EULER'S MET-
C                     HOD IS APPLIED:
C                         X(next)=X(current)+DINT*XT(current)
C                     WHERE X IS THE STATE VECTOR OF THE SYSTEM, XT IS
C                     THE FIRST DERIVATIVE OF THE STATE VECTOR, DINT
C                     IS THE INTEGRATION INTERVAL DEFINED BY:
C                         next instant= current instant +DINT
C                     SUBROUTINE COMPUTES THE FIRST DERIVATIVE OF THE
C                     STATE VECTOR (AT THE CURRENT INSTANT - ONCE AT
C                     EACH INTEGRATION INTERVAL) BY COMPUTATION OF
C                     THE RIGHT SIDE OF THE DIFFERENTIAL EQUATIONS OF
C                     THE ROBOT MODEL IN THE STATE SPACE USING THE
C                     CURRENT VALUE OF THE STATE VECTOR
C           NOTE: This is the simplest method for numerical inte-
C                 gration. It is easy to extend the programme to
C                 include other more reliable and more precise
C                 methods for numerical integration
C-------------------------------------------------------------------------
C       INPUT VARIABLES:
C               FILE(5) - Robot's name
C               N        - number of joints for the user's robot
C               NUK      - the order of the entire system
C               YSINT(18)-initial error of the system state vector
C                         (deviation of the state vector from the
C                         nominal state at the initial moment)
C               YS01(12)- nominal trajectory - values in the previous
C                         point: (1-6) joints anles (displacements)
C                         (7-12) joints velocities
C               YS02(12)- nominal trajectory - next point
C               QU01     - nominal accelerations at the previos point
C               QU02     -       "         "      at the next point
C               VR1,VR2 - time instants at the nominal trajectory in
C                         which the nominal trajectory is calculated
C                         and memorized in the file ***.ANG
C               NI(6)   - the orders of the actuators models
C               A(3,3,6)- matrices of the actuators models
C               AHM(3,3,6)    "       subsystems models-see INPUT_ACT
C                         (include moment of inertia of mechanism)
C               B(3,6)   - input distribution vectors of actuators
C                         models
C               BHM(3,6)-                 "            of subsystems
C                         models
C                         (include moment of inertia of mechanism)
C               F(3,6)   - load distribution vectors of actuators
C                         models
C               FHM(3,6)-                 "            of subsystems
C                         models
C                         (include moment of inertia of mechanism)
C               UMAX(2,6) amplitude constraints upon the actuators
C                         inputs
C               HMIN(6) - minimal moments of inertia of mechanism
C                         (see SM_MAX_IN)
```

```
C                   HH(6,6) - inertia matrix of the mechansm for the
C                             given value of joints angles
C                   H1(6)   - vector of centrifugal,Coriolis and gra-
C                             vity moments (forces) for  given joints
C                             angles and velocities
C                   OPTIONS: (for each joint)
C                   IOPCNN(6) apply nominal control (1), or not (0)

C                   IOPCNM(6) apply centralized nominal control(1),or
C                             apply decentralized nominal control(0)
C                   IOPTGG(6) global control is applied in the form of
C                             on-line computation of gravity, centrifu-
C                             gal and Coriolis moments around the i-th
C                             joint (0-no,1-yes)
C                   IOPTGH(6) global control is applied in the form of
C                             on-line computation of the inertia moments
C                             around the i-th joint (0-no, 1-yes)
C                   IOPTG(6)  =0 global control is not applied in
C                               the i-th joint
C                             =1 global control gain is in the form
C                               PKG(I)*(gradv*F)/(gradv*B)
C                             =2 global control gain is in the form
C                               PKG(I)*F(2,I)/B(3,I)
C                   TINTS   - time interval for printings at terminal
C                             display
C                   DINT    - integration interval
C                   PK1,PK2,PK3,PK4 - local feedback gains
C                                 (see SMLOC_E)
C                   PKG(6)  - global feedback gains
C-----------------------------------------------------------------------
C         OUTPUT VARIABLES:
C                   DY(18)  - Deviation of the state vector from the
C                             nominal trajectory (at the current
C                             sampling interval)
C                   P1(6)   - actual (simulated) driving torques in the
C                             joints developed during traking of the
C                             nominal trajectory
C                   U(6)    - actual (simulated) control (actuator
C                             input) signals
C                   TIME    - time moment along the simulation
C                   SPOWER(6) -actual required power in each joint
C                   SENERG(6) -actual (simulated) energy consumption
C-----------------------------------------------------------------------
C         LOCAL VARIABLES:
C                   YSP(18) -actual state vector at the current inte-
C                             gration interval
C                   YSOP(18) -nominal state vector at the current interval
C                   YS0      -      "            at the next interval
C                   POP(6)  -nominal driving torques at the current
C                             interval
C                   P0(6)   -      "            at the next interval
C                   Y0(12)  -nominal joints angles and velocities at the
C                             current integration interval
C                   QU0(6)  - robot accelerations
C                   DYS(18) - the first derivative of the actual state
C                             vector at the current integration interval
C-----------------------------------------------------------------------
C         SUBROUTINES REQUIRED:
C                   MODEL - computes the inertia matrix H and the vector h
C                             of gravity, centrifugal and Coriolis moments
C                             for the given joints angles SI and velocities
```

```
C                        SIDOT
C                SMOPTN - enables selection of a control law which will
C                         be simulated
C                SMCONT - computes local control signals for the given
C                         deviations of the state coordinates from
C                         the nominal trajectory DX(3)
C                SMCONG - computes the global gains for the given
C                         deviations of the state vector DX(3)
C-----------------------------------------------------------------------
          SUBROUTINE SMSIMD
C
          BYTE FILE(5)
          COMMON/FILE/FILE
C
          INCLUDE 'IN2:SMCOM.COM'
C
          INCLUDE 'IN2:MODELM.MOD'
C
          COMMON/MODOPC/ MOD,ITIP
          COMMON/NOVLON/AHMN(2,2,6),BHMN(2,6)
          COMMON/OPTNOM/IOPCNM(6),IOPCNN(6)
          COMMON/NKOORD/ YS0P(18),P0(6),U0(6)
          COMMON/SIMTIM/ TIME
          COMMON/POMINT/ POMINT(6)
C
          DIMENSION YSP(18),DY(18),DYS(18),P0P(6),P1(6)
          DIMENSION DX(3)
          DIMENSION YS0(18)
          DIMENSION U(6),PKGP(6)
          DIMENSION YS01(12),YS02(12)
          DIMENSION Y0POM(12)
          DIMENSION SENERG(6),SPOWER(6)
          DIMENSION DX1(3),DXP(6)
          DIMENSION QU01(6),QU02(6),QU0(6)
          DIMENSION H11(6),H10(6)
          DIMENSION Y0(12),HP(6),HPOM(6,6),HPOM3(6,6)
          DIMENSION LL(6),MM(6)
C
          BYTE FTOT(14),FTOT1(14),FTOT2(14)
          DATA FTOT(1)/'I'/,FTOT(2)/'N'/,FTOT(3)/'1'/,FTOT(4)/':'/
       1  ,    FTOT(10)/'.'/,FTOT(11)/'A'/,FTOT(12)/'N'/,
       2       FTOT(13)/'G'/,FTOT(14)/0/
          DATA FTOT1(1)/'I'/,FTOT1(2)/'N'/,FTOT1(3)/'1'/,
       1       FTOT1(4)/':'/,FTOT1(10)/'.'/,FTOT1(11)/'S'/
       2  ,    FTOT1(12)/'I'/,FTOT1(13)/'M'/,FTOT1(14)/0/
          DATA FTOT2(1)/'I'/,FTOT2(2)/'N'/,FTOT2(3)/'1'/,
       1       FTOT2(4)/':'/,FTOT2(10)/'.'/,FTOT2(11)/'L'/,
       2       FTOT2(12)/'O'/,FTOT2(13)/'C'/,FTOT2(14)/0/
C-----------------------------------------------------------------------
C
          DO 1 I=1,5
          FTOT(I+4)=FILE(I)
          FTOT2(I+4)=FILE(I)
       1  FTOT1(I+4)=FILE(I)
          TYPE 6699
    6699  FORMAT(/,' Simulation of tracking of the nominal trajectory')
          TYPE *,('-',I=1,54)
C
C         open file ***.ANG containing data on the nominal
C         joint trajectories
C
```

```
          OPEN(UNIT=1,NAME=FTOT,TYPE='OLD',ERR=999)
          GO TO 55
C
 999      TYPE 998,(FILE(I),I=1,5)
 998      FORMAT(3X,'DIAG*** FILE:',5A1,' .ANG cannot be found')
          RETURN
C
C         Options: the integration interval is set to 0.0001 [s]; user
C         might introduce variable integration interval and impose
C         it in the file ***.DAT
C
 55       DINT=0.0001
C
C         test whether the file with simulation results already exists
C
          OPEN(UNIT=2,NAME=FTOT1,TYPE='OLD',ERR=6691)
          TYPE 6697,FILE
 6697     FORMAT(/,' WARNING*** File ',5A1,'.SIM already exists')
C
          TYPE 6696
 6696     FORMAT($,' Want to form a file with new simulation [Y/N]?:')
          ACCEPT 6695,YES
 6695     FORMAT(A1)
          IF(YES.NE.'Y')GO TO 6694
C
          CLOSE(UNIT=2)
 6691     CALL ASSIGN(2,FTOT1,14)
C
          GO TO 6698
C
C         if the user doesnot want new simulation-exit from the routine
C
 6694     CLOSE(UNIT=2)
          CLOSE(UNIT=1)
          RETURN
C----------------------------------------------------------------
C
C         open file ***.LOC containing synthesized local feedback gains
C
 6698     OPEN(UNIT=3,NAME=FTOT2,TYPE='OLD',ERR=2000)
C
          DO 2002 I=1,N
          READ(3,2003)PK1(I)
          READ(3,2003)PK2(I)
          READ(3,2003)PK3(I)
 2002     READ(3,2003)PK4(I)
 2003     FORMAT(F12.5)
C
          CLOSE(UNIT=3)
          GO TO 2001
C
C         if there is no file ***.LOC with local feedback gains - exit
C         from the routine
C
 2000     TYPE *,'DIAG*** File ***.LOC cannot be found'
          CLOSE(UNIT=1)
          RETURN
C----------------------------------------------------------------
C
C         call subroutine to select control law which will be simulated
C
```

```
 2001    CALL SMOPTN
C
 2112    NN=N+N
10001    FORMAT($,' Want printings at the display during simulation
    >    [Y/N]?:')
 5597    TYPE 10001
         ACCEPT 6695,YES
         IF(YES.NE.'Y')TINTS=0.
         IF(YES.NE.'Y')GO TO 10005
         TYPE 10002
10002    FORMAT($,' How oten [s - F10.5]?:')
         READ(5,10003,ERR=5597)TINTS
10003    FORMAT(F10.5)
         IF(TINTS.GE.0.01)GO TO 10005
         TYPE *,' DIAG*** The values less than 0.01 are not allowed'
         GO TO 5597
C
C        Intialize time counter for printings
C
10005    TSTAMP=0.
C
C-------------------------------------------------------------------
C
C        initialization of the variables for simulation
C
         DO 9 I=1,N
         U(I)=0.
         SENERG(I)=0.
         POMINT(I)=0.
 9       SPOWER(I)=0.
C
C        read initial point at the nominal trajectory
C
         READ(1,100)VR1
         TIME=VR1
         READ(1,100)(YS01(I),I=1,N)
         READ(1,100)(YS01(I),I=N+1,NN)
         READ(1,100)QU01(I)
 100     FORMAT(6E13.5)
C
C        put nominal angles and velocities in the auxiliary vectors
C        SI and SIDOT to transfer data to the subroutine MODEL
C
         DO 40 I=1,N
         SI(I)=YS01(I)
 40      SIDOT(I)=YS01(I+N)
C
         ITIP=0
         MOD=0
         IPS=1
C
C        call subroutine MODEL to calculate inertia matrix H and vector
C        h for the nominal joints angles and velocities (initial point)
C
         CALL MODEL
C
C        in auxiliary vector H10 the nominal vector h is memorized
C        calculate the nominal driving torques (initail point)
C
 2010    DO 41 I=1,N
         POP(I)=H1(I)
```

```
        H10(I)=H1(I)
        DO 41 J=1,N
  41    POP(I)=POP(I)+HH(I,J)*QU01(J)
C
        II=2
C
        DO 43 I=1,N
        YSOP(II)=YS01(I+N)
        YSOP(II-1)=YS01(I)
        IF(NI(I).EQ.2)GO TO 44
C
C       if the actuator model is of order 3 then compute the nominal
C       value of the third state coordinate of the i-th actuator
C
        YSOP(II+1)=(QU01(I)-A(2,2,I)*YSOP(II)-F(2,I)*POP(I))/
     1  A(2,3,I)
C
C       test whether centralized or local nominal control is to be ap-
C       plied
C
  44 ·   IF(IOPCNM(I).EQ.1)GO TO 43
C
C       computation of the subsystem matrices to include minimal va-
C       lues of the moments of inertia of the mechanim around the
C       joints axes (necessary for computation of nominal local
C       control)
C
          FHM(2,I)=1./(HMIN(I)+1./F(2,I))
          POM=(1./F(2,I))*FHM(2,I)
          AHM(2,2,I)=A(2,2,I)*POM
          IF(NI(I).EQ.2)GO TO 43
          AHM(2,3,I)=A(2,3,I)*POM
C
        YSOP(II+1)=(QU0(I)-AHM(2,2,I)*YSOP(II))/AHM(2,3,I)
  43    II=II+NI(I)
C
C       computation of the actial initial state coordinates (since
C       initial errors of state coordinates are imposed in SMOPTN)
C
        DO 10 I=1,NUK
  10    YSP(I)=YSOP(I)+YSINT(I)
C********************************************************************
C       Start cycle for simulation
C
  2     CONTINUE
C
C       compute deviation (errors) between actual state coordinates
C       (YSP) and the nominal state coordinates (YSOP)
C
        DO 77 I=1,NUK
  77    DY(I)=YSP(I)-YSOP(I)
C
C       write data on simulated robot performance in file ***.SIM
C
        WRITE(2,100)TIME
        WRITE(2,100)(DY(I),I=1,NUK)
        WRITE(2,100)(P1(I),I=1,N)
        WRITE(2,100)(U(I),I=1,N)
        WRITE(2,100)(SPOWER(I),I=1,N)
        WRITE(2,100)(SENERG(I),I=1,N)
```

```
C
C         print at terminal display
C
          IF(TSTAMP.GT.TIME+0.001.OR.TINTS.EQ.0.)GO TO 119
          TSTAMP=TSTAMP+TINTS
          TYPE 101,TIME
  101     FORMAT(/,' Time:',F10.5,'[s]')
          TYPE 10006,(I,I=1,N)
10006     FORMAT(' No. of joint        ',6(4X,I1,5X))
          TYPE 10007
10007     FORMAT(' Deviations of the actuator state coordinates')
          TYPE 10008
10008     FORMAT(' Angles [rad] or [m]:',$)
          II=1
          DO 4097 I=1,N
          TYPE 4098,DY(II)
 4097     II=II+NI(I)
 4098     FORMAT('+',$,E10.3)
          TYPE 4049
 4049     FORMAT('+','   ')
          II=2
          TYPE 10009
          DO 4095 I=1,N
          TYPE 4098,DY(II)
 4095     II=II+NI(I)
10009     FORMAT(' Velocit.[rad/s-m/s]:',$)
          TYPE 4049
          TYPE 10010
          II=3
          DO 4099 I=1,N
          IF(NI(I).EQ.2)GO TO 4096
          TYPE 4098,DY(II)
          GO TO 4099
 4096     TYPE 4094
 4099     II=II+NI(I)
 4094     FORMAT('+',$,10X)
10010     FORMAT(' Current/pressure[A]:',$)
          TYPE 4049
          TYPE 103,(P1(I),I=1,N)
  103     FORMAT(' Driving torques[Nm]:',6E10.3)
          TYPE 104,(U(I),I=1,N)
  104     FORMAT(' Control signals [V]:',6E10.3)
  119     CONTINUE
C-------------------------------------------------------------------
C
C         read data on nominal trajectory from the file ***.ANG -
C         - at the instant VR2
C
          READ(1,100,END=9000)VR2
          READ(1,100)(YS02(I),I=1,N)
          READ(1,100)(YS02(I),I=N+1,NN)
          READ(1,100)(QU02(I),I=1,N)
C
          POM1=VR2-VR1
C-------------------------------------------------------------------
C
C         computation of nominal centralized or decentralized programmed
C         control; incement simulation time for DINT
C
 91       TIME=TIME+DINT
C
```

442

```
C          interpolation along the nominal trajectory at the integration
C          intervals between the moments VR2 in which the nominal
C          trajectory has been computed and memorized in the file ***.ANG
C
           POM=(TIME-VR1)/POM1
C
           DO 7777 I=1,NUK
 7777      DY(I)=YSP(I)-YSOP(I)
C
           DO 3 I=1,NN
    3      Y0(I)=(YS02(I)-YS01(I))*POM+YS01(I)
C
C          call MODEL to calculate matrix H and vector h for the
C          nominal values of joints angles and velocities
C
           DO 4 I=1,N
           QU0(I)=(QU02(I)-QU01(I))*POM+QU01(I)
           SI(I)=Y0(I)
    4      SIDOT(I)=Y0(I+N)
C
           CALL MODEL
C
C          computation of the nominal driving torques
C          and the nominal control
C
           DO 5 I=1,N
           P0(I)=H1(I)
           H11(I)=H1(I)
           DO 5 J=1,N
    5      P0(I)=P0(I)+HH(I,J)*QU0(J)
C
           II=2
           DO 6 I=1,N
           YS0(II-1)=Y0(I)
           YS0(II)=Y0(I+N)
           IF(NI(I).EQ.2)GO TO 67
           III=II+1
C
C          if the actuator order is 3
C
           IF(IOPCNM(I).EQ.0)GO TO 349
C
C          computation of the nominal value of the third state coordinate
C          of the i-th actuator-if the computation of the nominal driving
C          torque is included in the computation of the nominal control
C
           YS0(III)=(QU0(I)-A(2,2,I)*YS0(II)-F(2,I)*P0(I))
    1      /A(2,3,I)
C
C          computation of the centralized nominal control
C
           U0(I)=((YS0(III)-YSOP(III))/DINT -A(3,2,I)*YSOP(II)-
    1      A(3,3,I)*YSOP(III))/B(3,I)
           GO TO 6
C
C          if the nominal driving torque is not introduced in the nominal
C          control (if local nominal control is required)
C
  349      YS0(III)=(QU0(I)-AHM(2,2,I)*YS0(II))/AHM(2,3,I)
C
C          computation of the local nominal control
```

```
C
          U0(I)=((YS0(III)-YS0P(III))/DINT-AHM(3,2,I)*YS0P(II)-
     1    AHM(3,3,I)*YS0P(III))/B(3,I)
          GO TO 6
C
C         if the order of the i-th actuator model is 2
C
67        U0(I)=(QU0(I)-AHM(2,2,I)*YS0P(II)-IOPCNM(I)*FHM(2,I)*(P0P(I)-
     1    HMIN(I)*QU0(I)))/BHM(2,I)
 6        II=II+NI(I)
C-------------------------------------------------------------------
C
C         SIMULATION - computation of the right hand sides of the diff.
C         equations of the state nonlinear model of the robot
C
          II=1
          DO 7 I=1,N
          SI(I)=YSP(II)
          SIDOT(I)=YSP(II+1)
 7        II=II+NI(I)
C
C         call MODEL to compute matrix H and vector h for the actual
C         joints angles and velocities
C
          CALL MODEL
C-------------------------------------------------------------------
C
C         computation of control signals for all actuators according
C         to the selected control law (DX(3)-deviation of state coord.
C         of the i-th actuator from the nominal trajectories)
C
          II=1
          III=2
          I1=3
          DO 8 I=1,N
          DX(1)=DY(II)
          DX(2)=DY(III)
          IF(NI(I).EQ.3)DX(3)=DY(I1)
C
C         computation of local control signals
C
          CALL SMCONT(I,U1,DX)
          DYS(II)=A(1,2,I)*YSP(III)
C
C         computation of the selected global feedback gains
C
          CALL SMCONG(I,PKG1,U1,DX)
          U1=U1+IOPCNN(I)*U0(I)
C
C         PKGP(6) - vector of the global feedback gains computed
C         for the actual state of the actuator (susbsystem)
C
          U(I)=U1
          PKGP(I)=PKG1
C
C         HP - auxiliary vector for computation of the right sides
C         of differential equations of the robot model-the second
C         equations in the models of actuators in the state space
C
          HP(I)=A(2,2,I)*YSP(III)+F(2,I)*H1(I)+A(2,3,I)*YSP(I1)
          IF(NI(I).EQ.3)GO TO 68
```

```
C
C          if the order of actuator model is 2
C
           IF(IOPTG(I).EQ.0) GO TO 69
C
C          computation of the global control according to the
C          selected option
C
           U1=U1+PKG1*(IOPTGH(I)*(H10(I)-P0P(I))*IOPCNM(I)+
     *            IOPTGG(I)*(H1(I)-IOPCNM(I)*H10(I)))
C
C          amplitude constraints upon the actuator inputs
C
  69       IF(U1.GT.UMAX(1,I))U1=UMAX(1,I)
           IF(U1.LT.-UMAX(2,I))U1=-UMAX(2,I)
C
           U(I)=U1
           HP(I)=HP(I)+B(2,I)*U1
  68       II=II+NI(I)
           III=II+1
           I1=III+1
   8       CONTINUE
C
C          end of control computation
C          (at current integration interval)
C-------------------------------------------------------------
C
C          computation of the auxiliary matrix HPOM
C
           DO 70 I=1,N
           DO 70 J=1,N
  70       HPOM(I,J)=-HH(I,J)*(F(2,I)+B(2,I)*PKGP(I)*IOPTGH(I))
           DO 71 I=1,N
  71       HPOM(I,I)=HPOM(I,I)+1.
C
C          matrix inversion
C
           CALL SMMINV(HPOM,HPOM3)
C
C          multipliaction by the velocity dependent moments
C
           II=2
           DO 72 I=1,N
           DYS(II)=0.
           DO 73 J=1,N
  73       DYS(II)=DYS(II)+HPOM(I,J)*HP(J)
  72       II=II+NI(I)
C
C          computation of the actual joints torques
C
           DO 74 I=1,N
           II=2
           P1(I)=H1(I)
           DO 74 J=1,N
           P1(I)=P1(I)+HH(I,J)*DYS(II)
  74       II=II+NI(J)
C
           II=0
           DO 75 I=1,N
           II=II+NI(I)
C
```

```
C           computation of the actual power and energy consumptions
C
            SPOWER(I)=ABS(P1(I)*Y0(I+N))
            SENERG(I)=SENERG(I)+SPOWER(I)*DINT
            IF(NI(I).EQ.2)GO TO 75
C
C           if the order of the actuator is 3 -add the global control
C
            IF(IOPTGH(I).EQ.0) GO TO 76
            U(I)=U(I)+PKGP(I)*(IOPTGH(I)*(P1(I)-H1(I)-
     1          IOPCNM(I)*(POP(I)-H10(I)))+IOPTGG(I)*
     2          (H1(I)-H10(I)*IOPCNM(I)))
C
C           amplitude constraints upon the actuator inputs if the
C           order of the actuator model is 3
C
   76       IF(U(I).GT.UMAX(1,I))U(I)=UMAX(1,I)
            IF(U(I).LT.-UMAX(2,I))U(I)=-UMAX(2,I)
C
C           computation of the first derivative of the third state
C           coordinate of the i-th actuator model (if its order is 3)
C
            DYS(II)=A(3,2,I)*YSP(II-1)+A(3,3,I)*YSP(II)+B(3,I)*U(I)
   75       CONTINUE
C--------------------------------------------------------------------
C
            DO 90 I=1,NUK
C
C           memorize the nominal state vector at the previous interval
C
            YSOP(I)=YS0(I)
C
C           numerical integration by Euler's method
C
   90       YSP(I)=YSP(I)+DYS(I)*DINT
C--------------------------------------------------------------------
C
C           memorize nominal torques from the previous integration
C           interval
C
            DO 92 I=1,N
            H10(I)=H11(I)
            POP(I)=P0(I)
   92       CONTINUE
C
C           memorize the values of the nominal trajectory at the
C           previous point if the current time has reached the next
C           time instant in which nominal trajectory has been
C           imposed in file ***.ANG
C
            IF(TIME.LT.(VR2-0.5*DINT))GO TO 91
            DO 93 I=1,N
            QU01(I)=QU02(I)
            II=I+N
            YS01(I)=YS02(I)
   93       YS01(II)=YS02(II)
            VR1=VR2
C
            GO TO 2
C
C           end of simulation cycle
C*********************************************************************
```

```
C
 9000    CONTINUE
         CLOSE(UNIT=1)
         CLOSE(UNIT=2)
         TYPE *,' WARNING*** End of simulation'
         RETURN
         END




C--------------------------------------------------------------------
C       SUBROUTINE:     SMOPTN
C--------------------------------------------------------------------
C       FUNCTION:       INTERACTIVE SPECIFICATION OF THE CONTROL LAW
C                       OPTIONS FOR SIMULATION OF THE ROBOT DYNAMICS
C--------------------------------------------------------------------
C       INPUT VARIABLES:
C                       N - number of joints of the user's robot
C--------------------------------------------------------------------
C       OUTPUT VARIABLES:
C                       YSINT - deviation of the initial state from
C                               the nominal trajectory
C               OPTIONS:
C                       IOPCNN(I)- nominal control in the i-th joint
C                                  to be applied or not (1-yes,0-no)
C                       IOPCNM(i)- type of nominal control in the
C                                  i-th joint (1-centralized,0-local)
C--------------------------------------------------------------------
C       SUBROUTINES REQUIRED:
C                       SMINGL - enables specification of global
C                                control
C--------------------------------------------------------------------
        SUBROUTINE SMOPTN
C
        INCLUDE 'IN2:SMCOM.COM'
C
        COMMON/OPTNOM/ IOPCNM(6),IOPCNN(6)
C--------------------------------------------------------------------
C
 2000    FORMAT(A1)
         TYPE 101
  101    FORMAT(' Want simulation with the already adopted control law'
       1 ,/,$,' (otherwise you have to specify it) [Y/N]?:')
         ACCEPT 2000,YES
         IF(YES.EQ.'Y')GO TO 2
         TYPE 102
  102    FORMAT(' Want to apply nominal control in any joint'
       1 ,/,$,' (either centralized or decentralized) [Y/N]?:')
         ACCEPT 2000,YES
         IF(YES.EQ.'Y')GO TO 1
         DO 33 I=1,N
   33    IOPCNN(I)=0
         GO TO 5
    1    DO 34 I=1,N
         TYPE 103,I
  103    FORMAT($,' Want to apply nominal control in the ',I3,
       1 '-th joint [Y/N]?:')
```

```
        ACCEPT 2000,YES
        IF(YES.NE.'Y')GO TO 3
        IOPCNN(I)=1
        TYPE 104,I
104     FORMAT($,' Want centralized nominal control in the',I3,'-th
   1       joint [Y/N]?:')
        ACCEPT 2000,YES
        IOPCNM(I)=0
        IF(YES.EQ.'Y')IOPCNM(I)=1
        GO TO 34
  3     IOPCNN(I)=0
 34     CONTINUE
C
  5     TYPE 105
105     FORMAT($,' Want to apply global control [Y/N]?:')
        ACCEPT 2000,YES
        IF(YES.EQ.'Y')CALL SMINGL
        GO TO 6
C


  2     DO 35 I=1,N
        IOPCNN(I)=1
 35     IOPCNM(I)=1
C
  6     TYPE 106
106     FORMAT($,' Want to specify some specific initial conditions
   1       [Y/N]?:')
        ACCEPT 2000,YES
        IF(YES.NE.'Y')RETURN
        TYPE 107
107     FORMAT(' Specify initial conditions (for joint angles) [rad]
   1       or [m]:')
        DO 37 I=1,NUK
 37       YSINT(I)=0.
C
        II=1
        DO 36 I=1,N
        TYPE 108,I
108     FORMAT($,'  Joint',I3,'. ')
        ACCEPT 2001,YSINT(II)
        II=II+NI(I)
2001    FORMAT(F10.5)
 36     CONTINUE
        RETURN
        END
```

```
C-------------------------------------------------------------------
C          SUBROUTINE:       SMINGL
C-------------------------------------------------------------------
C          FUNCTION:         INTERACTIVE SPECIFICATION OF THE GLOBAL
C                            CONTROL LAW FOR SIMULATION - THE USER HAS
C                            TO SPECIFY A FORM OF THE GLOBAL CONTROL
C                            AND THE GLOBAL FEEDBACK GAIN
C-------------------------------------------------------------------
C          INPUT VARIABLE:
C                    N    -   number of joints of the user's robot
C-------------------------------------------------------------------
C          OUTPUT VARIABLES:
C                    PKG(I)  - attenuation of the global control in the
C                              i-th joint
C                    OPTIONS: (for each joint)
C                    IOPTG(i)  =0 global control is not to be applied in
C                                 the i-th joint
C                              =1 the global gain is in the form:
C                                 PKG(i)*(gradv*F)/(gradv*B)
C                              =2 the global gain is in the form:
C                                 PKG(i)*F(2)/B(2)
C                    IOPTGF(i) =0 force feedback is not to be applied
C                                 in the i-th joint
C                              =1 force feedback is applied
C                    IOPTGG(i) - the global control is in the form of
C                                "on line" computation of gravity, centri-
C                                fugal, and Coriolis moments around the
C                                i-th joint (0 -no, 1 -yes)
C                    IOPTGH(i) - the global control is in the form of
C                                "on-line" computation of inertial moments
C                                around the i-th joint (0 - no, 1 -yes)
C-------------------------------------------------------------------
          SUBROUTINE SMINGL
C
          COMMON/OPTION/ IOPTOR(6),IOPTV(6),IOPTI(6),
     1                   IOPTG(6),IOPTGF(6),IOPTGG(6),IOPTGH(6)
          COMMON/GAINS/ PK1(6),PK2(6),PK3(6),PKG(6)
          COMMON/ORDGLB/ NUK,N
C-------------------------------------------------------------------
C
          DO 666 I=1,N
          IOPTG(I)=0
          IOPTGF(I)=0
          IOPTGG(I)=0
  666     IOPTGH(I)=0
C
          TYPE 901
  901     FORMAT(' Want to apply equal global control law in all joints'
     1    ,/,$,' [Y/N]?:')
          ACCEPT 2000,YES
 2000     FORMAT(A1)
          IF(YES.EQ.'Y')GO TO 100
C
          DO 10 I=1,N
          TYPE 9000,I
 9000     FORMAT($,' Want global control in the ',I3,
     1    '-th joint [Y/N]?:')
          ACCEPT 2000,YES
          IF(YES.EQ.'Y')IOPTG(I)=1
```

```
C
         IF(IOPTG(I).EQ.0)GO TO 10
         TYPE 902,I
 902     FORMAT($,' Want force feedback in the ',I3,
       1 '-th joint [Y/N]?:')
         ACCEPT 2000,YES
         IF(YES.EQ.'Y')IOPTGF(I)=1

         IF(YES.EQ.'Y')GO TO 10
C
         TYPE 903,I
 903     FORMAT($,' Want "on-line" computation of gravity,',/,$
       1 ' centrifugal and  Coriolis moments in the ',I3,
       2 '-th joint [Y/N]?:')
         ACCEPT 2000,YES
         IF(YES.EQ.'Y')IOPTGG(I)=1
         IF(YES.EQ.'Y')GO TO 10
C
         TYPE 9004,I
 9004    FORMAT(' Want "on-line" computation of inertial moments',/,
       1   $,' in the ',I3,'-th joint [Y/N]?:')
         ACCEPT 2000,YES
         IF(YES.EQ.'Y')IOPTGH(I)=1
 10      CONTINUE
         GO TO 500
C
 100     TYPE 9006
 9006    FORMAT($,' Want force feedback in all joints [Y/N]?:')
         ACCEPT 2000,YES
         IF(YES.NE.'Y')GO TO 200
         DO 20 I=1,N
         IOPTG(I)=1
 20      IOPTGF(I)=1
         GO TO 500
C
 200     DO 25 I=1,N
 25      IOPTGF(I)=0
         TYPE 9016
 9016    FORMAT(' Want "on-line" computation of gravity,
       1 centrifugal and Coriolis moments ',/,
       2   $,' in all joints [Y/N]?:')
         ACCEPT 2000,YES
         IF(YES.NE.'Y')GO TO 300
         DO 30 I=1,N
         IOPTG(I)=1
 30      IOPTGG(I)=1
         GO TO 500
C
 300     DO 35 I=1,N
 35      IOPTGG(I)=0
C
         TYPE 9007
 9007    FORMAT(' Want "on-line" computation of inertial moments',
       1   /,$,' in all joints [Y/N]?:')
         ACCEPT 2000,YES
         IF(YES.NE.'Y')GO TO 700
         DO 40 I=1,N
         IOPTG(I)=1
 40      IOPTGH(I)=1
C
```

```
500      DO 501 I=1,N
         IF(IOPTGF(I).EQ.0) GO TO 501
         IOPTGG(I)=1
501      IOPTGH(I)=1
         TYPE 9008
9008     FORMAT($,' Want equal attenuation of global control in
       1 all joints [Y/N]?:')
         ACCEPT 2000,YES
         IF(YES.EQ.'Y')GO TO 600
         DO 60 I=1,N
         TYPE 9009,I
9009     FORMAT($,' Specify global attenuation for the ',I3,
       1 '-th joint [0.-1.]:')
  60     ACCEPT 3000,PKG(I)

3000     FORMAT(F10.5)
         GO TO 61
C
600      TYPE 9010
9010     FORMAT($,' Specify global attenuation for all joints
       1 [0.-1.]:')
         ACCEPT 3000,POM
         DO 70 I=1,N
  70     PKG(I)=POM
         GO TO 61
700      DO 80 I=1,N
         IOPTGH(I)=0
  80     IOPTG(I)=0
C
  61     TYPE 800
800      FORMAT($,' Want simplified form of the global control
       1 gain [Y/N]?:')
         ACCEPT 2000,YES
         IF(YES.NE.'Y')GO TO 6633
         DO 90 I=1,N
  90     IOPTG(I)=2
C
6633     DO 6644 I=1,N
         IF(IOPTOR(I).EQ.0) THEN
            IOPTG(I)=2
            TYPE 400, I
         END IF
         IF(PKG(I).NE.0)GO TO 6644
         IOPTG(I)=0
         IOPTGF(I)=0
         IOPTGG(I)=0
         IOPTGH(I)=0
6644     CONTINUE
C
400      FORMAT (' DIAG*** Since in the ',I2,'-th joint optimal',
       > /,'          regulator has not ben synthesized you must',
       > /,'          select simplified form of global gain')
         RETURN
         END
```

```
C------------------------------------------------------------------
C       SUBROUTINE:     SMCONT
C------------------------------------------------------------------
C       FUNCTION:       COMPUTES THE LOCAL CONTROL SIGNAL FOR THE
C                       I-TH ACTUATOR AND FOR GIVEN STATE VECTOR
C------------------------------------------------------------------
C       INPUT VARIABLES:
C                I    - the order number of the actuator(subsystem)
C                DX   - deviation of the actual state vector (during
C                       simulation) from the nominal trajectory for
C                       the i-th subsystem (DX(1) - position error,
C                       DX(2)-velocity error, DX(3)- error in rotor
C                       current, or pressure)
C                PK1(I)  position local feedback gain
C                PK2(I)  velocity          "
C                PK3(I)  gain in feedback loop by current/pressure
C                PK4(I)  integral feedback gain
C                DINT -  integration interval
C------------------------------------------------------------------
C       OUTPUT VARIABLE:
C                U1   -  local control signal for the i-th subsystem
C                        for the given DX
C------------------------------------------------------------------
        SUBROUTINE SMCONT(I,U1,DX)
C
        COMMON/GAINS/ PK1(6),PK2(6),PK3(6),PKG(6),PK4(6)
        COMMON/ORDGLB/ NUK,N
C
        DIMENSION DX(3)
        COMMON/ACTORD/ NI(6)
        COMMON/INTINT/ DINT
        COMMON/POMINT/ POMINT(6)
C
C       POMINT(I)-auxiliary vector for the numerical integration
C                 necessary to implement feedback loop by integral
C                 of the position error DX(1)
C
C------------------------------------------------------------------
C
        U1=-PK1(I)*DX(1)-
     1                  PK2(I)*DX(2)
     2                  -PK3(I)*DX(3)
     3                  -PK4(I)*POMINT(I)
C
        POMINT(I)=POMINT(I)+DX(1)*DINT
        RETURN
        END
```

```
C-----------------------------------------------------------------
C         SUBROUTINE:      SMCONG
C-----------------------------------------------------------------
C         FUNCTION:        COMPUTES GLOBAL CONTROL GAIN FOR THE I-TH
C                          JOINT - ACTUATOR
C-----------------------------------------------------------------
C         INPUT VARIABLES:
C                     I      - the order number of the actuator/joint
C                              (subsystem)
C                     DX(3) - deviation of the i-th subsystem state
C                              from the nominal trajectory
C                     U1     - local control signal for the i-th subsystem
C                     F(3,I)- load distribution vector in the i-th actu-
C                              ator model
C                     B(3,I)- input distribution vector in the state model
C                              of the i-th actuator
C                     NI(I) - order of the i-th subsystem
C                     V      - Liapunov's function value for the i-th sub-
C                              system for the given state vector DX(3)
C                     GRADV(3) - gradient of the Liapunov's function for
C                              the i-th subsystem for the given DX
C                   IOPTG(I) - type of the global gain for the i-th
C                              subsystem (see subroutine SMINGL)
C                     PKG(I)  - global control attenuation in the i-th
C                              joint - user's selection (see SMINGL)
C-----------------------------------------------------------------
C         OUTPUT VARIABLE:
C                     PKG1 -  global gain for the i-th subsystem
C-----------------------------------------------------------------
C         SUBROUTINES REQUIRED:
C                     LIAP1S - computes the value of the subsystem
C                              Liapunov's function for the given DX
C                     GRAD1S - computes the gradient of the subsystem
C                              Liapunov's function for the given DX
C-----------------------------------------------------------------
          SUBROUTINE SMCONG(I,PKG1,U1,DX)
C
          INCLUDE 'IN2:SMCOM.COM'
C
          DIMENSION DX(3),GRADV(3)
C-----------------------------------------------------------------
C
C         if IOPTG(i) is equal to 0 - the global control is not
C         applied in the i-th joint
C
          IF(IOPTG(I).NE.0)GO TO 1
          PKG1=0
          RETURN
C
    1     IF(IOPTG(I).EQ.2)GO TO 6
C
C         call subroutines to compute Liapunov's function and
C         its derivative for the i-th joint and given DX
C
          CALL LIAP1S(DX,V,I)
          CALL GRAD1S(DX,GRADV,I,V)
C
C         computation of GRADV*F and GRADV*B
C
```

```
          FG=0.
          BG=0.
          DO 2 J=1,NI(I)
          FG=FG+GRADV(J)*F(J,I)
    2     BG=BG+GRADV(J)*B(J,I)
    C
    C     if GRADV*B is smaller than 0.001- it is constrained to 0.001
    C
          IF(ABS(BG).GT.0.001)GO TO 50
          BG=SIGN(0.001,BG)
    C
    C     computation of the global gain according to:
    C     GRADV*F/(GRADV*B) multiplied by attenuation PKG(I)
    C     specified by the user
    C
    50    PKG1=-PKG(I)*FG/BG
          RETURN
    C
    C     if the user has selected the simplified form of
    C     global gain IOPTG(I)=2
    C
    6     IF(NI(I).EQ.2)PKG1=-PKG(I)*F(2,I)/B(2,I)
          IF(NI(I).EQ.3)PKG1=-PKG(I)*F(2,I)/B(3,I)
          RETURN
          END
```

```
C------------------------------------------------------------
C       SUBROUTINE:      LIAP1S
C------------------------------------------------------------
C       FUNCTION:        COMPUTES THE LIAPUNOV'S FUNCTION FOR THE
C                        I-th SUBSYSTEM FOR THE GIVEN VALUE OF THE
C                        SUBSYSTEM STATE VECTOR
C------------------------------------------------------------
C       INPUT VARIABLES:
C            I  - the order number of the subsystem
C            DX - the state of the i-th susbsystem (deviation of
C                 the state vector from the nominal)
C            NI(I) - the order of the subsystem model
C            RK(4,4,I)  - Liapunov's matrix for the i-th sub-
C                         system (solution of the Riccati's
C                         algebraic equation - see subroutine
C                         SMOREG)
C------------------------------------------------------------
C       OUTPUT VARIABLES:
C            V - the Liapunov's function for the i-th subsystem
C                for the given state vactor value DX
C------------------------------------------------------------
C       SUBROUTINES REQUIRED:
C            GMPRD - SSP subroutine for matrix multiplication
C------------------------------------------------------------
        SUBROUTINE LIAP1S(DX,V,I)
C
        DIMENSION DX(3),HPOM(3,3)
        DIMENSION DGX(3)
```

```
C
        COMMON/SUBLIA/ RK(4,4,6)
        COMMON/ACTORD/ NI(6)
C---------------------------------------------------------------
C
C       HPOM - auxiliary matrix
C
        DO 1 J=1,NI(I)
        DO 1 JJ=1,NI(I)
   1    HPOM(J,JJ)=RK(J,JJ,I)
        NN=NI(I)
C
C       multiplication of the Liapunov's matrix by the state DX
C
        CALL GMPRD(HPOM,DX,DGX,NN,NN,1)
C
C       multiplication by the transpose of DX
C
        V=0.
        DO 2 J=1,NI(I)
   2    V=V+DX(J)*DGX(J)
C
        IF(V.GE.0)GO TO 200
        TYPE 201
 201    FORMAT(' DIAG*** Local controller is not synthesized as',/,
       1        '           local optimal regulator, and',/,
       2        '           therefore simplified form of the global',/,
       3        '           gain must be selected')
        CALL EXIT
C
 200    V=SQRT(V)
        RETURN
        END
```

```
C-------------------------------------------------------------------
C       SUBROUTINE:      GRAD1S
C-------------------------------------------------------------------
C       FUNCTION:        COMPUTES GRADIENT OF THE GIVEN LIAPUNOV
C                        FUNCTION OF THE ij-th SUBSYSTEM AND
C                        FOR THE GIVEN SUBSYSTEM STATE VECTOR
C-------------------------------------------------------------------
C       INPUT VARIABLES:
C                IJ - the order number of the subsystem
C                DX - the subsystem state vector
C                V  - value of the subsystem Liapunov's function
C                     for the given state vector DX
C                NI(ij) - order of the ij-th subsystem
C                RK(3,3,IJ) -Liapunov's matrix for the ij-th
C                            subsystem
C-------------------------------------------------------------------
C       OUTPUT VARIABLES:
C                DGX - the gradient of the Liapunov's function
C                      of the ij-th subsystem for the given DX
C-------------------------------------------------------------------
C       SUBROUTINE REQUIRED:
C                GMPRD - SSP subroutine for matrix multiplication
```

```
C-----------------------------------------------------------------------
          SUBROUTINE GRAD1S(DX,DGX,IJ,V)
C
          DIMENSION DX(3),DGX(3),HL(3,3)
          COMMON/ACTORD/ NI(6)
          COMMON/SUBLIA/ RK(4,4,6)
C-----------------------------------------------------------------------
C
C         put the Liapunov's matrix in the auxiliary matrix HL
C
          NN=NI(IJ)
          DO 1 I=1,NN
          DO 1 J=1,NN
    1     HL(I,J)=RK(I,J,IJ)
C
C         multiplication of the Liapunov's matrix by the value
C         of the subsystem state vector DX
C
          CALL GMPRD(HL,DX,DGX,NN,NN,1)
C
C         multiplication of DGX by 1./V
C
          IF(V.LT.0.000001) THEN
              V=0.000001
          END IF
          DO 2 I=1,NN
    2     DGX(I)=DGX(I)/V
          RETURN
          END
```

```
C----------------------------------------------------------------------
C         SUBROUTINE:      SMMINV
C----------------------------------------------------------------------
C         FUNCTION:        AUXILIARY SUBROUTINE WHICH CALLS THE
C                          SSP SUBROUTINE MINV FOR THE MATRIX
C                          INVERSION (for various matrix dimen-
C                          sions)
C                          Note: This subroutine is not obligatory
C                          it could be omitted if matrix H is dimen-
C                          sioned in different way - depending
C                          on the FORTRAN compiler
C----------------------------------------------------------------------
C         INPUT VARIABLES:
C                          N - the matrix order
C                          HPOM - the matrix to be inverted
C----------------------------------------------------------------------
C         OUTPUT VARIABLES:
C                          HPOM3 - inverse of the matrix HPOM
C----------------------------------------------------------------------
C         SUBROUTINE REQUIRED:
C                MINV - SSP subroutine for the matrix inversion
C----------------------------------------------------------------------
          SUBROUTINE SMMINV(HPOM,HPOM3)
C
```

```
            COMMON/ORDGLB/ NUK,N
            DIMENSION HPOM3(N,N),HPOM(6,6)
            DIMENSION LL(6),MM(6)
C
            DO 1 I=1,N
            DO 1 J=1,N
    1       HPOM3(I,J)=HPOM(I,J)
C
            CALL MINV(HPOM3,N,D,LL,MM)
C
            DO 2 I=1,N
            DO 2 J=1,N
    2       HPOM(I,J)=HPOM3(I,J)
            RETURN
            END




C
C           INCLUDE FILE
C           'DL1:[107,21]SMCOM.COM'
C
C           Fail includes  COMMON - areas
C
            COMMON/SUBSYS/ A(3,3,6),B(3,6),F(3,6)
            COMMON/ORDGLB/NUK,N
            COMMON/ACTORD/ NI(6),KI(6)
C
            COMMON/SUBSHM/ AHM(3,3,6),BHM(3,6),FHM(3,6)
            COMMON/HMAX/ HMAX(6)
            COMMON/HMIN/ HMIN(6)
C
            COMMON/STDGIM/ ALFAI(6),ALFAIM
C
            COMMON/INTINT/ DINT
C
            COMMON/OPTION/ IOPTOR(6),IOPTV(6),IOPTI(6),IOPTG(6)
            COMMON/OPTION/IOPTGF(6),IOPTGG(6),IOPTGH(6),IOPTIN(6)
            COMMON/SUBCRT/ QQ(3,3,6),R1(6),QI(6)
            COMMON/GAINS/ PK1(6),PK2(6),PK3(6),PKG(6),PK4(6)
            COMMON/UMAX/ UMAX(2,6),UMAX1(6)
            COMMON/UM/ UQMAX(2,6),UDMAX(2,6)
            COMMON/SUBLIA/ RK(4,4,6)
            COMMON/CTIMIN/ TIMIN(60),ITMIN
C
            COMMON/STINIT/ YSINT(18)
```

```
        Fig. 1              ^ Im           Fig. 2              ^ Im
             :              |                   :              |
             :              |                   :              |
             :              |                   :              |
             :              |                   :              |
             :              |                   :              |
   S3        :              |            *S2    :              |
 -*-{}-------+-----*----+--> ---------{}-----+-----*----+-->
             :    S1=S2   | Re       *S3      :w0    S1  | Re
             :            |                   :          |
```

A. Auxiliary file CONF.DAT

```
6                                            ; defauld data
3
   .00000E 00      .10000E 01      .00000E 00  ;auxiliary data
   .00000E 00     -.11320E 02      .47500E 02  ;on actuators
   .00000E 00     -.11333E 04     -.66178E 02
   .00000E 00      .00000E 00      .74900E 02
   .00000E 00     -.03140E 02      .00000E 00
3
   .00000E 00      .10000E 01      .00000E 00
   .00000E 00     -.11320E 02      .47500E 02
   .00000E 00     -.11333E 04     -.66178E 02
   .00000E 00      .00000E 00      .74900E 02
   .00000E 00     -.18850E 01      .00000E 00
3
   .00000E 01      .10000E 01      .00000E 00
   .00000E 00     -.97700E 01      .41040E 02
   .00000E 00     -.85000E 03     -.49629E 02
   .00000E 00      .00000E 00      .56100E 02
   .00000E 00     -.32500E 00      .00000E 00
3
   .00000E 00      .10000E 01      .00000E 00
   .00000E 00     -.16600E 03      .80000E 03
   .00000E 00     -.30000E 03     -.80000E 02
   .00000E 00      .00000E 00      .80000E 02
   .00000E 00     -.33000E 03      .00000E 00
3
   .00000E 00      .10000E 01      .00000E 00
   .00000E 00     -.16600E 03      .80000E 03
   .00000E 00     -.30000E 03     -.80000E 02
   .00000E 00      .00000E 00      .80000E 02
   .00000E 00     -.33000E 03      .00000E 00
3
   .00000E 00      .10000E 01      .00000E 00
   .00000E 00     -.16600E 03      .80000E 03
   .00000E 00     -.30000E 03     -.80000E 02
   .00000E 00      .00000E 00      .80000E 02
   .00000E 03     -.33000E 03      .00000E 00
   .10000E 01                                  ;weighting matrix Q
                   .10000E 00                  ;for optimal regulator
                                   .10000E-01
   .10000E 00      .50000E 01      .50000E 01  ;weighting elements
   .50000E 01      .50000E 01      .50000E 01  ;by control for reg.
   .00000E-04      .00000E-02      .10000E-04  ;weighting elemnst
   .10000E-04      .10000E-04      .10000E-04  ;by integral coord.
1                                             ;options on feedback
1                                             ;loops
1
1
1
1
1
1
1
1
1
1
1
1
1
1
1
1
1                  A. Example - input - file UMSPR.DAT
1
```

3

```
 0.000000    ;type of joint- KSI2                         1
 0.000000    ;joint unit axis- eii (x)                    1
 0.000000    ;joint unit axis- eii (y)                    1
 1.000000    ;joint unit axis- eii (z)                    1
 0.000000    ;i+1 joint unit axis- ei,i+1 (x)             1
 0.000000    ;i+1 joint unit axis- ei,i+1 (y)             1
 1.000000    ;i+1 joint unit axis- ei,i+1 (z)             1
 0.300000    ;link vectors- Rii (x)                       1
 0.000000    ;link vectors- Rii (y)                       1
 0.000000    ;link vectors- Rii (z)                       1
-0.300000    ;link vectors- Ri,i+1 (x)                    1
 0.000000    ;link vectors- Ri,i+1 (y)                    1
 0.000000    ;link vectors- Ri,i+1 (z)                    1
 0.000000    ;spec. vect. for lin. j.- uii (x)            1
 0.000000    ;spec. vect. for lin. j.- uii (y)            1
 0.000000    ;spec. vect. for lin. j.- uii (z)            1
 0.000000    ;spec. vect. for lin. j.- ui,i+1 (x)         1
 0.000000    ;spec. vect. for lin. j.- ui,i+1 (y)         1
 0.000000    ;spec. vect. for lin. j.- ui,i+1 (z)         1
 0.000000    ;type of joint- KSI2                         2
 0.000000    ;joint unit axis- eii (x)                    2
 0.000000    ;joint unit axis- eii (y)                    2
 1.000000    ;joint unit axis- eii (z)                    2
 0.000000    ;i+1 joint unit axis- ei,i+1 (x)             2
 0.000000    ;i+1 joint unit axis- ei,i+1 (y)             2
 1.000000    ;i+1 joint unit axis- ei,i+1 (z)             2
 0.000000    ;link vectors- Rii (x)                       2
 0.000010    ;link vectors- Rii (y)                       2
 0.200000    ;link vectors- Rii (z)                       2
 0.000000    ;link vectors- Ri,i+1 (x)                    2
 0.000000    ;link vectors- Ri,i+1 (y)                    2
-0.300000    ;link vectors- Ri,i+1 (z)                    2
 0.000000    ;spec. vect. for lin. j.- uii (x)            2
 0.000000    ;spec. vect. for lin. j.- uii (y)            2
 0.000000    ;spec. vect. for lin. j.- uii (z)            2
 0.000000    ;spec. vect. for lin. j.- ui,i+1 (x)         2
 0.000000    ;spec. vect. for lin. j.- ui,i+1 (y)         2
 0.000000    ;spec. vect. for lin. j.- ui,i+1 (z)         2
 1.000000    ;type of joint- KSI2                         3
 0.000000    ;joint unit axis- eii (x)                    3
 0.000000    ;joint unit axis- eii (y)                    3
 1.000000    ;joint unit axis- eii (z)                    3
 0.000000    ;i+1 joint unit axis- ei,i+1 (x)             3
 0.000000    ;i+1 joint unit axis- ei,i+1 (y)             3
 0.000000    ;i+1 joint unit axis- ei,i+1 (z)             3
 0.000000    ;link vectors- Rii (x)                       3
 0.200000    ;link vectors- Rii (y)                       3
 0.300000    ;link vectors- Rii (z)                       3
 0.000000    ;link vectors- Ri,i+1 (x)                    3
-0.200000    ;link vectors- Ri,i+1 (y)                    3
 0.000000    ;link vectors- Ri,i+1 (z)                    3
 1.000000    ;spec. vect. for lin. j.- uii (x)            3
 0.000000    ;spec. vect. for lin. j.- uii (y)            3
 0.000000    ;spec. vect. for lin. j.- uii (z)            3
 1.000000    ;spec. vect. for lin. j.- ui,i+1 (x)         3
```

A. Example - input - file UMSPR.CNF

```
0.000000    ;spec. vect. for lin. j.- ui,i+1 (y)        3
0.000000    ;spec. vect. for lin. j.- ui,i+1 (z)        3
0.000000    ; First joint axis-ext. coordinate(x)
0.000000    ; First joint axis-ext. coordinate(y)
1.000000    ; First joint axis-ext. coordinate(z)
```

A. Example - input - file UMSPR. CNF (cont.)

```
0.000000    ;type of link                               1
7.000000    ;mass of link                               1
0.000000    ;moment of inertia Jxx/Js                   1
0.000000    ;moment of inertia Jyy/Jn                   1
0.300000    ;moment of inertia Jzz                      1
3.140000    ;qmax                                       1
-3.141000   ;qmin                                       1
0.000000    ;type of link                               2
0.000010    ;mass of link                               2
0.000000    ;moment of inertia Jxx/Js                   2
0.000000    ;moment of inertia Jyy/Jn                   2
0.000000    ;moment of inertia Jzz                      2
3.141000    ;qmax                                       2
0.000000    ;qmin                                       2
0.000000    ;type of link                               3
4.000000    ;mass  of link                              3
0.010000    ;moment of inertia Jxx/Js                   3
0.010000    ;moment of inertia Jyy/Jn                   3
0.000000    ;moment of inertia Jzz                      3
0.500000    ;qmax                                       3
0.000000    ;qmin                                       3
```

A. Example - input - file UMSPR.DNM

```
DC DC DC
    2.000000    ;Act. model order
    1.500000    ;Mechanical constant
    1.430000    ;Electromotor constant
    0.000030    ;Rotor moment of inertia
    0.005829    ;Viscous coefficient
    1.600000    ;Rotor resistance
   24.000000    ;Input amplitude constraint-upp. bound
   24.000000    ;Input amplitude constraint-low  bound
   31.000000    ;Gear speed ratio
   31.000000    ;Gear torque ratio
    0.002300    ;Rotor inductivity
   20.000000    ;Max. torque
   10.000000    ;Motor power
    0.000000    ;Auxiliary variable
    0.000000    ;Auxiliary variable
    2.000000    ;Act. model order
    1.500000    ;Mechanical constant
    1.430000    ;Electromotor constant
    0.000030    ;Rotor moment of inertia
    0.005800    ;Viscous coeficient
    1.600000    ;Rotor resistance
   24.000000    ;Input amplitude constraint-upp. bound
   24.000000    ;Input amplitude constraint-low  bound
   31.170000    ;Gear speed ratio
   31.170000    ;Gear torque ratio
```

A. Example - input - file UMSPR.ACT

```
    0.002300    ;Rotor inductivity
   20.000000    ;Max. torque
  100.000000    ;Motor power
    0.000000    ;Auxiliary variable
    0.000000    ;Auxiliary variable
    2.000000    ;Act. model order
  125.000000    ;Mechanical constant
  120.000000    ;Electromotor constant
    0.000030    ;Rotor moment of inertia
   40.000000    ;Viscous coeficient
    1.600000    ;Rotor resistance
   24.000000    ;Input amplitude constraint-upp. bound
   24.000000    ;Input amplitude constraint-low  bound
 2616.000000    ;Gear speed ratio
 2616.000000    ;Gear torque ratio
    0.002300    ;Rotor inductivity
   20.000000    ;Max. torque
  100.000000    ;Motor power
    0.000000    ;Auxiliary variable
    0.000000    ;Auxiliary variable
```

A. Example - input - file UMSPR.ACT (cont.)

```
0.0     0.4     0.               ;init.p.
0.0     0.6     0.               ;term. p.
0.01    0.6                      ;samp.in.
                                 ;duration
```

A. Example - input - file UMSPR.TRA

```
2                              ; number of points for test
 0.00000E+00                   ; instant at nom. traj.
 3.00000E+00                   ; prescribed stability degree
 0.60000E+00                   ; instant at nom. traj.
 3.00000E+00                   ; prescribed stability degree
```

A. Example - input - file UMSPR.INT

```
THE ROBOT NAME [max. length 5 characters]: UMSPR

DIAG*** File: UMSPR.CNF  is found

CHARACTERISTIC KINEMATIC VARIABLES
----------------------------------
        Joint coordinates:
0.00000E+00   0.00000E+00   0.00000E+00

Transformation matrix of the terminal link An:
   0.00000    1.00000    0.00000
  -1.00000    0.00000    0.00000
   0.00000    0.00000    1.00000

Joint axes with respect to absolute coord. frame

E(1)=      0.00000    0.00000    1.00000

E(2)=      0.00000    0.00000    1.00000

E(3)=      0.00000    0.00000    1.00000

Position vectors of the centers of masses and of    joints

R(1,1)=     0.300    0.000    0.000 R(1,2)=   -0.300    0.000    0.000

R(2,2)=     0.000    0.000    0.200 R(2,3)=    0.000    0.000   -0.300

R(3,3)=     0.200    0.000    0.300 RT0=    -0.200    0.000    0.000

DIAG*** Assembling of the mechanism:           SUCCESSFUL

 DIAG*** File: UMSPR.DNM  has been found

Want to compute maximal moments of inertia of the  mechanism
 (otherwise you have to directly specify  them) [Y/N]: Y

 DIAG*** Computation of maximal moments of    inertia

ax. mom.inert. H11max=         3.490068
in. mom.inert. H11min=         2.050324
....................
ax. mom.inert. H22max=         0.160016
in. mom.inert. H22min=         0.160016
....................
ax. mom.inert. H33max=         4.000003
in. mom.inert. H33min=         4.000001
....................
```

A. Example - interaction between user and package
(bold letters are user's answers)

```
WARNING*** File: UMSPR.ACT is found


Actuator model in the joint: 1
 Matrix A:
        0.000000        1.000000
        0.000000      -46.703224
 Vector b:
        0.000000
       32.518211
 Vector f:
        0.000000
      -34.686089

Actuator model in the joint: 2
 Matrix A:
        0.000000        1.000000

        0.000000      -46.194183
 Vector b:
        0.000000
       32.164471
 Vector f:
        0.000000
      -34.308769

Actuator model in the joint: 3
 Matrix A:
        0.000000        1.000000
        0.000000      -45.858898
 Vector b:
        0.000000
        0.380534
 Vector f:
        0.000000
       -0.004871
Do you want to print the matrices of subsystems models
(which include the maximal mom. of inertia)   [Y/N]: Y

 Subsystem: 1
Subsystem matrix A:
        0.00000         1.00000
        0.00000        -0.38264
Subsystem vector b:
        0.00000
        0.26642
Subsystem vector f:
        0.00000
       -0.28418
-----------------------------------------------------------------

 Subsystem: 2
Subsystem matrix A:
        0.00000        1.00000
        0.00000       -7.11780
Subsystem vector b:
        0.00000
        4.95604_
```

A. Example - interaction between user and package

```
Subsystem vector f:
        0.00000
       -5.28644
------------------------------------------------------------------

 Subsystem: 3
 Subsystem matrix A:
        0.00000         1.00000
        0.00000       -44.98249
 Subsystem vector b:
        0.00000
        0.37326
 Subsystem vector f:
        0.00000
       -0.00478
------------------------------------------------------------------


Synthesis of local controllers around robot joints


Do you want synthesis of local optimal regulator
for the  1th joint [Y/N]?: N
Specify the resonant structural frequency
of  the mechanism [rad/s]
for the servo in  1 -th joint [rad/s]: 12.
------------------------------------------------------------------

Local feedback servo gains:

Position  servo  gain   KP(1)= 0.13513E+03
Velocity  servo  gain   KV(1)= 0.43606E+02
------------------------------------------------------------------

Open-loop transfer function:

W(S)=     0.26642/(S**2+     0.38264*S)


------------------------------------------------------------------

Closed-loop transfer function:

G(S)=    36.00000/(S**2+    12.00000*S+    36.00000)


------------------------------------------------------------------

Do you want synthesis of local optimal regulator
for the  2th joint [Y/N]?: N
Specify the resonant structural frequency
of  the mechanism [rad/s]
for the servo in  2 -th joint [rad/s]: 50.
------------------------------------------------------------------

Local feedback servo gains:

Position  servo  gain   KP(2)= 0.12611E+03
Velocity  servo  gain   KV(2)= 0.86525E+01
------------------------------------------------------------------
```

          A. Example - interaction between user and package

Open-loop transfer function:

W(S)=   4.95604/(S**2+   7.11780*S)

Closed-loop transfer function:

G(S)=   625.00000/(S**2+   50.00000*S+   625.00000)

Do you want synthesis of local optimal regulator
for the 3th joint [Y/N]?: **N**
Specify the resonant structural frequency
of the mechanism [rad/s]
for the servo in 3 -th joint [rad/s]: **50.**

Local feedback servo gains:

Position servo gain   KP(3)= 0.16744E+04
Velocity servo gain   KV(3)= 0.13442E+02

Open-loop transfer function:

W(S)=   0.37326/(S**2+   44.98249*S)

Closed-loop transfer function:

G(S)=   625.00000/(S**2+   50.00000*S+   625.00000)

Want to write the feedback gains in the file [Y/N] ?: **Y**

 Synthesis of the joints nominal trajectories

Want to print the nominal trajectories [Y/N]?: **N**
Select the velocity profile-triangular or trapezoid [TA/TP]: **TA**

    Computation of the nominal dynamics of the robot
        - nominal driving torques and nominal
              programmed control -

Want centralized or local nominal programmed control? [C/L]: **C**
Want to print the nominal dynamics [Y/N]?: **N**

Do you want linear analysis [Y/N]?: **Y**

 Stability analysis

        A. Example - interaction between user and package

Eigen-values of the open-loop matrix of
linearized model of the robot in the time instant   0.00000 [s]

  Eigen-values:
 No.   Real part   Complex part Indic.

    1   0.00000E+00   0.00000E+00   2
    2   -0.16505E+02   0.00000E+00   2
    3   -0.43290E+00   0.00000E+00   2
    4   -0.53679E-08   0.00000E+00   2
    5   0.00000E+00   0.00000E+00   2
    6   -0.44982E+02   0.00000E+00   2

Eigen-values of the closed-loop matrix of
linearized model of the robot in the time instant   0.00000 [s]

  Eigen-values:
 No.   Real part   Complex part Indic.

    1   -0.23935E+02   0.00000E+00   2
    2   -0.34195E+01   0.00000E+00   2
    3   -0.98590E+02   0.00000E+00   2
    4   -0.14510E+02   0.00000E+00   2
    5   -0.25000E+02   0.00000E+00   2
    6   -0.25000E+02   0.00000E+00   2

Achieved stability degree of the robot -3.41955

Eigen-values of the open-loop matrix of
linearized model of the robot in the time instant   0.60000 [s]

  Eigen-values:
 No.   Real part   Complex part Indic.

    1   -0.16519E+02   0.00000E+00   2
    2   -0.43022E+00   0.00000E+00   2
    3   0.16400E-10   0.00000E+00   2
    4   0.27251E-08   0.00000E+00   2
    5   0.00000E+00   0.00000E+00   2
    6   -0.44982E+02   0.00000E+00   2

Eigen-values of the closed-loop matrix of
linearized model of the robot in the time instant   0.60000 [s]

  Eigen-values:
 No.   Real part   Complex part Indic.

    1   -0.23929E+02   0.00000E+00   2
    2   -0.34205E+01   0.00000E+00   2
    3   -0.98602E+02   0.00000E+00   2
    4   -0.14506E+02   0.00000E+00   2
    5   -0.25000E+02   0.00000E+00   2
    6   -0.25000E+02   0.00000E+00   2

Achieved stability degree of the robot -3.42049

A. Example - interaction between user and package

The linearized model is stabilized with
the prescribed exponential stability degree

Simulation of tracking of the nominal trajectory

WARNING*** File UMSPR.SIM already exists
Want to form a file with new simulation [Y/N]?: Y
Want simulation with the already adopted control law
(otherwise you have to specify it) [Y/N]?: Y
Want to specify some specific initial conditions [Y/N]?: Y
Specify initial conditions (for joint angles) [rad] or [m]: Y
 Joint 1. -0.03
 Joint 2. -0.03
 Joint 3. -0.03
Want printings at the display during simulation [Y/N]?: Y
How oten [s - F10.5]?: 0.1

Time: 0.00000[s]
No. of joint   1   2   3
Deviations of the actuator state coordinates
Angles [rad] or [m]:-0.300E-01-0.300E-01-0.300E-01
Velocit.[rad/s-m/s]: 0.000E+00 0.000E+00 0.000E+00
Current/pressure[A]:
Driving torques[Nm]: 0.000E+00 0.000E+00 0.000E+00
Control signals [V]: 0.000E+00 0.000E+00 0.000E+00

Time: 0.10000[s]
No. of joint   1   2   3
Deviations of the actuator state coordinates
Angles [rad] or [m]:-0.281E-01-0.114E-01-0.142E-01
Velocit.[rad/s-m/s]: 0.611E-01 0.135E+00 0.195E+00
Current/pressure[A]:
Driving torques[Nm]: 0.229E+01 0.463E+00-0.429E+02
Control signals [V]: 0.255E+01 0.104E+01 0.206E+02

Time: 0.20000[s]
No. of joint   1   2   3
Deviations of the actuator state coordinates
Angles [rad] or [m]:-0.208E-01-0.272E-02-0.246E-02
Velocit.[rad/s-m/s]: 0.766E-01 0.497E-01 0.485E-01
Current/pressure[A]:
Driving torques[Nm]: 0.675E+00 0.223E+00-0.428E+02
Control signals [V]: 0.828E+00 0.999E+00 0.297E+01

Time: 0.30000[s]
No. of joint   1   2   3
Deviations of the actuator state coordinates
Angles [rad] or [m]:-0.138E-01 0.134E-03-0.308E-03
Velocit.[rad/s-m/s]: 0.599E-01-0.194E-02 0.663E-02
Current/pressure[A]:

Driving torques[Nm]:-0.214E+01-0.391E+00-0.398E+02
Control signals [V]:-0.221E+01 0.514E+00-0.751E-01

A. Example - interaction between user and package

```
Time:    0.39998[s]
No. of joint              1         2         3
Deviations of the actuator state coordinates
Angles [rad] or [m]:-0.889E-02 0.672E-03-0.339E-04
Velocit.[rad/s-m/s]: 0.404E-01 0.206E-02 0.761E-03
Current/pressure[A]:
Driving torques[Nm]:-0.189E+01-0.454E+00-0.393E+02
Control signals [V]:-0.196E+01 0.876E-01-0.456E+00

Time:    0.49997[s]
No. of joint              1         2         3
Deviations of the actuator state coordinates
Angles [rad] or [m]:-0.558E-02 0.682E-03-0.349E-05
Velocit.[rad/s-m/s]: 0.264E-01-0.144E-02 0.801E-04
Current/pressure[A]:
Driving torques[Nm]:-0.166E+01-0.423E+00-0.392E+02
Control signals [V]:-0.174E+01-0.202E+00-0.497E+00

Time:    0.59999[s]
No. of joint              1         2         3
Deviations of the actuator state coordinates
Angles [rad] or [m]:-0.346E-02 0.508E-03-0.344E-06
Velocit.[rad/s-m/s]: 0.166E-01-0.195E-02 0.801E-05
Current/pressure[A]:
Driving torques[Nm]:-0.150E+01-0.398E+00-0.392E+02
Control signals [V]:-0.158E+01-0.495E+00-0.502E+00
 WARNING*** End of simulation


Want another simulation of robot [Y/N]?: N
```

A. Example - interaction between user and package

```
 135.12567
  43.60567
   0.00000
   0.00000
 126.10879
   8.65252
   0.00000
   0.00000
1674.42932
  13.44234
   0.00000
   0.00000
```

A. Example - output - file UMSPR.LOC

Subject Index

M. Vukobratović (Ed.)

Introduction to Robotics

With contributions by M. Djurović,
D. Hristić, B. Karan, M. Kirćanski,
N. Kirćanski, D. Stokić, D. Vuijić,
M. Vukobratović

1989. 228 figures. XIV, 301 pp.
Hardcover ISBN 3-540-17452-4

Contents: Preface. – General Introduction to
Robotics. – Manipulator Kinematic Model.
– Dynamics and Dynamic Analysis of Mani-
pulation Robots. – Hierarchical Control of
Robots. – Microprocessor Implementation of
Control Algorithms. – Industrial Robot
Programming Systems. – Sensors in Robot-
ics. – Elements, Structures and Application
of Industrial Robots. – Robotics and Flexible
Automation Systems. – Appendix.

This book provides a general introduction to
robot technology with an emphasis on robot
mechanisms and kinematics. It is conceived
as a reference book for students in the field
of robotics.

Springer-Verlag
Berlin Heidelberg New York
London Paris
Tokyo Hong Kong

Springer

M. Vukobratović

Applied Dynamics of Manipulation Robots

Modelling, Analysis and Examples

1989. Approx. 495 pp. 176 figs. Hardcover
ISBN 3-540-51468-6

This book is devoted to the study of manipulation robot dynamics and its applications. It contains a computational procedure for the automatic generation of mathematical models of robot dynamics, comprising linearized models of robot dynamics and parameter sensitivity models. The presentation is complemented by a selection of problems and solutions presenting mathematical models of different types of drives and examples of dynamic models of characteristic manipulation systems.

Springer-Verlag
Berlin Heidelberg New York
London Paris
Tokyo Hong Kong

Springer